Flywheel Energy Storage

Armin Buchroithner

Flywheel Energy Storage

in Automotive Engineering

Springer

Armin Buchroithner
Institute of Electrical Measurement and Sensor Systems
Graz University of Technology
Graz, Steiermark, Austria

ISBN 978-3-658-35341-4 ISBN 978-3-658-35342-1 (eBook)
https://doi.org/10.1007/978-3-658-35342-1

This book is a translation of the original German edition “Schwungradspeicher in der Fahrzeugtechnik” by Buchroithner, Armin, published by Springer Fachmedien Wiesbaden GmbH in 2019. The translation was done with the help of an artificial intelligence machine translation tool. A subsequent human revision was done primarily in terms of content, so that the book will read stylistically differently from a conventional translation. Springer Nature works continuously to further the development of tools for the production of books and on the related technologies to support the authors.

This Springer imprint is published by the registered company Springer Fachmedien Wiesbaden GmbH, part of Springer Nature.
The registered company address is: Abraham-Lincoln-Str. 46, 65189 Wiesbaden, Germany

Formula Symbols

Symbol	Designation	Unit
A	Area	m^2
c_w	Air drag coefficient	–
E	Modulus of elasticity	N/m^2
E_k	Kinetic energy	Joule
E_p	Potential energy	Joule
F_a	Axial force	N
F_{req}	Required tractive force	N
F_{air}	Vehicle aerodynamic drag	N
f_R	Rolling resistance coefficient	–
F_r	Radial force	N
F_{roll}	Wheel rolling resistance force	N
F_{climb}	Slope resistance of the vehicle	N
g	Gravitational constant	m/s^2
G	Balance quality class	Mm/s
G_b	Thermal conductivity value	W/K
H	Angular momentum	$kg{\cdot}m^2/s$
ΔH	Height difference	m
h_0	Initial height	m
i	Gear ratio of a gearbox	
I	Moment of inertia	$kg{\cdot}m^2$
K_{dyn}	Dynamic index of driving cycle	–
K_{shape}	Form factor of the flywheel	–
L	Length	m
m	Mass	kg
m^*	Generalized mass	kg
M_K	Gyroscopic torque	Nm

(continued)

Symbol	Designation	Unit
M_R	Torque loss of rolling bearing	Nm
n_{max}	Maximum rotational speed limit	RPM
P	Equivalent bearing load	kN
r	Radius (general)	m
R	Outer radius	m
$R_{p0,2}$	Yield strength	N/mm^2
R_m	Tensile strength	N/mm^2
s^*	Wheel slip	–
u	Unbalance mass	g
U	Unbalance	g·mm
v	Speed	m/s
v_0	Initial speed	m/s
w	Deflection	m
W_{req}	Energy demand	Joule
W_{rec}	Recoverable energy	Joule
x	Path length	m
α	Rotation around the X-axis of the vehicle (rolling)	rad
α_s	Slope angle	°
β	Rotation around the Y-axis of the vehicle (pitching)	rad
γ	Rotation around the Z-axis of the vehicle (yawing)	rad
γ_k	Angle around the rotation axis of the FESS	rad
λ	Rotational mass factor	–
λ_j	Eigenvalue of the transcendental equation	–
λ_w, λ_{spoke}	Specific thermal conductivity	W/m*K
μ	Poisson's ratio	–
$\mu(x)$	Mass distribution	kg/m^2
ρ_L	Air density	kg/m^3
σ_{max}	Maximum stress	N/mm^2
σ_r	Radial stress	N/mm^2
σ_t	Tangential stress	N/mm^2
Φ_{ish}	Correction factor lubricant film thickness	–
Φ_{rs}	Influence factor lubricant displacement	–
ω_j	Natural frequency	Hz

Acknowledgments

This book would not have been possible without the cooperation of the following persons:

- *Gunter Jürgens*, who pointed out the importance of the systematic analysis of existing systems already several years ago when I was working on my master's thesis and later on encouraged me to write this book.
- *Michael Bader*, without whose extensive support and permanent input of know-how, the execution of the numerous empirical studies that form the core of this book would not have been possible.
- *Hannes Wegleiter* and *Bernhard Schweighofer,* whose excellent and meanwhile long-lasting friendly cooperation cannot be appreciated enough and resulted in the foundation of the Energy Aware Systems work group.
- *Peter Haidl*, who, with boundless idealism, contributed valuable input regarding the investigation of rolling bearings at the limits of what is technically possible.
- *Andreas Brandstätter* and *Manes Recheis,* who used numerical and empirical methods to make the gyro kinematics of flywheels easier to understand.
- *Clemens Voglhuber,* who through his committed cooperation helped to gain new insights in the investigation of the torque loss of rolling bearings for flywheel energy storage systems.
- *Christoph Birgel* and *Rupert Preßmair*, who showed particular dedication to the investigation of rotors and safety concepts for flywheel applications.
- *Thomas Murauer* and *Martin Simonyi*, who have contributed important results concerning the thermal behavior of rolling bearings in vacuum.

Special thanks also go to *Melanie Kaiser* and my entire family, who always gave me the freedom and support to dedicate myself to projects like this book.

Summary

Energy storage must be regarded as the greatest technological challenge of the beginning twenty-first century and plays a central role in the decarbonization of our society. Efficient energy storage is not only essential for the transition to renewable, volatile energy sources but is also a key element of all mobile applications and must be given particular importance in the context of sustainable transportation.

This book deals with the design and optimization of *flywheel energy storage systems* (FESSs) in vehicles as an alternative to conventional solutions such as chemical batteries or capacitors. A possible vehicle topology with FESS is shown in Fig. 1. Despite the supposed simplicity of the physical principle, namely storing energy in kinetic form in a rotating mass, only very few successful, production-ready solutions are available on the market to date.

In the first part of the book, the *supersystem analysis*, FESSs are evaluated in a global context by using a holistic approach. External influences such as a vehicle, driver, operating strategy, and environment all the way to socio-psychological aspects are analyzed with regard to their interaction with the actual energy storage unit. Based on these findings, not only optimal application scenarios for FESS are derived but also the development goals relevant for market success are defined. The *supersystem analysis* is also based on a detailed investigation of more than 50 historical flywheel hybrid vehicle concepts that were identified in the course of extensive literature research to evaluate the state of the art.

On the basis of the technical-energetic specific target properties of flywheel energy storage, which were determined in the course of the *supersystem analysis,* the second part of this book will follow with a detailed consideration of the FESS *subsystem*. Those critical components within FESS, which are responsible for achieving the desired specifications and key performance indicators, are identified. While strictly pursuing maximum cost reduction, concrete technical solutions for key components are discussed and their suitability is validated by empirical studies. The focus is clearly set on the optimization of FESS housing, bearings, and rotor, and practical case studies based on actual prototypes are given for each of the three components.

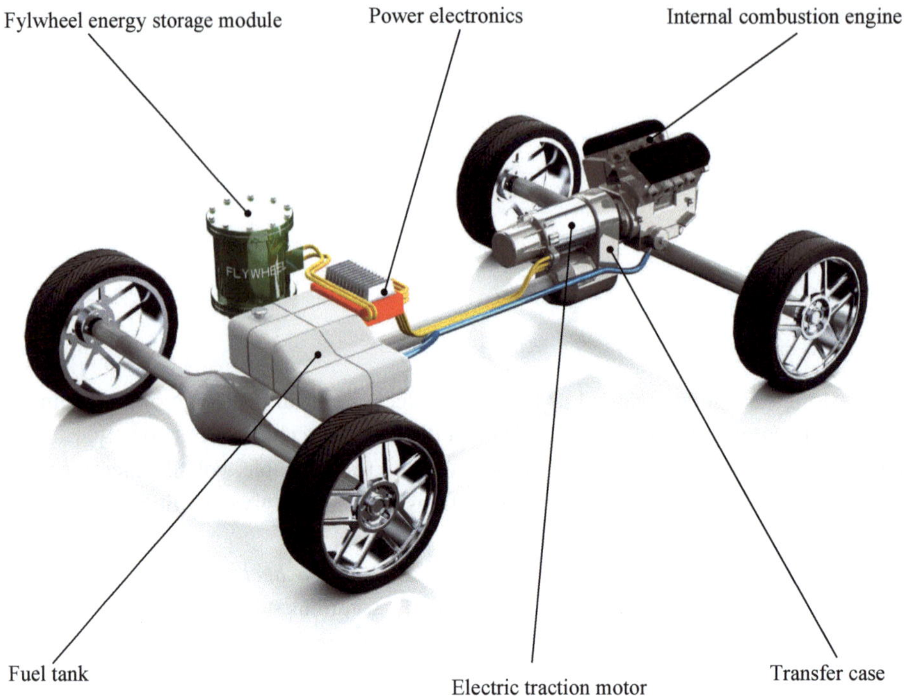

Fig. 1 Typical topology of a hybrid powertrain with flywheel energy storage system (FESS). The energy storage device allows load point shifting, brake energy recuperation, and "boosting," e.g., during overtaking maneuvers

Finally, an alternative, stationary FESS concept is presented, which largely circumvents the specific problems of mobile flywheel energy storage systems, but is nevertheless able to make a significant contribution to sustainable, electric mobility.

Contents

Abbreviations

ASM	Asynchronous machine
ATTB	Advanced Technology Transit Bus
CFRP	Carbon fiber reinforced plastic
CMO	Clean Motion Offensive
CVT	Continuously Variable Transmission
E3oN	Efficient electrical energy storage for public transport
EMS	Institute for Electrical Measurement and Sensor Systems
EV	Electric vehicle
FESS	Flywheel Energy Storage System
FFG	(Austrian) Research Promotion Agency
FIMD	Fully integrated multi-disc design
FTP	Federal Test Procedure
ICE	Internal combustion engine
IME	Institute for Machine Elements and Method of Development
KERS	Kinetic energy recovery system
LESS	Lebensdauer-Erhöhung bei Schwungrad-Speichern [Increased service life of flywheel storage systems]
NASA	National Aeronautics and Space Administration
NEDC	New European Driving Cycle
PMSM	Permanent-magnet synchronous
PTO	Power take-off
PV	Photovoltaics
Ref.	Reference (source of literature)
SRM	Switched reluctance machine
SynRM	Synchronous reluctance machine
TUG	Graz University of Technology
UPS	Uninterruptible power supply
WLTP	Worldwide harmonized Light vehicles Test Procedure

1 Introduction

In 1973, the Western industrialized countries felt the danger of unrestricted dependence on Middle Eastern oil imports for the first time. The US American involvement in the fourth Arab-Israeli war had led to an oil boycott that took the entire world by surprise. However, it subsequently also triggered a multitude of initiatives to investigate and promote alternative and renewable energy sources. In November 1974, the *International Energy Agency* (IEA) was founded under US President Jimmy Carter, which received a start-up budget of 25 billion US dollars [1]. But only shortly after, in 1979, when Ayatollah Khomeini called for the Iranian revolution and thus caused the second oil embargo, the Western world seemed as surprised and ill-prepared as it had been only 5 years earlier.

More than 40 years have passed since then, and despite all the climate conferences and declarations by the industrial nations (such as the Kyoto Protocol), the global transport system and our energy consumption behavior has hardly changed at all. On the contrary, the increasingly powerful emerging economies such as India, and especially China, are striving to gain access to the largest remaining oil and gas reserves on the planet with an enormous hunger for energy. But it does not matter whether peak oil has already been reached, or the analysts are wrong and the reserves will still last for several more decades: The burning of fossil fuels and the associated CO_2 emissions have already led to a noticeable climate change, which has devastating consequences for the entire planet [2].

But can technical science solve these problems simply by increasing energy efficiency? To what extent could efficient energy storage systems like the flywheels, which are discussed in detail in this book, contribute to improving this threatening situation?

"The first drink from the cup of science makes one an atheist, but at the bottom of the cup God is waiting" is a quote attributed to the German quantum physicist Werner Heisenberg. Somewhat less dramatic, but a similarly misleading experience, is the first contact with flywheel energy storage (FESS) for many scientists or technology aficionados. In the 1960s and 1970s, the charming simplicity of the FESS's underlying physical

A. Buchroithner, *Flywheel Energy Storage*,
https://doi.org/10.1007/978-3-658-35342-1_1

Fig. 1.1 An early flywheel hybrid vehicle: The VW T2 of RWTH Aachen University, 1977, in Germany. (Image rights: Institute of Automotive Engineering and Piston Engines, RWTH Aachen University)

principle and the long list of theoretical advantages compared to chemical energy storage devices led many scientists to almost prophetically announce the flywheel as a "panacea" against rising energy prices, proposing a revolution of mobile energy storage devices. In this "golden era" of flywheel energy storage, which reached its peak during the two abovementioned oil crises in 1973 and 1979, numerous popular science journals propagated sometimes questionable applications of this storage technology. Everything should be powered by flywheels: from the hand drill to the motorboat [3]. Nevertheless, even at that time, considerable energy-saving potentials of 25% and more were demonstrated by drivetrain concepts incorporating flywheel energy storage. A successful example is the *VW T2 Hybrid* shown in Fig. 1.1, which was developed by the *Institute of Automotive Engineering and Piston Engines* (*Institut für Kraftfahrwesen und Kolbenmaschinen*) at *RWTH Aachen University* in Germany.

But what scientific goals and technical properties must be achieved in order to really help this technology achieve a final breakthrough and actual market success? Which application is the most promising and rewarding one? Does the flywheel energy storage system find its home in racing, where in 2008 renowned racing teams already achieved a number of successes in the *Formula 1* and *Les Mans* classes, as shown in Figs. 1.2 and 1.3? Will it become a luxurious add-on for high-powered luxury class SUVs, hypocritically putting the "green coat of brake energy recuperation" on the shoulders of the wealthy

Fig. 1.2 Les Mans racing car with flywheel energy storage system by *PUNCH Flybrid* in 2011. The associated flywheel module is shown in Fig. 1.3. (Image rights: PUNCH Flybrid)

Fig. 1.3 Mechanical flywheel energy storage module for a Les Mans racing car by *PUNCH Flybrid*, which was used as a kinetic energy recovery system (KERS). (Image rights: PUNCH Flybrid)

Fig. 1.4 The legendary *MFO Gyrobus*. Chassis of the bus during assembly (*left*) and charging the flywheel energy storage system via pantographs (*right*). (Image rights: Historical Archive ABB Switzerland, N.3.1.53232 and N.3.1.54566)

buyer? Or will the much-cited *Gyrobus* of *Maschinenfabrik Oerlikon* (MFO) from 1955 (see Fig. 1.4), which has recently experienced a renaissance in the form of a modernized edition by the *PUNCH Flybrid* company, remain the prototype of an ideal application?

In order to answer all these questions, an isolated technical view of flywheel energy storage is not sufficient in itself, which is why this book takes a holistic approach that goes far beyond the optimization of a few sub-components of the actual technological device.

1.1 Structure of this Book

As shown in Fig. 1.5, the reader's view will expand from the flywheel energy storage system per se to an analysis of the *supersystem*, which attempts to examine the complex relationships between the energy storage system, the vehicle, and the environment and consequently leads to the determination of desirable specifications and target properties of

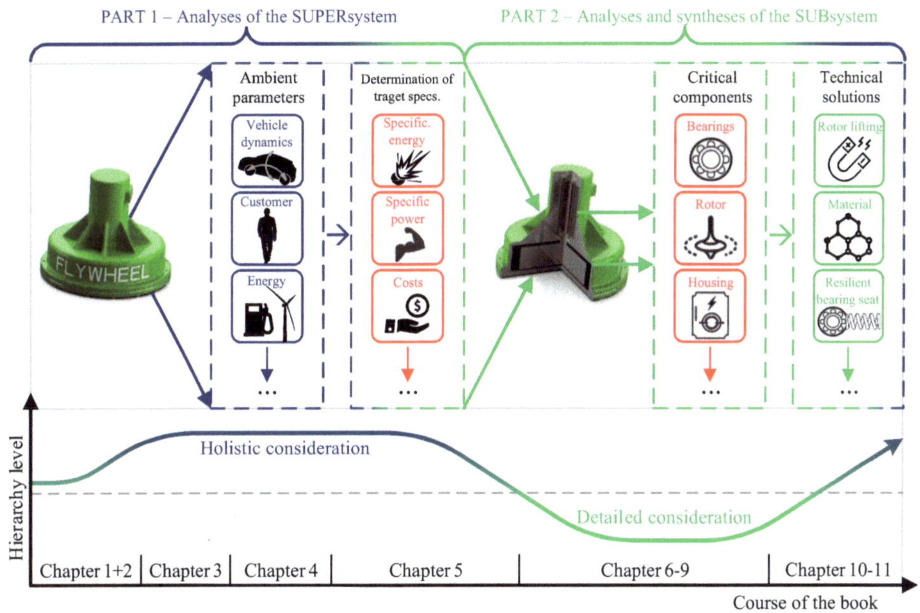

Fig. 1.5 Structure and course of the book from the reader's perspective

the storage system itself. The reader is then led from the top hierarchical level down to the *subsystem*, to highly detailed considerations where critical components and their system-internal interdependencies are identified. The scientific core of the book is the empirical investigation and detailed description of practical solutions in the field of rotor design, bearing concept, and burst containment.

In order to ensure high practical relevance, this book often contains application examples. Particularly important key statements are highlighted and allow the reader to grasp the essential aspects of the respective chapter, even when just browsing the pages. Pictograms are often used in tables for a better overview.

All graphics, which do not have an explicit source reference, were created by the author himself.

1.2 Motivation for a Holistic Assessment of the System "Energy Storage—Vehicle—Environment"

As stressed in the introduction, our mobility behavior, which (as will be described in more detail in Sect. 2.1) is mainly based on road transport, has profound economic, social, and political implications. Similarly, the issue of transport-related energy consumption is not an isolated technical one, as the efficiency of mobility depends on many factors, may they be personal or environmental. This system dependency of efficiency applies equally to a holistic mobility concept, which can be composed of different means of transport, as it

Fig. 1.6 Russian pin with representation of the "Kulibin tricycle," the first flywheel-powered vehicle from 1791

does to the vehicle itself. Even a highly optimized internal combustion engine cannot compensate for the deficits of poor driving style of the driver. In some cases, the energy-saving potential through driver training may even be many times higher than what can be achieved through advanced engine technology measures. No technical background knowledge is required to understand that doubling the number of vehicle occupants also doubles the efficiency of passenger transport—an increase in efficiency that no currently available technical measure can achieve.

But the mere availability of even the most efficient technologies is not enough. For example, despite its proven high efficiency, the heat pump for domestic heating, which has been available for more than 100 years, could not have achieved economic success without appropriate marketing [6].

The idea of using a flywheel to store energy in a vehicle was first documented in 1791. A tricycle, developed by the Russian engineer *Ivan Petrovich Kulibin* (1735–1818), was driven by muscle power fed energy into a steel flywheel when descending and decelerating and released it again when accelerating (compare Fig. 1.6). The maximum stored energy of the flywheel made it possible to cover a horizontal distance of about 400 m autonomously [8]. The principle was therefore already known at least since this time, but for many years, no attention was paid to it. Since then, the effectiveness of the system has been demonstrated by several prototypes and small series. Simple, forged steel flywheels were transformed into high-performance energy recovery systems for race car applications. Figure 1.7 shows a carbon fiber rotor from *Ricardo PLC* during dynamic balancing.

Table 1.1 shows a selection of more than 50 vehicles that had flywheel energy storage implemented so far. Considering the large number of developments, the reader is justified

Fig. 1.7 Modern flywheels combine high-tech materials with precision mechanical engineering: dynamic fine balancing of a flywheel rotor from *Ricardo PLC*. (Image rights: Ricardo PLC)

in asking why these vehicles have not long since left their mark on our roads. However, the social and economic-political relations mentioned above affect flywheel hybrid vehicles just as much as all other aspects of mobility. External parameters such as legislation, global market situation, road network and infrastructure, energy prices, and the complex socio-psychological behavior of the end customer determine the profitability of flywheel energy storage systems in hybrid vehicles. A detailed investigation of these aspects in a historical context was published by the author in [8].

To give an example of a rail vehicle successfully incorporating flywheel energy storage, Fig. 1.8 shows the *Parry People Movers* (PPM) *Class 139 Railcar*, which is in service in Stourbridge, England. Figure 1.9 shows the corresponding chassis of the vehicle, with the flywheel unit in the center clearly visible.

The durability and longevity of the flywheel energy storage technology were impressively demonstrated by the *MFO Gyrolokomotive*, among others. The rail vehicle, designed for use in the Gonzen iron mine in Switzerland, was in service underground for many years and has been used for visitor trains and guided tours since 1994—still in perfect working condition. Figure 1.10 shows the "Gyrolok" in a 1956 photograph taken by Maschinenfabrik Oerlikon.

In the course of a techno-economic assessment of the *Advanced Technology Transit Bus* (ATTB) (see Fig. 1.11 or also Sect. 2.2.2.2), Larry Hawkins, founder and director of technology of the company *Calnetix* wrote in an e-mail correspondence with the author:

> "The program [of the ATTB] was a technical success, but didn't get follow-on funding to commercialize."

Table 1.1 Overview of flywheel vehicles in the commercial vehicle, personal car, and rail vehicle sectors

	Year	Name	Manufacturer/developer	Country	Energy content of flywheel *kWh*	Max. rot. speed *RPM*	Rotor mass *kg*
Commercial vehicles	1953	Gyrobus	Oerlikon machine factory	CH	9.15	3000	1500.0
	1981	M.A.N. test bus	M.A.N.	D	1.50	12,000	104.0
	1985	New York bus system	Garrett Corp.	USA	16.00	16,000	340.0
	1988	Munich city buses	MAN/Neoplan/Magnent engine	D	2 × 2.75	11,000	181.0
	2002	ATTB	Center for Electromechanics	USA	2.0	40,000	59.0
	2006	AutoTram	Fraunhofer Institute	D	4.00	23,000	300.0
	2012	GKN	GKN Hybrid Power	AR	0.5	36,000	55
Road cars	1792	Kulibin tricycle	Lt. I.P. Kulibin	RUS	0.011	500	50.0
	1978	Garrett 4 passenger sedan	Garrett Corp.	USA	1.0	25,000	22.7
	1993	Chrysler Patriot	Chrysler Motors	USA	1.0	58,000	60.0
	1996	Hybrid III	ETH Zurich	CH	0.070	6000	48.0
	2000	Zero Inertia - VW Bora	TU Eindhoven/Van Doorne	NL	0.040	8000	12.2
	2009	Porsche 911 GT3 R Hybrid	Porsche	D	0.20	40,000	14.0
	2010	Jaguar XF	Torotrak/Xtrac CVT	AR	0.120	60,000	5.0
Rail vehicles	1860	Schuberski locomotive	Lieutenant Z. Schuberski	RUS	31.670	–	5000
	1950	Gyro Tractor	Oerlikon machine factory	CH	–	–	–
	1954	Gyro Lok	Oerlikon machine factory	CH	5.6	3000	1530
	1974	New York Subway	Garrett Corp.	USA	1.6	14,000	4 × 68
	1975	Advanced concept train	Boeing Vertol	USA	4.500	11,000	–
	1992	PPM Class 139 Railcar	Parry People Movers	AR	3.750	2600	720
	2001	ULEV TAP I	CCM	NL	4.000	1500	–
	2004	Lirex MDS K5	Alstom/magnet motor	D	2 × 2	12,000	600
	2006	Lirex MDS K6	Alstom/WTZ Rosslau	D	2 × 6	25,000	–

Fig. 1.8 *Parry People Moves Class 139 Railcar* with flywheel energy storage in winter operation in Stourbridge, England. (Image rights: Parry People Movers Ltd)

Fig. 1.9 Chassis of the *PPM Class 139 Railcar* and its flywheel energy storage system. (Image rights: Parry People Movers Ltd)

Fig. 1.10 A proof of the unbeatable service life of flywheel energy storage systems: The *MFO gyro locomotive*, here on a photo taken in 1956 at the Gonzenwerk Sargans, Switzerland, is still in operation today. (Image rights: Historical Archive ABB Switzerland N.3.1.64734)

Fig. 1.11 The *Northrop Grumman Advanced Technology Transit Bus* with flywheel energy storage [4, 5]. (Image rights: Center for Electromechanics, University of Texas)

So what is the point of further isolated technological development of the flywheel energy storage technology if even solutions that have been available for years and have been proven to work have not been able to establish themselves so far?

The *ATTB* is only one of many technically successful concepts, which were never produced in series. But why is that so? Which characteristics of the flywheel technology have to be improved to ensure a successful market entry? Which technological solutions can serve in order to achieve this? This book attempts to provide answers to these and other

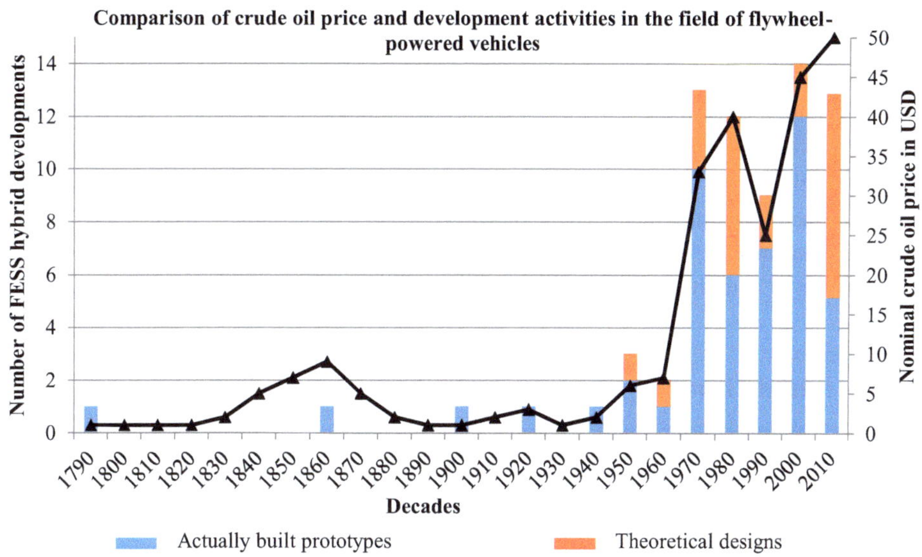

Fig. 1.12 Comparison of crude oil price and development activities in the field of flywheel-powered vehicles [8]

questions arising in the course of research and development work on the subject of flywheel energy storage.

But one thing seems to be clear: Even if the advantages of FESS technology in terms of reducing CO_2 emissions were less than initial results suggest, in the end, it is always about economic profit. And the situation was no different 30 years ago. *Major Richard Cope*, of the *Advanced Research Projects Agency* (USA), said in a 1994 interview about flywheel energy storage [7]:

> "The vision, the technology and the payoff are all clear. But three problems stand in the way: Costs, costs and costs!"

A good example of the circumstance that economics and politics—i.e., aspects of *the flywheel energy storage supersystem*—influence the profitability and thus also the research activity of this technology sector is illustrated in Fig. 1.12. The diagram shows an amazingly strong correlation between the nominal crude oil price and the number of developments in the field of flywheel-powered (hybrid) vehicles.

Therefore, when talking about the "optimization of a flywheel energy storage systems for automotive use," these *external factors*, some of which have a much greater impact than all the possible measures available in the technical field, cannot be ignored.

However, complex interdependencies also occur within the *subsystem* of the energy storage system, which can only be adequately captured by a holistic view of the energy storage system itself (see Sect. 6.2). As Chap. 9 will show in more detail, the bearing system is one of the most critical components in FESS. However, determining the load spectrum for bearing design inevitably involves an examination of external influences such as forces, accelerations, etc., i.e., an examination of the *supersystem*.

The traditional technological problem-solving approach, which does not go beyond determining the direct bearing load (or component load in general), must be expanded and applied to all system components of the FESS. In doing so, external influences that cannot be determined and quantified strictly physically—such as economic factors, trends, and customer psychology—must also be taken into account and analyzed in the context of system-internal and cross-system interdependencies.

1.3 Initial Situation: Europe in the Energy Revolution

Central Europe's economically and politically motivated striving for independence from oil as an imported commodity, propagated by public figures such as Angela Merkel, has led to a boom in renewable, predominantly *volatile energy sources* such as wind and solar energy. Here, one of the main challenges lies in the smoothing of supply and demand through the storage of (predominantly electrical) energy. It is not without reason that the lack of storability is described as one of the "seven paradigms of the energy industry" [9]. In addition, in order to keep in particular CO_2 emissions within reasonable limits, it is necessary to further reduce primary energy consumption. The political efforts of the global powers USA, Russia, and China to gain access to the valuable resource oil have shaped the history of modern times in a sometimes dramatic way. Wars have been waged, leading powers overthrown and borders redrawn for the sake of conquering fossil fuels. Thousands of kilometers of pipelines, supertankers, and multibillion dollar corporate businesses enable the upper middle classes of the Western industrialized countries to refuel their vehicles at reasonable prices. For the people of many oil-rich countries, especially non-OPEC countries, the exploitation of their natural resources results in abject poverty [1]. It must therefore be regarded as the moral duty of the modern automotive engineer to work on reducing the import of oil, or fossil fuels in general.

Taking a look at Fig. 1.13, the reader can see that *mobility* and *building heating* represent the highest energy consumption and therefore also offer the greatest potential for savings. The CO_2 emissions of the different sectors in the EU 27 are shown in Fig. 1.14. However, low-emission limits can be achieved neither by political regulation alone nor by simply increasing the efficiency of vehicles, but depend above all on the awareness and willingness of each participant [10]. However, intensive cooperation between politicians, legislators, industry, and end users can only take place if there is a *coherent understanding of the term energy*. Surveys conducted by the *Institute of Electricity Economics and Energy Innovation* (*Institut für Elektrizitätswirtschaft und Energieinnovation*) at *Graz University of Technology* have shown that energy is a physical quantity that is extremely difficult to perceive and assess by us humans [6]. This energy-psychological circumstance, which is

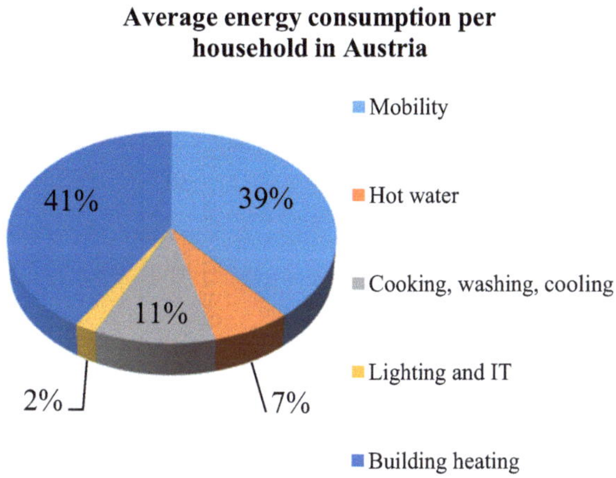

Fig. 1.13 Average energy consumption per household in Austria, Europe, including mobility in 2012 [11]

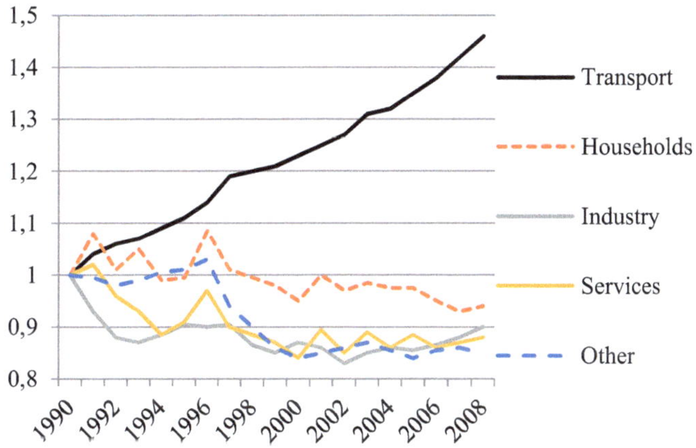

Fig. 1.14 Relative CO_2 emissions of the various sectors in the EU 27 standardized to the 1990 consumption value [12]

explained in more detail in Sect. 3.5, inevitably leads to partial misconduct of the population in dealing with our precious energy resources.

1.4 The Role of the Transport Sector

Nowhere else does the difficulty of storing energy have such a striking effect as in mobile applications. This applies not only to portable devices, but above all to vehicles. The key to vehicle hybridization, electrification, or the *zero-emission vehicle* in general, therefore, lies not, as a layman might assume, in the improvement of the electric traction motor, but in the development of efficient, mobile energy storage systems.

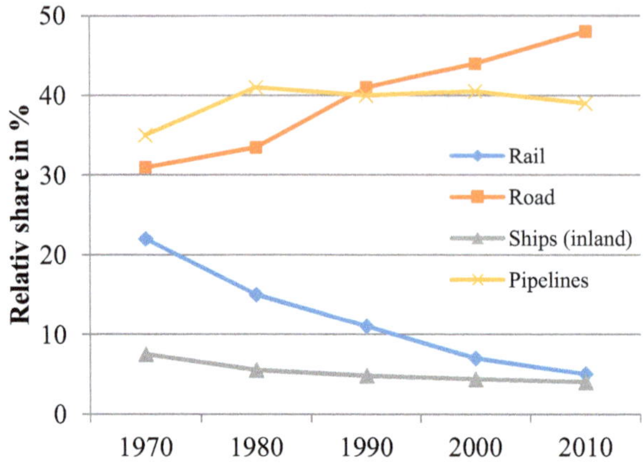

Fig. 1.15 Shares of energy consumption by transport sector in the EU. (data from [10] and [14])

However, the topic of "sustainable mobility" has long since ceased to be a purely technological one. It has rather become a question of the *interaction of external parameters* such as politics, marketing, economy, and also the psychology of the end customer. If the efficiency of mobility were to be viewed strictly from a techno-energetic point of view, a shift of commercial road vehicle to rail transport would appear to be an obvious first solution [13]. Theoretically, the existing rail infrastructure could be used, and the share of pure "e-mobility" in transport could be multiplied at a stroke. (This applies at least to countries with a high share of electrified rail tracks). In addition, the fact that rail vehicles in Central Europe are predominantly grid-connected means that electricity generated from renewable energies can be used directly and recuperated braking energy (e.g., in pumped-storage power stations) can be stored for later use. Paradoxically, the share of rail transport in Europe in the mobility mix has consistently declined since the 1970s. In 1970, the share of energy required for the transport of passengers and goods by rail was 22%, and in 2010, it was only about 5%; it has thus fallen to about 1/4 of its original value (Fig. 1.15). In addition, the transport sector—compared to the other energy consumers in the EU—is the one showing the strongest growth (see Fig. 1.16).

However, the share of electric vehicles in registration number remains modest in absolute terms. In 2010, only 0.07% of vehicles in the EU average were electrically powered, with Norway leading the way with 1.23% [16]. In 2015, 0.15% of European cars were electric [17]. Although the share of EVs in new registrations is increasing and in 2017 there were 882,000 electric and hybrid vehicles on Europe's roads [18], this percentage is also rather low at around 0.29% in comparison with the almost 300 million total cars in the EU.

Three main statements can be derived from this:

1. It is obvious that the development of sustainable technologies alone is not enough, since some of them already exist (e.g., rail transport), but are not or not sufficiently used.

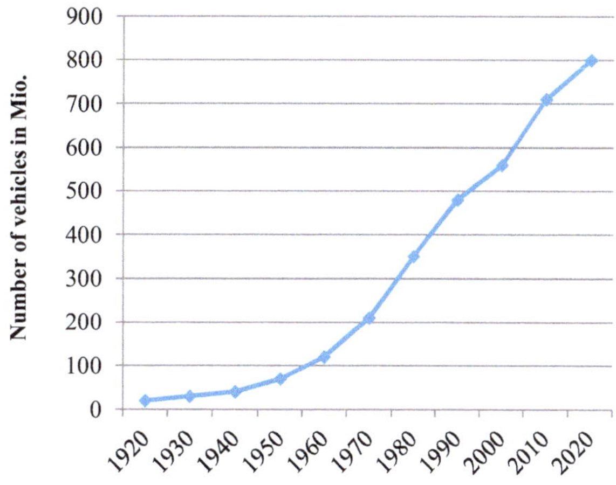

Fig. 1.16 Progress of the global vehicle population, based on data from [15]

2. If we do not want to rely on an intrinsic change in the awareness of the population, technologies must be developed, which can take over *sustainable actions* for people without noticeably restricting their freedom or comfort.
3. Economic motives alone are not enough to increase the use and acceptance of efficient vehicles and sustainable mobility concepts. In addition to political and legislative measures, customer benefits above all must be maximized by adding value (time savings, comfort, image, etc.).

1.5 The Future of Transportation

The scenarios described in Sects. 1.2 and 1.3 illustrate that it is not without reason that words such as "energy revolution," "electrification," and "sustainability" are increasingly appearing on the front pages of daily newspapers. While heat insulation, combined heat and power generation, and alternative heating systems are gaining ground in the building sector, the automotive industry *does not* seem to be able or willing to agree on a *uniform solution*. The former common denominator and source of hope for the major vehicle corporations, electromobility, seemed to have been in a crisis for some time. As so often in a product cycle, the initial euphoria was followed by a "valley of disappointments." This reflects not only the limited sales figures (as already described in Sect. 1.3) and the lack of customer satisfaction[1] in some cases with battery electric vehicles but also the unisonous echo from vehicle engineers, for example, at the *VDI Congress for Innovative Vehicle Drives 2012*. At this event, *Prof. Günter Hohenberg* described the e-mobility hype as "having reached a phase of disillusionment" [19]. Although battery electric vehicles seem

[1] Problems such as the significantly reduced range of EVs during winter, long charging times, and the warranty and disposal issue of the batteries could not be completely solved to date.

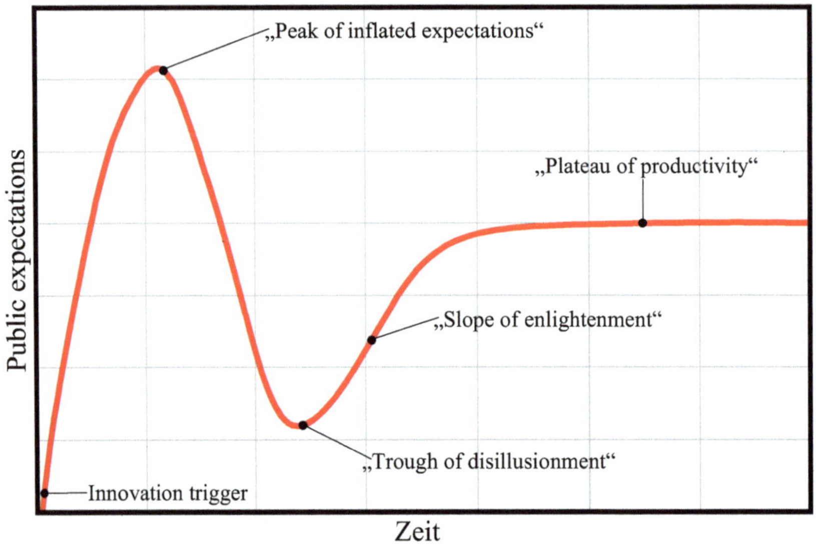

Fig. 1.17 So-called hype cycle, which shows the phases of public attention when a new technology is introduced

to be establishing themselves more and more in the passenger car sector, the commercial vehicle sector is still struggling with the insufficient range of EVs and is looking for alternatives such as hydrogen. At higher penetration rates, however, aspects of resource procurement (lithium), excessive loads on the electricity grid (see Chap. 10), and life cycle assessment (LCA) issues are also having significant impact. In this context, it must be mentioned that approximately 300 kWh of energy are required to produce a battery with a storage capacity of 1 kWh (Fig. 1.17).

So what will the future of modern transportation look like? How does the automotive industry see it and what is the customer's take on it? Will new battery concepts such as the zinc-air battery help the electric vehicle to rapidly gain lasting popularity, or will a hybrid vehicle concept win the race? And if so, which one? *Mild*, *Micro*, or *Full*? Or is *Daimler* right, and the fuel cell will prevail? If the current results of economic transformation research are to be believed, sustainable technologies are not enough. What is needed is a completely new mobility concept based on so-called voluntary simplicity [20].

The different approaches and objectives of the automotive giants (e.g., Volkswagen group, General Motors, etc.) and the efforts of each corporation to establish their own standard[2] result in highly divergent and isolated research and development activities in the field of low-emission and zero-emission mobility. This explains why some technologies,

[2]Discussions between car manufacturers and politicians concerning standard plugs for charging electric vehicles and a uniform battery system that allows battery swapping seem to have been fruitless so far, but are still ongoing.

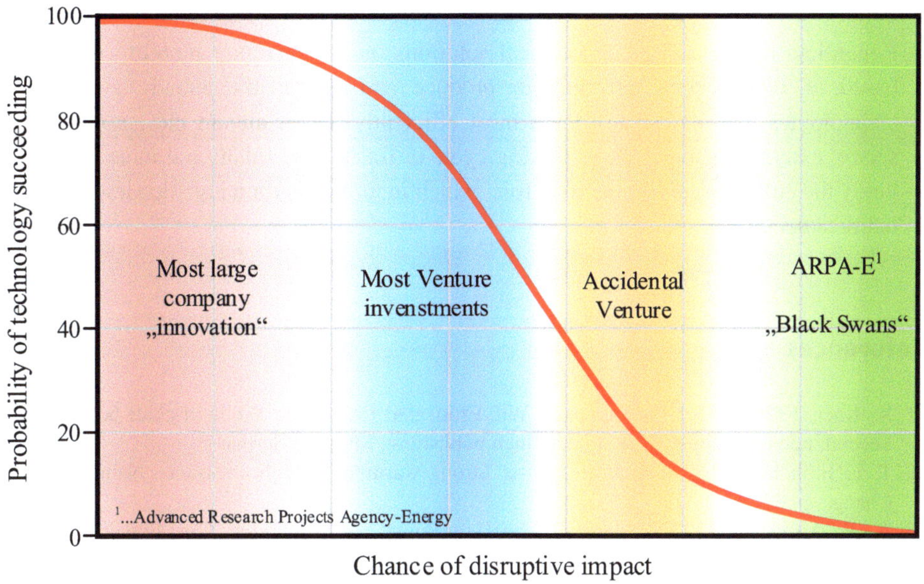

Fig. 1.18 Probability of realization and implementation of innovation content versus impact of new technologies [22]. (Image rights: Khosla Ventures)

such as the flywheels or compressed air storage, have been condemned to exist in the shadows to date. In contrast, the idea that the automotive industry is pursuing a common goal "with joined forces" sounds tempting, but carries the danger of overlooking truly revolutionary and disruptive technologies. The well-known financial mathematician and philosopher *Nassim Nicholas Taleb* describes rare but truly groundbreaking innovations as *black swans* [21]. Based on Taleb's works, the US businessman and major investor in environmental technology *Vinod Khosla* wrote the *Black Swan Thesis of Energy Transformation* [22]. Khosla believes that small, incremental improvements are not the way to achieve monumental, life-changing advances in energy and mobility technology. The highly innovative "black swans," on the other hand, have a low probability of successful technological implementation, as shown in Fig. 1.18.

If, however, one of these unlikely *black swan approaches* were to be implemented and achieve market success, this would be a technological quantum leap.

A gradual, incremental increase in efficiency, especially in energy production, can even lead to increased consumption by the population due to the price reduction often associated with it. If vehicle operating costs fall, more people can afford a car and the CO_2 might increase despite the lower specific consumption. This phenomenon is known as the *rebound effect* [23].

Regardless of the exact technological solutions on which future mobility will be based, it is important not only to produce and master the already available technologies, but also to use them most efficiently. Especially in the automotive sector, external factors such as driving cycle or traffic, and finally but most importantly the customer, play a decisive role, which in turn calls for a holistic view of the entire *supersystem*.

References

1. K. Kneissl (2006) Der Energiepoker: Wie Erdöl und Erdgas die Weltwirtschaft beeinflussen. Second, updated edition 2008. FinanzBuch publishing, Munich, Germany.
2. T. L. Frölicher (2016) Climate response: Strong warming at high emissions. Nature Climate Change, p. 823–824.
3. A. P. Armagnac (1970) Super Flywheel to Power Zero-Emission Car. Popular Science, pp. 41–43, Issue August 1970.
4. BMP Center of Excellence / Northrop Grumman (1997) Advanced Technology Transit Bus. Northrop Grumman, Military Aircraft Systems Division, El Segundo, California, USA. http://www.bmpcoe.org/bestpractices/internal/north/north_22.html. [Accessed July 2nd 2011].
5. R.J. Hayes, J.P. Kajs, R.C. Thompson and J.H. Beno (1999) Design and Testing of a Flywheel Battery for a Transit Bus. SAE International Congress and Exposition, Detroit, Michigan, USA.
6. H. Stiegler and U. Bachhiesl (2013) Grundlagen der Energieinnovation. Graz University of Technology, Austria.
7. M. DiChristina (1994) Emerging Technologies for the Supercar. Popular Science, p. 99, Issue June 1994.
8. A. Buchroithner and M. Bader (2011) History and development trends of flywheel-powered vehicles as part of a systematic concept analysis. European Electric Vehicle Congress (EEVC), November 2011, Brussels, Belgium.
9. H. Stiegler (1999) Rahmen, Methoden und Instrumente für die Energieplanung in der neuen Wirtschaftsorganisation der Elektrizitätswirtschaft. Graz University of Technology, Austria.
10. P. L. Schiller, E. C. Brunn and J. R. Kenworthy (2010) An Introduction to Sustainable Transportation – Policy, Planning and Implementation. EARTHSCAN, Washington DC, USA
11. W. Pölz (2001) Kohlendioxid-Reduktionspotentiale der Klimabündnisgemeinde Mistelbach, Institut für Land-, Umwelt- und Energietechnik der Universität für Bodenkultur, Vienna, Austria.
12. C. Sessa and R. Enei (2010) EU transport demand: Trends and drivers. European Commission (ISIS).
13. J. Pluy (2012) Energieeffiziente und kostengünstige Elektromobilität mit der Bahn. EnInnov – 12. Symposium Energy Innovation 2012, Graz, Austria.
14. European Commission (2012) Transport in Figures – Statistical Pocketbook 2012. Publications Office of the European Union, Luxembourg.
15. M. Fish (2006) Where Global Warming Comes From. http://www.globaltrees.co.uk/facts_.php. [Accessed April 12th 2011].
16. J. Bates (2011) Incentives Fail to Stimulate European Electric Vehicle Sales. JATO Dynamics GmbH, Limburg, Germany.

17. European Environment Agency (2016) Electric vehicles in Europe. Publications Office of the European Union, Luxembourg.
18. A Tsakalidis and C. Thiel (2018) Electric vehicles in Europe from 2010 to 2017: is full-scale commercialisation beginning? JRC Science for Policy Report, EUR 29401 EN, European Commission.
19. G. Hohenberg et al (2012) Range Extended E-Mobility. VDI-Berichte 2183: 8. VDI-Tagung mit Fachausstellung – Innovative Fahrzeugantriebe, November 2012, pp. 129–143, Dresden, Germany.
20. S. Alexander and S. Ussher (2011) The Voluntary Simplicity Movement: A Multi-National Survey Analysis in Theoretical Context. Journal of Consumer Culture. https://doi.org/10.1177/1469540512444019.
21. N. N. Taleb (2008) The Black Swan: The Impact of the Highly Improbable. Random House Publishing Group, New York, USA.
22. V. Khosla (2011) The Black Swan Thesis of Energy Transformation. Khosla Ventures, Menlo Park, California, USA.
23. J. Jenkins, T. Nordhaus and M. Shellenberger (2011) Energy Emergence – Rebound & Backfire as Emergent Phenomena. Breakthrough Institute, Oakland, California, USA.

References

2 Complexity, Importance, and Overall System Dependency of the Vehicle Operating Strategy

2.1 A Holistic View: Vehicle, Driver, and Environment

In particular since the establishment of standardized emission test cycles, such as the *New European Driving Cycle* (*NEDC*), there is a widespread belief that the responsibility for reducing vehicles' fuel consumption lies primarily with vehicle developers and is, therefore, a purely technical issue. Optimization of the internal combustion engine (ICE), lightweight design, and "smart" energy management can undoubtedly contribute to reducing fuel consumption, but external influences such as the choice of vehicle itself[1] and its usage profile often hold much greater potential. Since in some cases it is not possible to precisely quantify and determine the most important influencing parameter, the vehicle must from now on be viewed as a holistic, system-dependent optimization problem. This also means that all the parties that participate in and hence define the *supersystem* of the vehicle are involved in an interdisciplinary process and must take responsibility for its joint development. Figure 2.1 shows the interaction between *hybrid vehicle*, *subsystem* (consisting of the essential technical components of the vehicle), and *supersystem*, which describes the external influences that have an impact on vehicle topology, operating strategy, and ultimately energy consumption. The actual *subsystem* of the energy storage system and the various hierarchical levels of the systematic approach are dealt with in Chap. 4.

[1] While in the past a continuous reduction of fuel consumption was observed in vehicle fleets, CO_2 emissions from new cars in Germany are rising again due to the increasing sales figures of SUVs despite more efficient engines [1].

A. Buchroithner, *Flywheel Energy Storage*,
https://doi.org/10.1007/978-3-658-35342-1_2

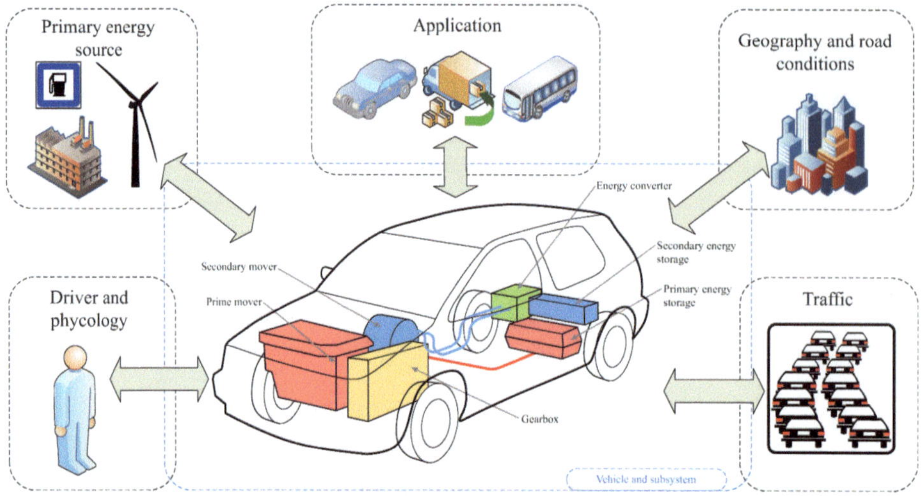

Fig. 2.1 Interaction between hybrid vehicle and its environment

2.2 *Subsystem* of Flywheel Energy Storage

If one simulates—as it is a common practice in today's modern development process—a *complete virtual vehicle* in order to determine the specifications of the energy storage system in the hybrid powertrain, the virtual vehicle "sees" only the terminals of an *idealized reference energy storage*. Thus, the desired behavior of this "black box" is defined externally, but in fact ignores internal processes. However, when it comes to the actual, technical realization of the energy storage device, these "inner processes" are of great importance, as they are responsible for the relevant properties appearing in the *supersystem*.

In the case of a chemical battery, these internal processes, that is, processes within the *subsystem* of the storage, are, for example, diffusion processes, which depend on the chemical structure of the battery and substantially influence fundamental properties such as power density, service life, etc. The concrete course of these processes, however, is again influenced by external parameters such as temperature. (The interdependencies between *sub*- and *supersystem* are described in greater detail in Sect. 6.2.) The processes, which take place in the subsystem of a flywheel storage device and are relevant for the energetic specifications, are primarily of mechanical and electrical nature. The components required for this and the typical structure of an FESS are briefly outlined in the following chapters.

2.2.1 Fundamentals of Kinetic Energy Storage

If a mass m moves on a straight path at a given speed v, its kinetic energy can be calculated:

$$E_k = \frac{1}{2} m v^2 \tag{2.1}$$

If the mass now moves on a circular track, its speed is v proportional to the angular velocity ω and the radius r, and can be expressed as

$$v = r * \omega \tag{2.2}$$

$$E_k = \frac{1}{2} m v^2 = \frac{1}{2} m (r\omega)^2 = mr^2 \ \frac{\omega^2}{2} \tag{2.3}$$

An analogy to longitudinal motion becomes immediately obvious. The angular velocity ω corresponds to the speed v and the expression mr^2 is equivalent to the mass m during movement on a straight path. It is called the moment of inertia I:

$$E_k = \frac{1}{2} I \omega^2 \tag{2.4}$$

For rotation around a spatially fixed axis, the moment of inertia is a scalar quantity:

$$\mathrm{I} = \int_0^{\mathrm{R}} 2\rho\pi h r^3 \ dr = 2\rho\pi h \frac{R^4}{4} \tag{2.5}$$

whereby the introduction and subsequent factoring out of the density ρ is only valid for homogeneous bodies with a constant density distribution. Since many flywheels can be approximated as a cylindrical disk, the moment of inertia for the entire cylinder also follows as:

$$\mathrm{I} = \int_0^{\mathrm{R}} 2\rho\pi h r^3 \ dr = 2\rho\pi h \frac{R^4}{4} \tag{2.6}$$

The mass of a cylinder with a constant density changes to $\rho\pi h R^2$, allowing the moment of inertia to be calculated:

$$I_{Zyl} = \frac{1}{2} m R^2 \tag{2.7}$$

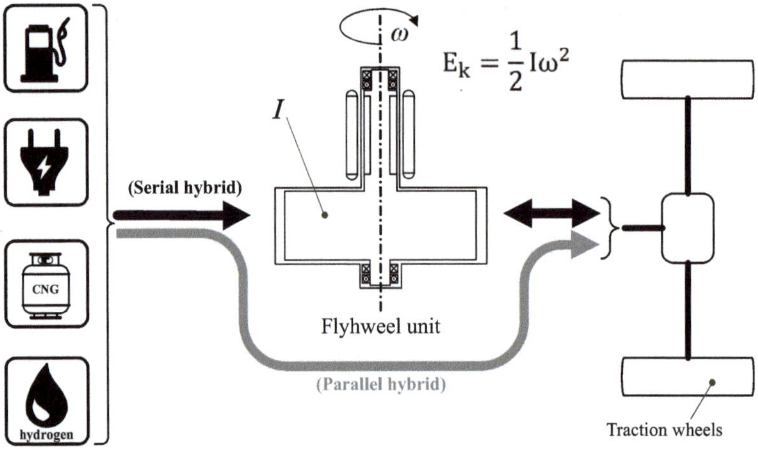

Fig. 2.2 Principle of a flywheel energy storage system in a hybrid vehicle

Since an increase in flywheel mass results in only a linear increase in energy content and the increase in radius is limited by design criteria, increasing the speed must be regarded as the most elegant method of achieving high-energy contents of FESS.

The energy stored in the flywheel can thus be precisely quantified by Eq. 2.4 and can be directed to the traction wheels of a vehicle in several different ways. In this regard, mechanical and electrical energy transfers are the most popular solutions. The basic equation of dynamic rotation allows a reversible energy conversion process, so power can also flow from the vehicle's wheels to the energy storage device. Furthermore, this vehicle topology can be combined with different primary energy and power sources, as shown schematically in Fig. 2.2.

In this book, the description of the history of the world's oldest energy storage principle is limited to mentioning the fact that pottery wheels were already being used in Mesopotamia more than 6000 years ago and the first flywheel-powered vehicle was documented in Russia in 1792. Further historical concepts are described in detail in [4]. A comprehensive analysis of historic mobile FESS applications in vehicles was carried out in 2011 in [5] and published in [6, 7].

2.2.2 Differentiation According to Transmission of the Stored Energy

As mentioned in Sect. 2.2.1, the kinetic energy stored in the flywheel can be transmitted in different ways to the traction wheels of the vehicle. Theoretically, even a compressed air turbine or a hydraulic motor could be used; however, for reasons of efficiency, only two

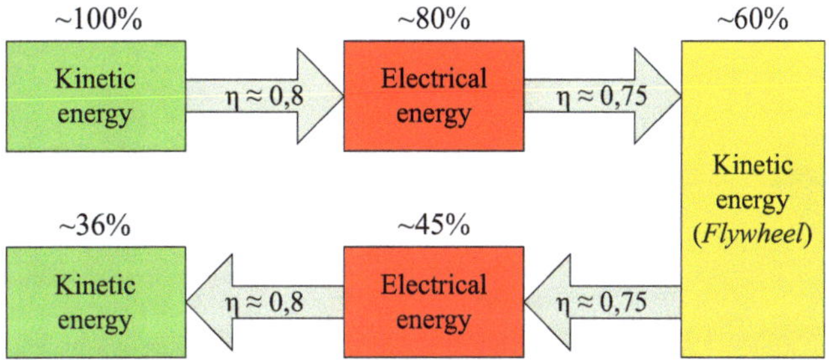

Fig. 2.3 Unfavorable efficiency chain of multiple energy conversion of an electromechanical FESS

concepts of energy transmission—namely, *electrical* and *mechanical*—have been predominantly used so far. A description of the two methods follows in Sects. 2.2.2.1 and 2.2.2.2.

2.2.2.1 Purely Mechanical FESS

Mechanical transmissions for power transmission and torque conversion in vehicles generally have relatively high efficiencies, easily reaching values above 90% [8]. If the energy stored in the flywheel is transmitted by means of mechanical elements, the advantage is that the energy always "remains mechanical," that is, does not have to be converted into another form of energy. This "shortening" of the efficiency chain (compare Fig. 2.3) allows theoretically high-overall efficiencies when storing and reusing energy (*round-trip efficiencies*).

However, the problem lies in the large transmission gear ratio required, since flywheels usually operate at very high speeds (10,000–60,000 rpm) and the traction wheel speed of passenger cars lies between 0 and about 600 rpm. The continuously variable transmission (CVT)[2] used should ideally have a spread of 100 or more, whereas in reality values of only around 8 are achieved. The remaining differential speed when engaging or disengaging the flywheel must be compensated by slip in the required clutches. The process of energy storage in the flywheel in the case of regenerative braking can be outlined as follows:

1. The clutch between the drive wheels and CVT to the flywheel is open during constant driving of the vehicle.
2. The braking process is initiated and the clutch to the CVT is closed while the CVT is in the lowest gear ratio. (Energy is lost in the form of heat until the synchronous speed of the clutch disks is reached.)
3. When the synchronous speed is reached and the clutch is no longer slipping, the CVT's gear ratio is shifted from the lowest to the highest value, which increases the flywheel speed and reduces the speed of the drive wheels.

---

[2]CVT: *continuously variable transmission*, a transmission with a continuous gear ratio and without discrete shift points. Often also called *variomatic*

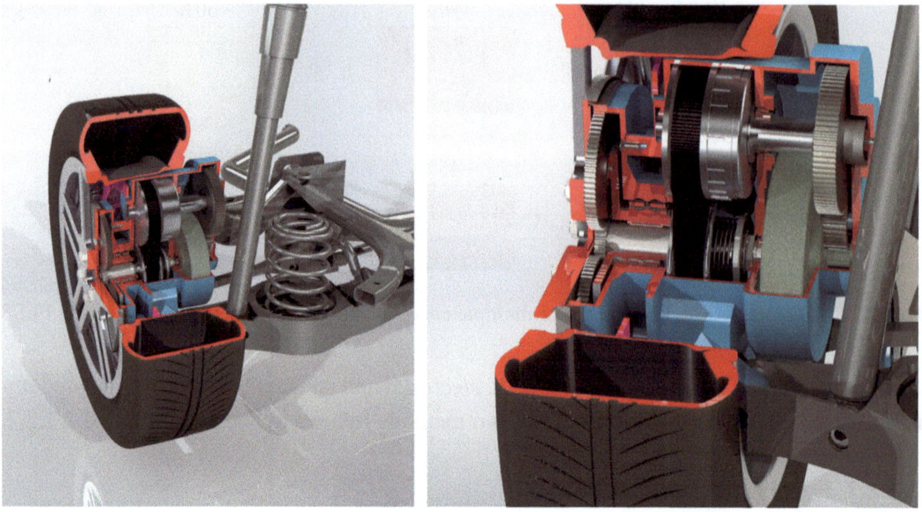

Fig. 2.4 Purely mechanical concept of a flywheel energy storage system (research project "HEuV" [9]). (Image rights: Advanced Mechatronic System Development KG (AMSD))

4. When the maximum transmission ratio of the CVT is reached, the clutch is disengaged again. The traction wheels can be brought to a complete stop by the vehicle's friction brake. The kinetic energy is now stored in the rotation of the mechanically decoupled flywheel.

In the course of the research project "Highly integrated energy storage systems for urban traffic (HEuV)" at Graz University of Technology (TU Graz), a feasibility study on a purely mechanical FESS was carried out. The designed concept, which tries to use the space available inside the rim of the car to allow recuperation of braking energy in urban traffic, is shown in Fig. 2.4.

Volvo V40 Flywheel Hybrid

This is a concept, which elegantly solves the problem of the large required gear ratio by using a so-called *geared neutral* architecture and has already been successfully implemented in practice as shown in Fig. 2.5. It is a test vehicle, type *Volvo S40* which was converted to a flywheel hybrid by the *Eindhoven University of Technology (Technische Universität Eindhoven—TU Eindhoven)* in the Netherlands. The concept is based on purely mechanical energy transfer, whereby the flywheel made of a fiberglass composite is connected to a CVT. In this case a *Van Doorne push belt* is used, with the CVT operating in a so-called i^2 mode. Four hydraulically operated multi-plate clutches allow a variety of operating modes, such as conventional CVT mode with ICE at its most efficient operating point, or local emission-free operation, in which the ICE is switched off and the energy is taken only from the flywheel [10].

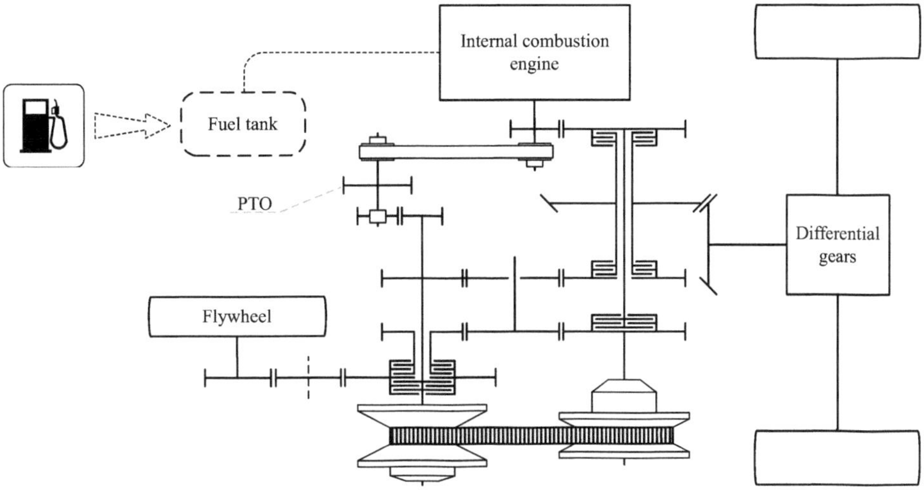

Fig. 2.5 Schematics of the *Volvo S40* flywheel hybrid of the *TU Eindhoven* of 1996, based on [10]. (Image rights: VDI)

Although this purely mechanical system achieves a high degree of efficiency (also in regenerative braking), the concept encounters problems when the ICE is operated in an on-off mode, as the auxiliary units such as the brake booster, servo pump, and oil pumps can no longer be supplied with the mechanical energy of the crankshaft. Mechanical decoupling of the auxiliary units and their electrification is currently being pushed in the vehicle industry, also for reasons of reduced fuel consumption.

VW T2 Flywheel Hybrid

A solution combining a mechanical coupling of the FESS with a hybrid electric powertrain was developed in the mid-1970s at the *Institute for Automotive Engineering and Piston Engines (Institut für Kraftfahrwesen und Kolbenmaschinen)* at the *RWTH Aachen* in Germany and led to the so called VW T2 Hybrid (see Fig. 2.6).

The hybrid powertrain, consisting of an internal combustion engine, an energy storage flywheel, and an electric motor, had a power-split architecture, an approach later used in one of the best-selling hybrid vehicles, the *Toyota Prius*.

The concept, which at that time was still designed based on the calculation results of analog computers, was tested on the chassis dynamometer as well as in real traffic and driving operation. In addition to the confirmation that the hybrid drive can indeed achieve substantial fuel savings (~ 25% in city traffic), some weaknesses in the area of driving comfort and manual vehicle control were revealed.

When designing the drive unit, attention was not only paid to energy aspects, but an attempt was also made to keep adaptations to the series vehicle as low as possible. In other

Fig. 2.6 The test vehicle *VW T2 flywheel hybrid* during test bench investigations on the *RWTH Aachen University* in Germany. *(Image rights: Institute of Automotive Engineering and Piston Engines, RWTH Aachen University)*

words, a compact design was pursued in order to make do with the existing space. Figure 2.7 shows the drive unit of the prototype.

The flywheel, which for reasons of simplicity was made of steel and not of a fiber composite material, is connected to the internal combustion engine via an angular gear with a constant ratio, which means that its speed is proportional to the engine shaft speed, and therefore causes a strong phlegmatization of the ICE.

In order to keep the inertial forces and thus the loads on the flywheel bearings low, an extremely well-balanced and smooth running internal combustion engine had to be used. The choice was therefore first made in favor of a Wankel engine, which was later replaced by a four-stroke engine from Fiat due to the Wankel's low-power output and poor efficiency. An essential characteristic of the design is that the power of the ICE, the electric motor, and the flywheel are superimposed in an epicyclical gearbox (planetary gear). Since the flywheel is connected in parallel to the ICE with a fixed transmission ratio, the torques of the respective power sources are in a ratio determined by the planetary gear, as can be seen in Fig. 2.8.

The electric motor therefore determines the output torque and output power, and is essentially controlled by the accelerator pedal. Figure 2.9 shows an abstraction of the entire drive system.

Fig. 2.7 Image of the assembled drive unit of the *VW T2 flywheel hybrid. (Image rights: Institute of Automotive Engineering and Piston Engines, RWTH Aachen*, Germany*)*

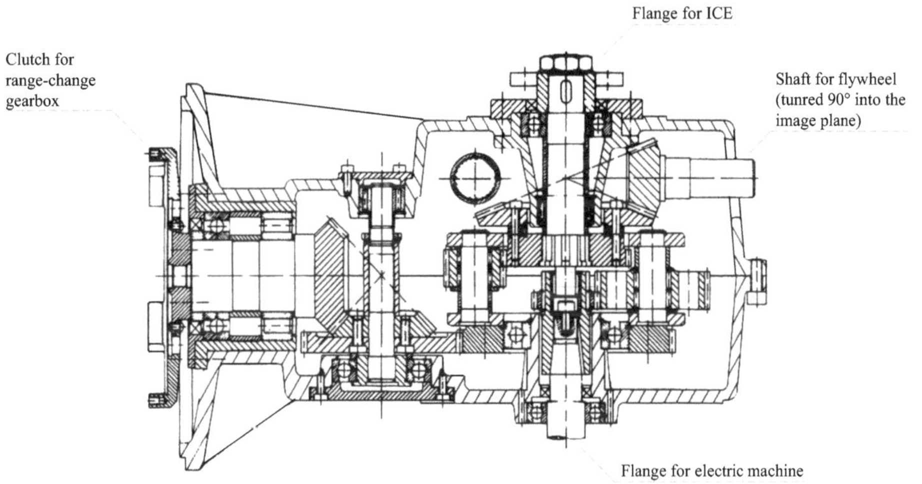

Fig. 2.8 Sectional view of the planetary gears of the *VW T2* flywheel hybrid by RWTH Aachen, Germany. (Image rights: Institute of Automotive Engineering and Piston Engines, RWTH Aachen University)

A list of the advantages and disadvantages of *purely mechanical* flywheel energy storage concepts is listed in Table. 2.1.

2.2.2.2 Electromechanical FESS

Due to the disadvantages of mechanical FESS mentioned in Table 2.1 and the rapid development of electric/electronic components as well as the trend toward increasing

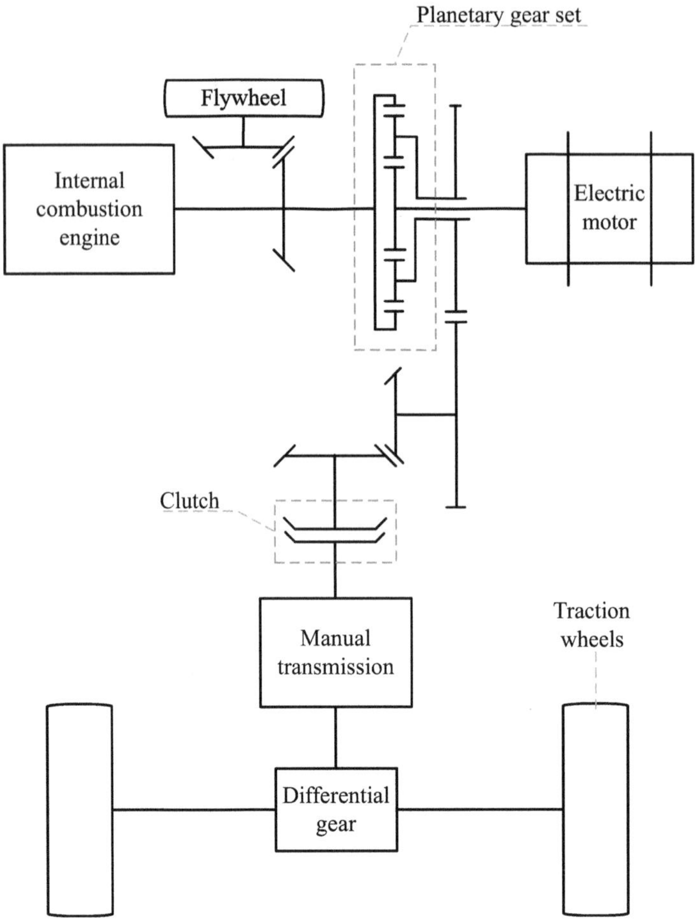

Fig. 2.9 Basic sketch of the VW T2 of the RWTH Aachen University, Germany. (Image rights: Institute for Automotive Engineering and Piston Engines, RWTH Aachen University)

electrification of the powertrain, this book focuses primarily on electromechanical flywheel energy storage systems, as these also have the greatest potential for future improvement. The ever-increasing demand for electrical energy in the vehicle[3], which is caused by—among many reasons—the increased electrical auxiliary units such as servo pumps, air-conditioning compressors, or even turbochargers, also speaks in favor of electromechanical FESS in vehicles. (Furthermore, most stationary and grid-connected FESS applications also require electrical energy.)

Typically, electromechanical FESS in vehicles are not used as the sole (primary) energy storage device due to their relatively low-specific energy, but in *parallel* or *hybrid serial*

[3]This is also confirmed by the introduction of a 48 V on-board network for passenger cars.

Table 2.1 Main advantages and disadvantages of *purely mechanical flywheel energy storage* for vehicles

Advantages		Disadvantages	
Designation	Description	Designation	Description
High efficiency	Kinetic energy is transferred/stored in mechanical form and not converted several times	Rigidity and spatial constraints	Energy transmission via shafts and rigid machine elements requires close proximity to *power takeoff* (PTO), that is, vehicle transmission
Thermal behavior	Heat losses that may occur in clutches or transmission components are easy to dissipate	Self-discharge	A vacuum feedthrough for the shaft is required, or high windage losses occur if the flywheel is operated at ambient pressure
High-power density	Maximum power is limited only by the maximum transmittable torque of the components	Wear	Slippage in clutches and CVT belt drives results in mechanical wear, which limits the service life of the system
No aging of power electronics	The service life of electronic components (MOSFETs, IGBTs) in electrical FESS is highly dependent on environmental conditions. Predictions are difficult [11]	Low Potential for improvement	The cost reduction potential and efficiency of gear components relevant to FESS are almost exhausted. A much more rapid development can be observed in the area of electric/electronic components
Low-cost, low-performance system possible	If efficiency plays only a minor role, a system can be designed consisting of only one clutch and one flywheel rotor	Limited response and control times	Electromechanical FESS can respond within milliseconds, whereas the reaction speed of mechanical systems is limited by inertia

arrangements (see also Sect. 3.1). The systematic design of the successful test vehicle *Advanced Technology Transit Bus (ATTB)* is shown in Fig. 2.10.

The *ATTB* was developed by *Northrop Grumman* and the *Center for Electromechanics* at the *University of Texas*. The vehicle features purely electrical energy transfer in the powertrain. The basic platform, a 12-m-long 13 ton bus, was built as a serial hybrid and the body parts were made of lightweight composite materials (sandwich panels with a light foam core). A natural gas-powered V8 internal combustion engine serves as the primary power unit and operates a 360 V alternator. The permanent availability of electrical energy allows complete electrification of all auxiliary units [12, 13].

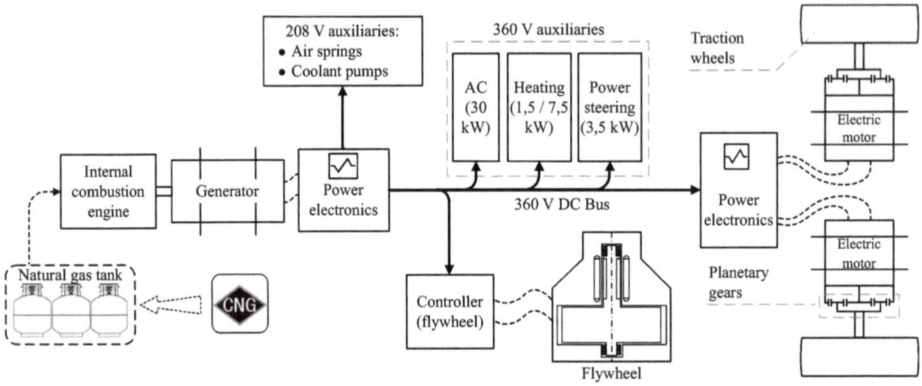

Fig. 2.10 Abstraction of the *Advanced Technology Transit Bus*, *Center for Electromechanics*, USA, in 2002, based on [12]. (Image rights: Center for Electromechanics, University of Texas)

The combination of a consequent lightweight design and a composite flywheel rotor for regenerative braking and load point averaging, applied to a vehicle in public transport, is a promising concept that holds great potential, as both the systematic analyses (e.g., in [5]) and the success of the *GKN Gyrodrive* [14] confirm.

When the term "flywheel energy storage system" or "FESS" is used in this book, an *electromechanical solution* is always assumed, unless a purely mechanical system is explicitly mentioned.

2.2.3 System Components of a FESS

The technical design of an electromechanical FESS appears trivial at first glance, as it contains only a few essential system components. Basically, it consists of the elements listed in Table 2.2 that make up the entire energy storage system. A possible and also typical structure of a flywheel energy storage system including peripheral components is shown in Fig. 2.11.

The optimization of auxiliary components such as *power electronics*, *cooling*, and *vacuum technology* will not be dealt with in this book, since these components—as will be shown—are not key elements in the development of FESS and can therefore be regarded as sufficiently well solved. The *housing suspension*, on the other hand, is in strong interaction with the rolling bearing of the flywheel energy storage system due to the rotor dynamics and will hence be examined in more detail in Sect. 9.2.1.

The alleged triviality of the system is deceptive, as the special operating conditions of the flywheel (extremely high-speed, vacuum, gyroscopic reactions, to name but a few aspects) lead to complex interactions and interdependencies between the components. Different arrangements (topologies) of the essential components show a strongly different

Table 2.2 Essential components of a mobile FESS divided into *sub-* and *supersystem*

	Component	Task	Typical realization
Subsystem	**Flywheel rotor** 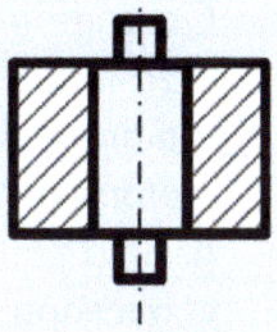	Increasing the moment of inertia/ content of energy in kinetic form	Solid steel disk, sometimes in sheet metal design or wound synthetic fiber structures (see Chap. 7)
	Electric motor/ generator 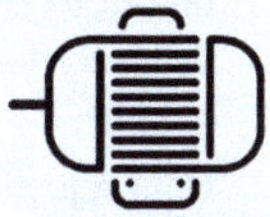	Conversion of the kinetic energy of the flywheel to electrical energy for the drive and vice versa	Asynchronous or reluctance machine, in some cases permanent magnet synchronous motor
	Bearing system 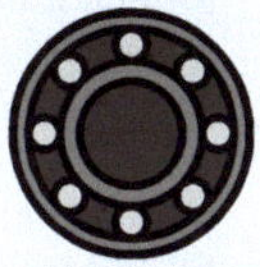	Low-friction positioning and guiding of rotating parts	Magnetic bearing or roller bearing
	Housing 	Positioning of all components and protection in case of rotor failure, vacuum-tight housing	Steel or aluminum construction with mostly cylindrical contours
	Condition monitoring measurement technology 	Monitoring the state of health of critical components, and the operating condition of the FESS in general	Optical contrast sensors for speed measurement, Pt100 sensors for temperature measurement, piezoelectric accelerometers for monitoring of rolling bearings, Pirani element for pressure monitoring
	Optional: Lifting magnet 	Magnetic compensation of rotor weight (axial bearing load)	Neodymium-iron-boron magnet in ring shape, or electromagnets
Supersystem	**Vacuum components** 	Maintaining a low-atmospheric pressure to reduce aerodynamic losses	One- to two-stage rotary vane pumps, turbo pumps and standard ISO-K or ISO-F components

(continued)

Table 2.2 (continued)

Component		Task	Typical realization
Power electronics		Generation of a high-frequency rotating field for the motor/generator, and desired voltage conversion for output to an AC or DC intermediate circuit	Frequency converters with power switching elements such as MOSFETs or IGBTS
Cooling		Heat rejection of the losses of the motor/ generator (stator windings) and bearings	Standard water cooling circuits with centrifugal pumps and heat exchangers
Housing suspension	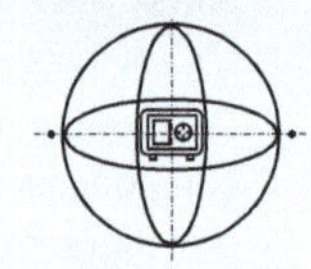	Connection of the FESS housing to the vehicle	Elastomer elements, rope springs, or fully cardanic suspension

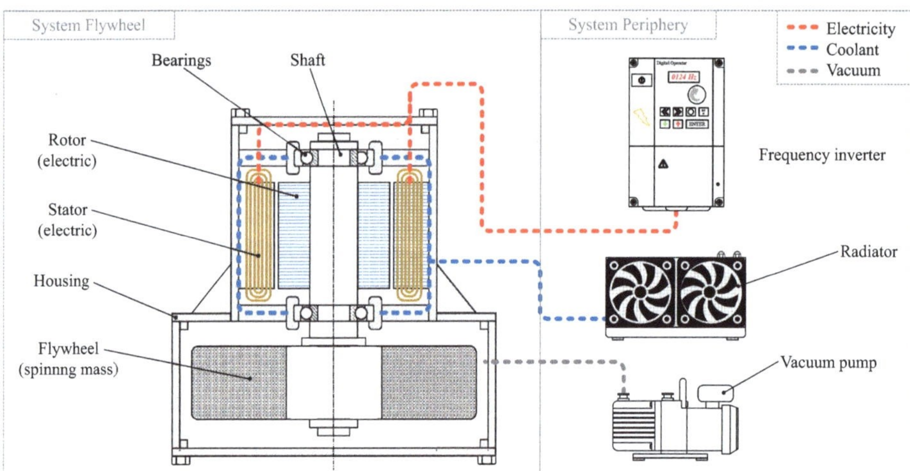

Fig. 2.11 Typical structure of an electromechanical flywheel energy storage device in a nonintegrated design including peripheral components

system behavior and must therefore be considered individually in each concrete application. An overview of the common topologies is given in Fig. 2.15, and *system-internal interdependencies* are described in more detail in Sect. 6.2 (Fig. 2.12).

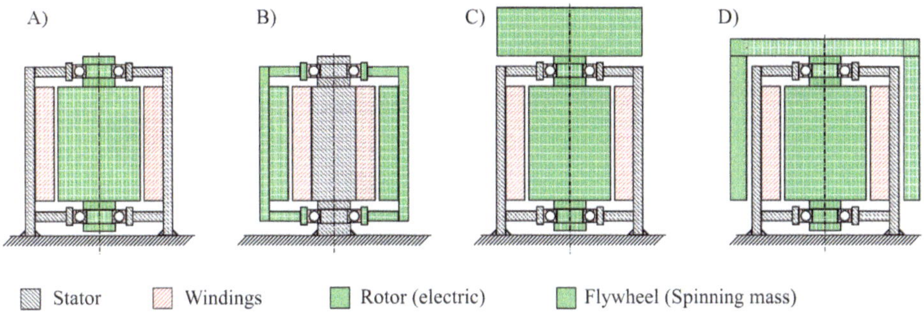

Fig. 2.12 Different topologies of electromechanical flywheel energy storage devices. (**a**) *Internal rotor, fully integrated.* (**b**) *External rotor.* (**c**) *Internal rotor, nonintegrated.* (**d**) *Internal rotor with bell rotor*

2.3 State of the Art in the Field of Flywheel Energy Storage Systems

Already the brilliant pioneer in the field of methods of development in engineering, *Gustav Niemann*[4] propagated the analysis of existing systems as the first step in the development process of a new product. The history of flywheel energy storage systems in general was presented in [4] and a specific analysis of historic systems in mobile applications in [5]. For reasons of a better overview, a short summary of these publications is given in Table 3.1 in Chap. 3.

2.3.1 Existing Systems: Stationary FESS

Although the typical energetic properties of today's electromechanical flywheel energy storage devices (i.e., their position in the *Ragone plot*; see Sect. 3.4.1) clearly indicate their primary suitability for a highly dynamic load cycle—just like it predominantly occurs in mobile applications—more manufacturers are devoting themselves to the production of flywheels for stationary use. The use of these "stationary flywheels" usually focuses on two applications:

(a) **Short-term storage** to mitigate voltage fluctuations in the electrical supply network, which is called "power quality management." This involves maintaining a perfect, sinusoidal AC voltage and its frequency within narrow tolerance limits, which is crucial for production machines in semiconductor manufacturing, for example.

---

[4] *Gustav Niemann*, 1899–1982, Professor for Machine Elements at the *TU Braunschweig* and *TU Munich*, Germany, is considered one of the most important scientists in the field of methods of development in mechanical engineering.

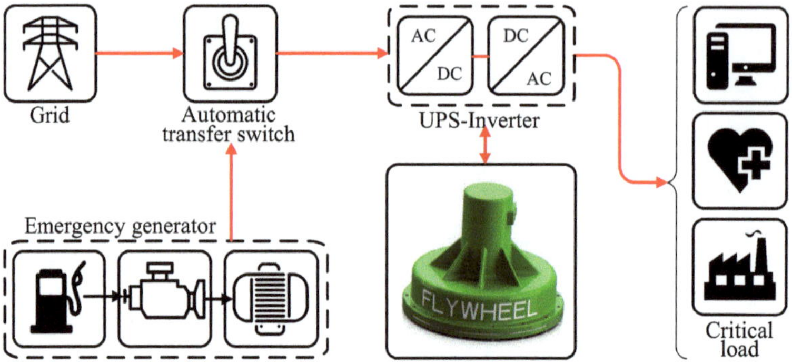

Fig. 2.13 Diagram of a flywheel energy storage system for uninterruptible power supply

(b) **UPS systems** (*uninterruptible power supply*) are another popular application. Here, an electromechanical flywheel energy storage system is used to bridge the time gap between the actual power failure and the start of the emergency power generator, thus ensuring unrestricted operation of a "critical load" (e.g., computer centers, medical instruments, or similar).

The schematic of a UPS flywheel (application b) is shown in the following figure (Fig. 2.13). However, as the title suggests, this book is mainly devoted to mobile applications.

In addition to the actual application, the following reasons can be given for the current predominant use of stationary FESS:

- High demand for "buffer capacity" in the electrical power grid due to increasing integration of volatile energy sources as part of the energy revolution.
- "24/7 operation" allows higher profitability and faster payback than in most vehicle applications.
- No occurrence of gyroscopic reactions due to vehicle dynamics.
- Low gravimetric energy density (specific energy content of the FESS) or high weight is of less consequence.
- Problems of "automotive certification" and crash safety are avoided.

An overview of several manufacturers of stationary flywheel energy storage systems including their products and properties is given in Table 2.3.

Figure 2.14 shows different stationary flywheel systems offered by the Piller company.

2.3.2 Mobile Flywheel Energy Storage Systems for Vehicles

In vehicles, FESS are used almost exclusively in a parallel hybrid configuration as secondary energy storage, as the application examples in Figs. 2.5 and 2.10 have shown. After the "flywheel euphoria" of the 1970s subsided, FESS technology only experienced a

Table 2.3 Selection of relevant commercial and noncommercial stationary flywheel energy storage systems

Manufacturer	Country	Year	Type	Energy	Performance	Speed	Service life	Ref.
(model)				*kWh*	*kW*	*RPM*		
Siemens	D	2014	Prototype	0.5	120	10,000	n.a.	[15]
Velkess	USA	2011	Prototype	5–15	9	9000	10 years	[16]
Active power (CleanSource® 750HD UPS)	USA	2015	Series	2.9	675	7700	20 years	[17]
Temporal power	Can	2016	Series	50	1000	12,000	20 years	[18]
Beacon power	USA	2008	Series	25	100	16,000	n.a.	[19]
Quantum power	USA	2015	Prototype	360	150	6000	20 years	[20]
Kinetic traction systems	USA	2015	Series	1.5	333	36,000	10 million cycles	[21]
PowerTHRU	USA	2014	Series	0.53	225	52,000	400 years (with a 6-year Service interval)	[22]
Calnetix Vycon (VDC-XXE)	USA	2015	Series	1.67	300	36,750	20 years	[23]
Rosseta (T4)	D	2011	Small series	1.5	250	50,000	20 years	[24]
Piller Powerbridge (Uniblock UBT)	D	2006	Series	4.6	500	3600	20 years	[25]
Boeing phantom	USA	2007	Prototype	5	100	15,000	n.a.	[26]
Dynastore	D	2006	Prototype	11	2000	10,000	n.a.	[27]
Amber kinetics (GEN-2)	USA	2015	Series	25	6.25	8500	30 years	[28]
Gerotor GmbH	D	2015	Series	0.0375	50	50,000 rpm	>ten million cycles	[29]

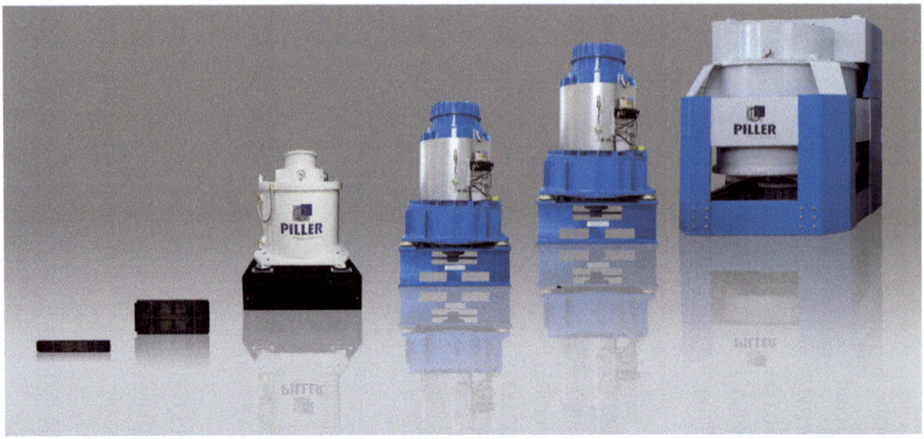

Fig. 2.14 Product range of the Piller Group GmbH in the segment of stationary flywheel energy storage systems. (Image rights: Piller Group GmbH)

Fig. 2.15 Flywheel energy storage in a racing car: cross section of the Porsche GT3 Hybrid [30]. (Image rights: Porsche AG)

renewed upswing many years later when racing (primarily *Formula 1*) was in search of powerful short-term energy storage systems for regenerative braking applications due to a change in regulations in 2008. Other applications followed, some of which were quite successful, such as the *Porsche GT3 Hybrid* (2010) shown in Fig. 2.15 or the *Audi R18 e-tron Quattro* (2012). They all have the following characteristics in common (Table 2.4):

- Very high specific power.
- Mostly fiber composite rotors.
- Short required service life.
- High costs.
- Relatively high self-discharge.
- Low absolute energy content.
- Higher energy density than stationary FESS.

Table 2.4 Selection of relevant commercial and noncommercial mobile flywheel energy storage systems

	Name/model	Manufacturer/ developer	Country	Year	Type	Energy content and type	Max. speed	Rotor mass	Reference
						kWh	*RPM*	*Kg*	
Commercial vehicles	*Gyrobus*	*Oerlikon plants*	CH	1953	Small series	9.15 (electric)	3000	1500.0	[31]
	Gyreacta	*Robert clerk*	GB	1961	Prototype	- (mechanical)	15,000	–	[32]
	M.A.N. test bus	*M.A.N.*	D	1981	Prototype	1.50 (electric)	12,000	104	[33]
	Stockholm city bus	*Volvo*	SWE	1982	Small series	- (electric)	10,000	329	[34]
	New York bus system	*Garrett Corp.*	USA	1985	Small series	16.00 (electric)	16,000	340	[35]
	Munich city buses	*MAN/Neoplan/magnet engine*	D	1988	Small series	2 x 2.75 (electric)	11,000	181	[36]
	ATTB	*Center for Electromechanics, Austin*	USA	2002	Prototype	2.0 (electric)	40,000	59	[3]
	AutoTram	*Fraunhofer Institute*	D	2006	Prototype	4.00 (electric)	23,000	300	[37]
	Hydrogen hybrid shuttle bus	*Center for Electromechanics, Austin*	USA	2008	Prototype	1.87 (electric)	18,000	155	[38]
	Flybus	*Ricardo/Torotrac*	GB	2009	Small series	0.28 (mechanical)	60,000	15	[39]
	GKN Gyrodrive	*GKN hybrid power*	GB	2012	Series	0.5	36,000	55	[14]

(continued)

Table 2.4 (continued)

	Name/model	Manufacturer/ developer	Country	Year	Type	Energy content and type	Max. speed	Rotor mass	Reference
Passenger cars	*Kulibin tricycle*	*Lieutenant I.P. Kulibin*	RUS	1792	Prototype	0.011 (mechanical)	500	50	[40]
	VW T2 flywheel hybrid	*RWTH Aachen*	D	1974	Prototype	0.211 (mechanical)	13,400	58	[41]
	K-wagon	*N.V. Gulia*	RUS	1976	Prototype	- (mechanical)	–	–	[40]
	FESS pinto	*Prof. Andy Frank, University of Wisconsin*	USA	1976	Prototype	- (mechanical)	8000	90.7	[42]
	Garrett postal wagon	*Garrett Corp.*	USA	1978	Prototype	- (mechanical)	36,000	12.2	[43]
	Garrett 4 passenger sedan	*Garrett Corp.*	USA	1978	Prototype	1.0 (mechanical)	25,000	22.7	[44]
	GM FX 85	*General motors*	USA	1982	Prototype	- (mechanical)	–	–	[45]
	Volvo 240 FESS	*Volvo*	SWE	1983	Prototype	- (mechanical)	–	–	[46]
	Chrysler patriot	*Chrysler motors*	USA	1993	Prototype	1.0 (electric)	58,000	60	[47]
	AFS 20	*American flywheel systems*	USA	1994	Prototype	20 × 2.0 (electric)	20,0000	–	[48]
	Hybrid III	*ETH Zurich*	CH	1996	Prototype	0.07 (mechanical)	6000	48	[49]
	Volvo S40	*TU Eindhoven*	NL	1998	Prototype	0.12 (mechanical)	17,000	–	[10]

	Zero inertia VW bora	*TU Eindhoven/Van Doorne*	NL	2000	Prototype	0.04 (mechanical)	8000	12.2	[50]
	Porsche 911 GT3 R hybrid	*Porsche/Williams hybrid power*	D/GB	2009	Prototype	0.2 (electric)	40,000	14	[51]
	Porsche 918 RSR	*Porsche/Williams hybrid power*	D/GB	2011	Prototype	- (electric)	36,000	–	[52]
	Audi e-tron	*Audi/Williams hybrid power*	D/GB	2013	Small series	0.1 (electric)	45,000	19	[53]
	Volvo S60 Flybrid	*Volvo/PUNCH Flybrid*	SWE/GB	2014	Prototype	0.17 (mechanical)	60,000	6	[54]
	Jaguar XF	*Jaguar/PUNCH Flybrid*	GB	2010	Prototype	0.12	60,000	5	[55]
Rail vehicles	*Schuberski locomotive*	*Lieutenant Z. Schuberski*	RUS	1860	Prototype	31.67	–	5000	[40]
	British rail class 70	*Southern railway (S.R.)*	GB	1948	Small series	- (electric)	–	1500	[56]
	Gyro locomotive "MFO EG 120–17"	*Oerlikon machine factory*	CH	1955	Prototype	5.6 (mechanical)	3000	3500	[57]
	New York Subway	*Garrett Corp.*	USA	1974	Prototype	1.6 (electric)	14,000	4 × 68	[2]
	Advanced concept train	*Boeing Vertol comp*	USA	1975	Prototype	4.5 (electric)	11,000	–	[58]
	PPM 50 railcar	*Parry people movers*	GB	1992	Small series	3.75 (electric)	2600	720	[59]
	ULEV TAP I	*Alstom DDF, CCM*	NL	2001	Prototype	4.0 (electric)	1500	–	[60]
	Lirex MDS K5	*Alstom/magnet motor*	D	2004	Prototype	2 x 2.0 (electric)	12,000	600	[61]

(continued)

Table 2.4 (continued)

	Name/model	Manufacturer/ developer	Country	Year	Type	Energy content and type	Max. speed	Rotor mass	Reference
	ULEV-TAP II	*Alstom DDF, CCM*	NL	2005	Prototype	4.0 (electric)	22,000	375	[62]
	Lirex MDS K6 "Flytrain"	*Alstom/WTZ Rosslau*	D	2006	Prototype	2 × 6 (electric)	25,000	–	[63]
Flywheel energy storage modules	*Dynastore*	*Compact dynamics*	D	2008	Small series	4 × 0.053 (electric)	80,000	4 × 1.5	[64]
	HyKinesys—FlyCylinder	*PowerBeam Imperial College London*	GB	2008	Prototype	0.44 (mechanical)	71,000	12	[65]
	F1-KERS	*Flybrid systems/ Torotrac*	GB	2009	Series	0.11 (mechanical)	64,500	5	[66]
	Williams F1/GT3 FESS	*Williams hybrid power*	GB	2009	Small series	0.375 (electric)	45,000	47	[53]
	MK 1	*Flywheel energy systems Inc.*	CAN	2010	Prototype	0.77 (electric)	42,500	–	[67]
	e-KERS	*Enstor technologies (bankrupt in 2014)*	D	2011	Prototype	- (electric)	–	5	[68]
	DBS flywheel	*Dynamic boosting systems ltd.*	GB	2019	Prototype	0.14 (electric)	50,000	35	[65]
Construction machinery	*TorqStor*	*Ricardo*	GB	2014	Prototype	0.056	44,000	–	[69]
	Rubber-tired gantry (RTG) crane	*Center for Electromechanics, Austin/Vycon*	USA	2007	Prototype	2 × 1.27 (electrical)	36,000	–	[70]
	High-efficiency excavator (HFX)	*Ricardo*	GB	2013	Prototype	0.056	44,000	–	[69]

Figure 2.16 shows the *Jaguar XF* converted by *PUNCH Flybrid* to a flywheel-KERS vehicle and the installation of the flywheel module on the underbody. The components of the system are shown in Fig. 2.17.

Fig. 2.16 *Jaguar XF* with flywheel energy storage of *PUNCH Flybrid. (Image rights: PUNCH Flybrid)*

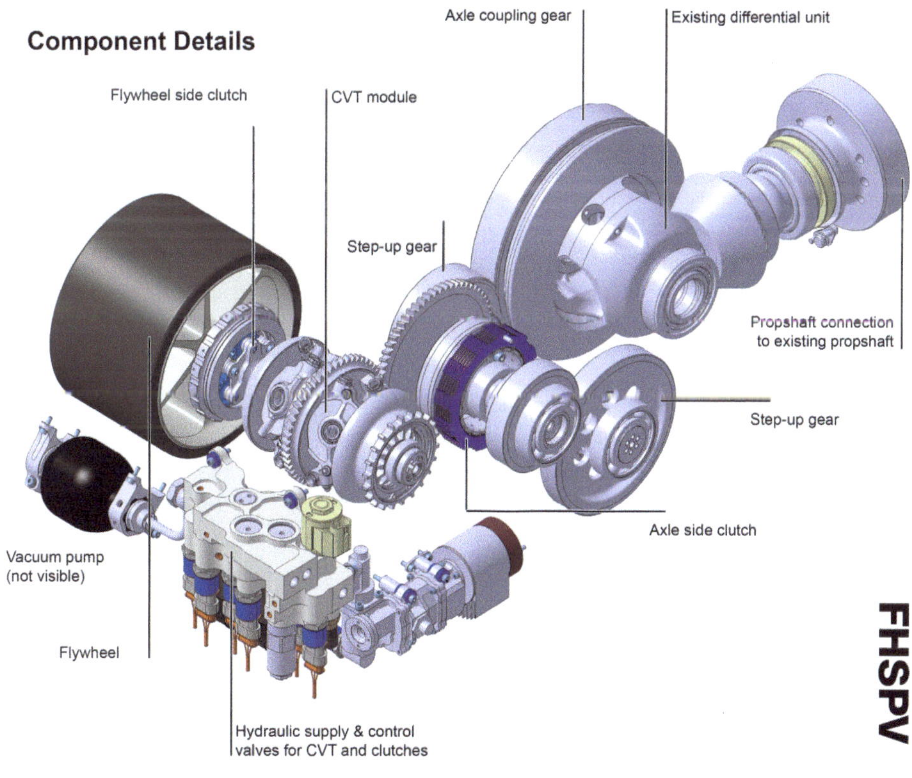

Fig. 2.17 Components of the *PUNCH Flybrid* flywheel module for the *Jaguar XF* demonstrator vehicle. (Image rights: PUNCH Flybrid)

To once again underline the versatility of flywheel technology, an off-highway application (construction machine) is shown in Fig. 2.18. It is a 17-ton-class excavator, the so-called high-efficiency excavator (HFX), which was developed by *Ricardo* in England. The *TorqStor* mechanical flywheel module with magnetic clutch (see Fig. 2.19) has been integrated into the machine's hydraulic system and recuperates energy, especially when lowering loads lifted by the excavator bucket. According to the manufacturer, the modular flywheel system shall be built with energy contents of 200 kJ to 4 MJ and can be used in various applications such as road and rail vehicles.

Fig. 2.18 The *Ricardo high-efficiency excavator* (HFX), which was practically tested in 2013. (Image rights: Ricardo PLC)

Fig. 2.19 The flywheel energy storage module *TorqStor* by *Ricardo*, which was realized in a transparent housing for the purpose of the *CONEXPO 2014* exhibition. (Image rights: Ricardo PLC)

References

1. C. Seyerlein (2018) CO_2-Ausstoß: So schnitten die Autobauer 2017 ab. Vogel Communications Group GmbH & Co. KG. https://www.kfz-betrieb.vogel.de/co2-ausstoss-so-schnitten-die-autobauer-2017-ab-a-687790/. [Accessed May 20th 2019].
2. A. P. Armagnac (1974) Flywheel Brakes Store New Train's Energy for Electricity-Saving Starts. Popular Science, pp. 41–43, Issue February 1974.
3. R.J. Hayes, J.P. Kajs, R.C. Thompson and J.H. Beno (1999) Design and Testing of a Flywheel Battery for a Transit Bus. SAE International Congress and Exposition 1999, Detroit, Michigan, USA.
4. G. Genta (1985) Kinetic Energy Storage: Theroy and Practice of Advanced Flywheel Systems. Dipartimento di Meccanica/Politecnico di Torino, Butterworths, London, UK.
5. A. Buchroithner (2011) Systematische Analyse von Hybridfahrzeugen mit Schwungradspeicher unter Erfassung von Entwicklungstendenzen. Institut für Maschinenelemente und Entwicklungsmethodik, Graz University of Technology, Austria.
6. A. Buchroithner and M. Bader (2011) History and development trends of flywheel-powered vehicles as part of a systematic concept analysis. European Electric Vehicle Congress (EEVC), November 2011, Brussels, Belgium.

7. A. Buchroithner and M. Bader (2012) Systematische Analyse von Hybridfahrzeugen mit Schwungradspeicher unter Erfassung von Entwicklungstendenzen. 8. VDI Wissensforum für innovative Fahrzeugantriebe, Dresden, Germany.
8. H. Naunheimer, B. Bertsche and G. Lechner (2007) Fahrzeuggetriebe, Springer-Verlag Berlin Heidelberg, Germany. DOI: https://doi.org/10.1007/978-3-662-07179-3.
9. G. Kelz, C. Nussbaumer, M. Bader and P. Haidl (2015) HEuV – Hochintegrierte Energiespeicher für den urbanen Verkehr. FFG (Österreichische Forschungsförderungsgesellschaft), Vienna, Austria.
10. R. Van der Graaf, D.B. Kok and E. Spijker (1999) Integration of Drive system, Subsystem and Auxiliary Systems of a Flywheel Hybrid Driveline with Respect to Design Aspects and Fuel Economy. VDI Berichte 1459 – Hybridantriebe, Helmond, NL, Verein Deutscher Ingenieure.
11. C. Kulkarni, J. Celaya, G. Biswas and K. Goebel (2012) Prognostics of Power Electronics, methods and validation experiments. IEEE Autotestcon, Anaheim, CA, USA. DOI: https://doi.org/10.1109/AUTEST.2012.6334578
12. H. Reutter (1999) Advanced Technology Transit Bus – Final Test Report for the ATTB Prototypes. U.S. Department of Transit – Federal Transit Administration, Springfield, VA, USA.
13. R.J. Hayes, J.P. Kajs, R.C. Thompson and J.H. Beno (1999) Design and Testing of a Flywheel Battery for a Transit Bus. SAE International Congress and Exposition, Detroit, Michigan, USA.
14. GKN Hybrid Power (2014) GYRODRIVE by GKN Hybrid Power – Driving Efficient Transport. Unit 1 Pentagon South, Abingdon Science Park, Barton Lane, Abingdon, Oxford, UK.
15. K. Schrein (2014) Smart Grids and Energy Storage – Fears of Power Loss Fade. Siemens Aktiengesellschaft, Werner-von-Siemens-Straße 1, 80333 München, Germany. https://www.siemens.com/innovation/en/home/pictures-of-the-future/energy-and-efficiency/smart-grids-and-energy-storage-flywheel-energy-storage.html.
16. J. Hewitt (2013) The Velkess Flywheel: A more flexible energy storage technology. Phys.org, 12. April 2013. https://phys.org/news/2013-04-velkess-flywheel-flexible-energy-storage.html. [Accessed January 20th 2017].
17. EMEA Active Power Solutions Ltd (2015) CleanSource® 750HD UPS. Lauriston Business Park, Pitchill, Evesham, UK.
18. T. Biggs (2016) A Flywheel like No Other. Temporal Power Ltd., 2-3750A Laird Rd, Mississauga, ON, Canada.
19. J. Arseneaux (2013) 20 MW Flywheel Energy Storage Plant. Beacon Power LLC., Wilmington, Massachusetts, USA.
20. Quantum Energy Storage Inc. (2015) Advanced Kinetic Energy Storage System. http://www.qestorage.com/technology
21. Kinetic Traction Systems, Inc. (2015) Flywheel Energy Storage UPS & Power Quality Applications (Produktinformation). Kinetic Traction Systems, Inc., 20360 Plummer Street, Chatsworth, CA 91311, USA.
22. PowerThru Inc. (2014) Cleann Flywheel Energy Storage – The Battery-Free Solution for Your UPS System. 11825 Mayfield, Livonia, Michigan 48150, USA.
23. Calnetix (2015) VYCON® Direct Connect (VDC®) – The Optimal UPS Energy Storage Solution for Mission-Critical Power Protection. Calnetix Technologies, 16323 Shoemaker Avenue, Cerritos, CA 90703 USA.
24. F. Täubner (2014) Schwungradspeicher in Vision und Realität. Konstruktionsbüro Frank Täubner, Ueckerstr. 4, 38895 Derenburg, Germany.
25. Piller Power Systems (2017) UNIBLOCK UBT+ Rotary UPS from 500 kW up to 40 MW. Langley Holdings plc., Retford, Nottinghamshire, UK. http://www.piller.com/en-GB/257/uniblock-ubt-rotary-ups-from-500-kw-up-to-40-mw.
26. M. Strasik et al (2007) Design, Fabrication, and Test of a 5-kWh/100-kW Flywheel Energy Storage Utilizing a High-Temperature Superconducting Bearing. IEEE Transactions on Applied Superconductivity, pp. 2133–2137.

27. W. Canders, H. May, J. Hoffmann, P. Hoffmann, F. Hinrichsen and I. Koch (2006) Flywheel Mass Energy Storage with HTS Bearing – Development Status. WCRE/Eurosolar International Conference on Renewable Energy Storage, Gelsenkirchen, Germany.
28. S. Sanders, M. Senesky, M. He and E. Chiao (2015) Low-Cost Flywheel Energy Storage Demonstration, California Energy Commission, USA.
29. Gerotor GmbH, "Der Gerotor HPS – Die Basis für ein aktives Energiemanagement!", http://gerotor.tech/gerotor-hps-schwungmassenspeicherenergiemanagement/
30. Porsche Cars North America, Inc. (2011) 911 GT3 R Hybrid Celebrates World Debut in Geneva. Press release on November 2nd 2011. http://www.porsche.com/usa/aboutporsche/pressreleases/pag/?pool=international-de&id=2010-02-11. [Accessed June 20th 2011].
31. E. B. Leutwiler (1991) Gyrobus. Tram, Nr. 1/91.
32. D. Scott (1961) Fith Wheel Runs Bus... Stops it Too! Popular Science, pp. 98–102, Issue May 1961.
33. D. Scott (1980) Hydrobus, gyrobus use brake-generated energy. Popular Science, pp. 76–77, Issue April 1980.
34. D. Scott (1985) Brake-power buses. Popular Science, p. 59, Issue January 1985.
35. S. Renner-Smith (1981) The coming era of flywheel buses. Popular Science, pp. 62–63, Issue August 1981.
36. J.B. Crawley (1992) Flywheel Trolley. Popular Science – Special Anniversary Issue, p. 15, Issue August 1992.
37. M. Klingner (2006) Fahrzeugtechnik im ÖPNV – Migrationspfade der Elektromobilität. Fraunhofer-Institut für Verkehrs und Infrastruktursysteme, Zeunerstraße 38, 01069 Dresden, Germany.
38. C.S Hearn et.al. (2007) Low Cost Flywheel Energy Storage for a Fuel Cell Powered Transit Bus. IEEE Vehicle Power and Propulsion Conference, Arlington, TX, USA. DOI: https://doi.org/10.1109/VPPC.2007.4544239
39. J. Wheals, J. Taylor and W. Lanoe (2016) Rail Hybrid using Flywheel. Den Danske Banekonference, Tivoli Congress Centre, Kopenhagen, Denmark.
40. N. Gulia (1986) Der Energiekonserve auf der Spur. Verlag Harri Deutsch, Thun, Germany.
41. H. Schreck (1977) Konzeptuntersuchung, Realisierung und Vergleich eines Hybrid-Antriebes mit Schwungrad mit einem konventionellen Antrieb, Aachen: Fakultät für Maschinenwesen der Rheinisch-Westfälischen Technischen Hochschule, Aachen, Germany.
42. A. Frank (1981) Engine never idles as steel flywheel spins out savings. Popular Mechanics, pp. 98–99, Issue June 1981.
43. R. F. Dempewolff (1978) Flywheels: New Boost for Engine Power. Popular Mechanics, pp. 98–102, Issue February 1978.
44. S. Renner-Smith (1981) Battery-saving flywheel gives electric car freeway zip. Popular Science, Issue October 1980.
45. General Motors Heritage Center (2012) GM's Flywheel Hybrid Vehicles. https://history.gmheritagecenter.com/wiki/index.php/GM%27s_Flywheel_Hybrid_Vehicles. [Accessed am March 18th 2017].
46. K. Pudenz (2011) Kraftstoff-Einsparpotenzial bis zu 20 Prozent – Volvo testet Schungradspeicher. ATZ Online, Issue 26. 05. 2011.
47. D. McCosh (1995) Seeing the Forest Instead of the Trees. Popular Science, Issue January 1995.
48. D. Stower (1995) Flywheel Power. Popular Science, Issue January 1995.
49. P. Dietrich (1999) Gesamtenergetische Bewertung verschiedener Betriebsarten eines Parallel-Hybridantriebes mit Schwungradkomponente und stufenlosem Weitbereichsgetriebe für einen Personenwagen (Dissertation) p. 86. ETH Zürich, Switzerland.
50. R.M. van Druten, P. van Tilborg, P. Rosielle and M. Schoutem (2000) Design and Construction Aspects of a Zero Inertia CVT for Passenger Cars. Seoul 2000 FISITA World Automotive Congress, Seoul, Korea.

51. R. Meaden (2010) Porsche 911 GT3 R Hybrid Review. Telegraph Media Group Limited, UK.
52. Porsche AG (2011) Porsche 918 RSR – racing laboratory with even higher-performance hybrid drive. http://www.porsche.com/usa/aboutporsche/pressreleases/pag/?pool=international-de&id=2011-01-10.
53. I. Foley (2013) William Hybrid Power – Flywheel Energy Storage. Williams F1, Grove, Oxfordshire, OX12 0DQ, UK.
54. S. Birch (2010) Volvo spins up flywheel technology research. SAE International, Issue June 19th 2011.
55. J. Rendell (2010) Jaguar's advanced XF ‚flybrid',“ AUTOCAR – First for Car News and Reviews. http://www.autocar.co.uk/car-news/concept-cars/jaguars-advanced-xf-flybrid. [Accessed March 18th 2017].
56. M. Bowman (2016) The C-C Booster Electric Class 70 Locomotive. https://www.slideshare.net/MarkBowman11/br-class-70-64046099
57. P. von Burg (1998) Moderne Schwungmassenspeicher – eine alte Technik in neuem Aufschwung. VDI Fachtagung Energiespeicherung für elektrische Netze, Gelesenkirchen, Verein Deutscher Ingenieure.
58. Der Spiegel (1974) Wucht im Kreisel. Der Spiegel, pp. 157–158, Issue 11 1974.
59. Parry People Movers Ltd. (2009) PPM Technology. Parry People Movers Ltd., Overend Road, Cradley Heath, West Midlands, B647DD, UK. http://www.parrypeoplemovers.com/technology.htm. [Accessed August 20th 2016].
60. Centrum voor Constructie en Mechatronica (2000) ULEV-TAP Newsletter August 2000, Issue No. 2. http://www.ulev-tap.org/ulev1/index.html. [Accessed May 14th 2011].
61. R. Benger (2007) Fachpraktikum Energiesystemtechnik: Elektrische Energiespeicher für dynamische Anforderungen, Institut für Elektrische Energietechnik, TU Clausthal, Germany.
62. U. Henning, F. Thoolen, M. Lampérth, J. Berndt, A. Lohner and N. Jäning (2005) Ultra Low Emission Vehicle – Transport Advanced Propulsion. http://www.railway-research.org/IMG/pdf/480-2.pdf
63. Deutsche Bahn AG (2002) Applications for energy storage flywheels in vehicles of Deutsche Bahn AG. Deutsche Bahn AG, Research & Technology Centre, Witthuhn, Germany.
64. W. Novy (2008) Start-Stopp – aber mit Schwung! Kietische Energiespeicher als Alternative zu Akkumulatoren und Kondensatoren. AUTOMOTIVE, pp. 64–66, Issue 11 2008.
65. K.R. Pullen, S. Shah and C. Ellis (2006) Kinetic Energy Storage for Vehicles. Department of Mechanical Engineering, Imperial College, London, UK.
66. C. Brockbank and C. Greenwood (2008) Full-Toroidal Variable Drive Transmission Systems in Mechanical Hybrid Systems – From Formula 1 to Road Vehicles. Torotrak (Development) Ltd. 1 Aston Way Leyland, PR26 7UX, UK.
67. Flywheel Energy Systems Inc. (2011) Performance Verification of a Flywheel Energy Storage System for Heavy Hybrid Vehicles. Flywheel Energy Systems Inc., 25C Northside Road, Ottawa, K2H 8S1, Canada.
68. M. Schmich (2011) Mit einer Augsburger Innovation Sprit sparen. Augsburger Allgemeine, Nr. 2014, p. 30, Issue September 16th 2011.
69. Ricardo plc. (2014) Ricardo to showcase 'TorqStor' high efficiency flywheel energy storage at CONEXPO. Press Release, 24th of February 2014. Ricardo plc. Shoreham-by-Sea, UK.
70. M. M. Flynn, P. McMullen and O. Solis (2007) High-Speed Flywheel and Motor Drive Operation for Energy Recovery in a Mobile Gantry Crane. APEC 07 – Twenty-Second Annual IEEE Applied Power Electronics Conference and Exposition, Anaheim, USA.DOI: 10.1109/APEX.2007.357660

Supersystem of Mobile Flywheel Energy Storage

3

The *supersystem* of the flywheel energy storage system (FESS) comprises all aspects and components, which are *outside the energy storage system* itself, but which interact directly or indirectly with the flywheel. These hierarchically superordinate components or influencing parameters can form their own system and are often summarized and considered a separate unit in a holistic view. In some cases, all relevant components of the vehicle are combined to form the *supersystem* "vehicle and vehicle topology" to set appropriate system limits and interfaces and allow for adequate system consideration.

3.1 Vehicle and Vehicle Topology

For some years now, customers have been offered an increasing number of drivetrain concepts and vehicle topologies. It is not only a matter of choosing between conventional and electric drive systems, as the various "intermediate stages," i.e., hybrid variants (*serial* vs. *parallel*, *micro* vs. *full hybrid*) in particular have a wide variety of characteristics (compare Figs. 3.1 and 3.2). Only rarely do car customers succeed in sufficiently and objectively defining their own requirement profile and selecting a correspondingly optimal drivetrain concept. Apart from this, economic or energetic criteria are rarely the decisive factor when buying a car, as described in Section 3.5. In the commercial vehicle sector (see Sect. 5.1) on the other hand, it is primarily the monetary payback time that determines the choice of vehicle and its topology.

The driving cycle (which is analyzed in more detail in Sect. 4.2.1) and vehicle topology mainly influence the energy recovery potential of a vehicle. Conversely, the technical properties of the regenerative braking system also influence the driving style that should be

A. Buchroithner, *Flywheel Energy Storage*,
https://doi.org/10.1007/978-3-658-35342-1_3

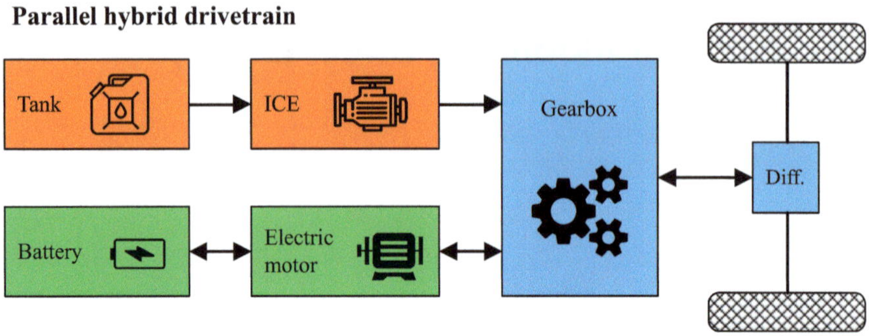

Fig. 3.1 Topology of a parallel hybrid drive system

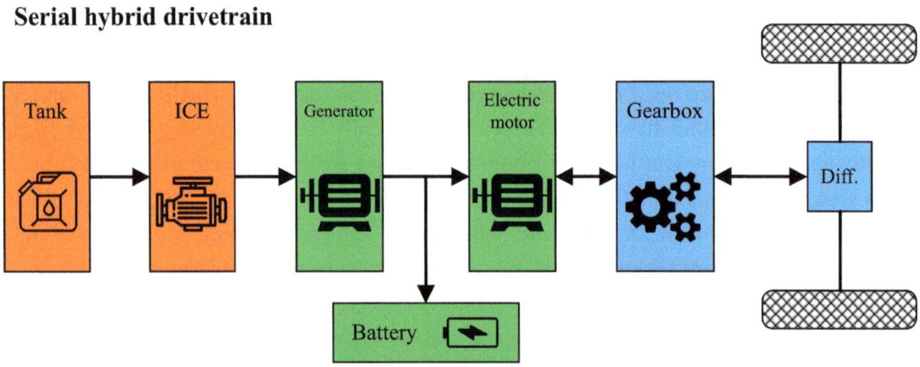

Fig. 3.2 Serial hybrid drive system as often used in heavy commercial vehicles

chosen for highest efficiency. The *Volvo S60 Flywheel KERS*[1] *Hybrid* is briefly discussed here as an example. As Fig. 3.3 shows, a flywheel energy storage system with mechanical energy transfer acts on the rear axle of the vehicle, while the front axle is equipped with a conventional drivetrain. When the vehicle is decelerated, the kinetic energy is transferred to the flywheel and its rotational speed is increased. Later, the energy of the flywheel can be used again to accelerate the vehicle. A detail of the KERS unit is shown in Fig. 3.4.

However, due to the dynamic axle load distribution, the braking power of the rear wheels must be considerably lower than that of the front wheels (usual front to rear brake force distribution is around 80/20 to 60/40), which means that only part of the theoretically available kinetic energy can be recuperated. If the regenerative braking system were to act on the front axle, this would already be an improvement. However, in the sense of exploiting the entire recuperation potential, four-wheel hub motors, which allow regenerative operation, would be ideal. But in this case, not only would the unsprang mass of the

[1] *KERS* stands for *Kinetic Energy Recovery System.*

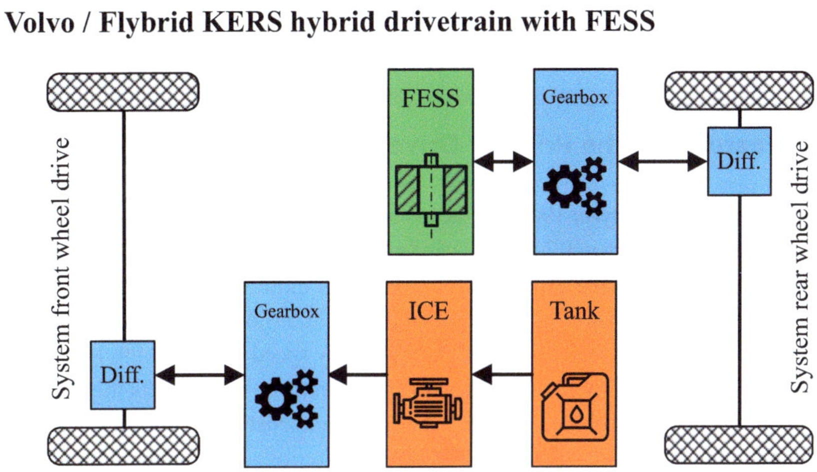

Fig. 3.3 Concept of the Volvo S60 Flywheel KERS (kinetic energy recovery system)

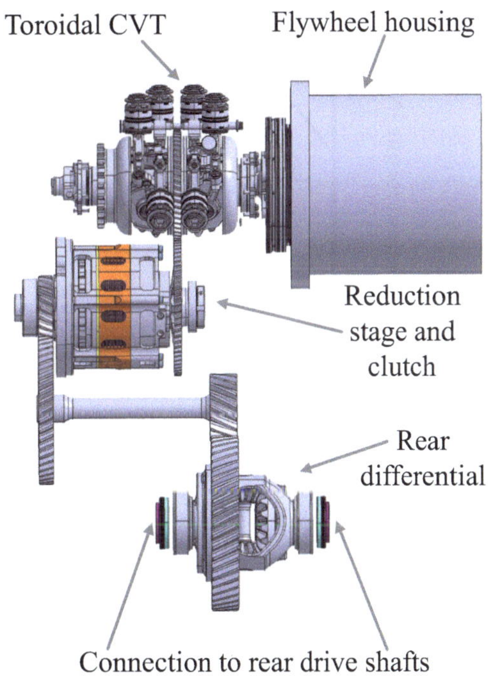

Fig. 3.4 Detail of Volvo rear axle drive with PUNCH Flybrid KERS unit [8]. (Image rights: PUNCH Flybrid/Volvo)

vehicle increase but an *electromechanical* FESS with correspondingly high power and energy content would also be required.

3.2 Features of the Primary Drive

In principle, the ideal, *perfect VKM*[2] achieves an efficiency η of at most ~67%; in practice, this value fluctuates between 0% (idling) and about 40% (best point) [8]. In general, it can be assumed that the average efficiency of a VKM is quite low when operated predominantly in a transient way (*stop-and-go* traffic). Figure 3.5 shows that the highest efficiency (here $\eta = 0.29$) is only achieved in a relatively narrow speed range at high loads. The load point shift made possible by hybridization of the powertrain counteracts this phenomenon but may cause another problem, namely, that engine acoustics and driving dynamics are

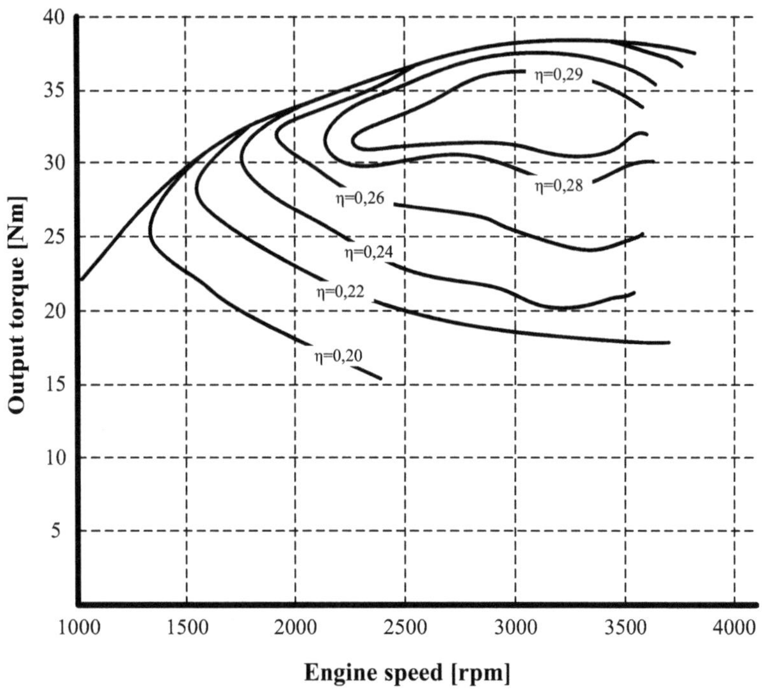

Fig. 3.5 Brake-specific fuel consumption (BSFC) plot of a typical gasoline internal combustion engine (Renault Clio) based on the data from [5]. (Image rights: Institute for Automotive Engineering and Piston Engines, RWTH Aachen University (Institut für Kraftfahrwesen und Kolbenmaschinen, RWTH Aachen))

[2]ICE assuming a *frictionless* and *adiabatic combustion process.*

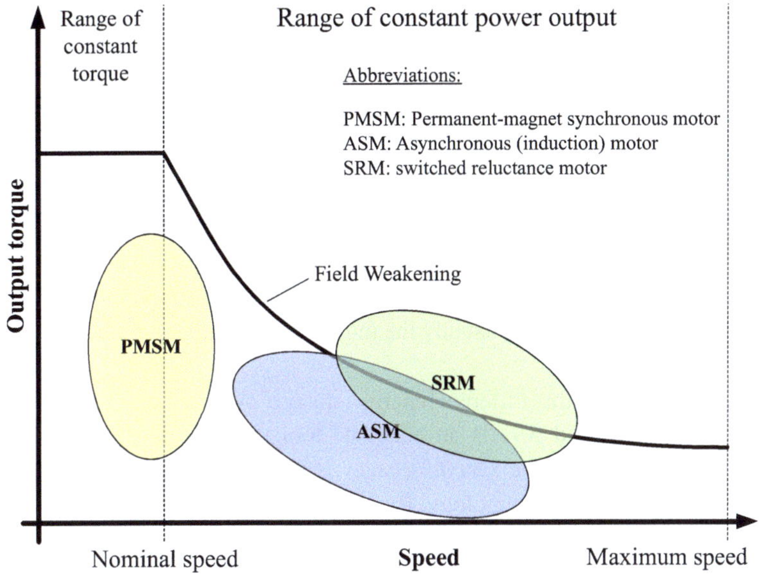

Fig. 3.6 Operating ranges of the different electric machines

decoupled from each other. This behavior is often considered unusual or even unpleasant by the driver. These *psychoacoustic aspects* usually also frequently prevent efficient operation of the conventional car by driving at extremely low engine speeds. This so-called overdrive characteristic is rejected by many drivers because they interpret the low "humming" of the drivetrain as "bad for the engine" [9] (compare also Section 3.5.).

An electric motor, on the other hand, has an acoustic behavior that is predominantly proportional to speed and also offers strikingly better part load efficiency. Compared to the ICE, the asynchronous motor (ASM or induction motor) and switched reluctance motor (SRM) offer a relatively wide speed range of constant power, while the permanent magnet synchronous motor (PMSM) is capable of delivering very high starting torques (compare Fig. 3.6). But these and other advantages of hybridization can only be fully exploited if appropriate electrical energy storage devices—such as an electromechanical flywheel or chemical batteries—are available.

3.3 Properties of Mobile Energy Storage Devices

In addition to the important task described in Sect. 2.1 of securing a clean and sustainable grid-based energy supply from volatile sources, the various energy storage technologies are naturally also becoming increasingly important in the automotive sector. Functions such as recuperation and load point shifting as a product of powertrain hybridization interact closely with the energetic properties of the energy storage devices. The exact design of an energy storage system for hybrid vehicles can only be based on a *known load profile* or predetermined driving cycle. In general, the design of energy storage systems is based on the pursuit of high specific energy and performance at low costs and good environmental compatibility. Unfortunately, these properties often stand in an unavoidable physical trade-off.

Table 3.1 compares some mobile energy storage systems in the course of an approximate, qualitative assessment [6]. It can be seen that chemical and electrical energy storage devices have high-energy densities, but perform worse than their mechanical counterparts in terms of recycling and manufacturing costs. At this point, it is important to note that cost and energy and power density are usually the most highly weighted criteria for automotive applications.

The representation of the different energy storage systems in a so-called *Ragone diagram* (see Figs. 3.5 and 3.7) is an adequate tool to compare the optimal areas of application. In this diagram, the *specific energy* is plotted against the *specific power* of the respective energy storage technology. By choosing a double-logarithmic scale, energy storage devices of different types and with strongly diverging properties can be compared in a clearly arranged way. The ordinate, which represents the specific energy, thus illustrates how much energy is available per weight, while the power density on the abscissa is a measure of how fast this energy can be released.

If a flywheel is strategically used as an "overtaking booster" in a car, as is the case with the KERS in Formula 1, for example, the release of all its energy within an extremely short time is important. In the case of a flywheel energy storage system, which is designed to allow a transit bus station to station in a local transport network, more attention will of course be paid to the specific energy content, i.e., subsequently "range/kg."

Traditionally, the technical research and development work has focused on the points discussed in Sects. 3.1, 3.2, and 3.3 (i.e., the elements of the vehicle subsystem). However, the influence of these sub-elements on the overall system performance and the physically possible development potential is limited by the laws of nature.

3.4 Geography, Infrastructure, and Intended Use of the Vehicle

3.4.1 Geography and Infrastructure

It is important to recognize that the influence of the *supersystem* on the design of the vehicle, and consequently its *subsystem* as well, is stronger than vice versa. This is also due to historical reasons since the automobile is of course many times younger than the environment in which it is used. The layout of many Central European cities was defined

Table 3.1 Characteristics of common energy storage systems for hybrid vehicles including qualitative assessment

	Lithium-ion battery	Lead-acid battery	Ni-Cd battery	Super-/UltraCap	Compressed air storage	Hydraulic energy storage	Flywheel energy storage
Energy density	***	**	***	*	*	*	*
Power density	**	**	***	***	***	***	***
Lifetime	**	**	*	**	***	***	***
Costs	**	**	**	*	***	**	*
Efficiency	***	***	***	***	*	*	**
Charging time	*	*	**	***	**	**	***
Recycling	*	**	*	*	***	***	***
Production effort	*	**	*	*	***	**	**

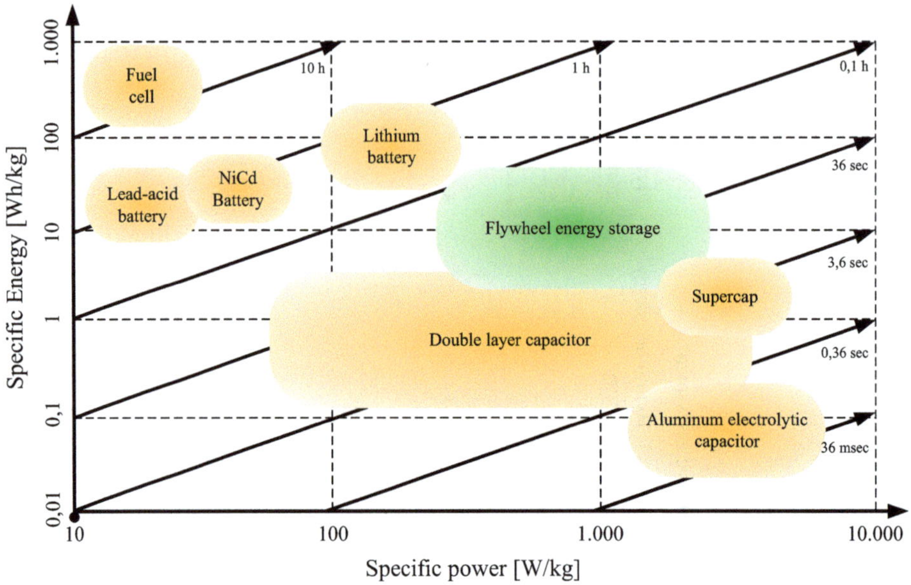

Fig. 3.7 Ragone plot of different energy storage systems

in medieval times, or even earlier, based on strategic, military, or agricultural and supply considerations. Traffic routes that were optimized for high travel speeds or the energy supply of vehicles did not play a role at that time, and consequently, from today's "energetic" point of view, some streets, districts, or even entire regions are not always optimally suited for the use of modern vehicles. Avoiding this problem by creating a subway network and hence designing a new transport system from scratch, for example, is not always a viable option. Younger cities, as can be found in many places in the USA, have a road network that has been "optimized" for use by road vehicles. Two examples are shown in Figs. 3.8 and 3.9.

However, this "optimization" usually only affects spatial aspects (e.g., parking), and problems such as adapting the energy supply network to alternative vehicle technologies remain mostly unsolved to date. The following examples illustrate the extent to which geography and infrastructure interact directly with the vehicle and its energy storage system:

1. The number of start-stop cycles caused by traffic lights or intersections as well as geodetic altitude difference and "curvyness" of a route determines the potential for recuperation and consequently energy savings for hybrid vehicles. In principle, the more dynamic the driving cycle, the more likely the advantages of a hybrid vehicle with regenerative braking can be exploited (see Sect. 4.2 and the entire Chapter 5).
2. Plug-in hybrids as well as pure EVs depend on the availability of a correspondingly powerful power grid. While a well-developed network of gas stations ensures the

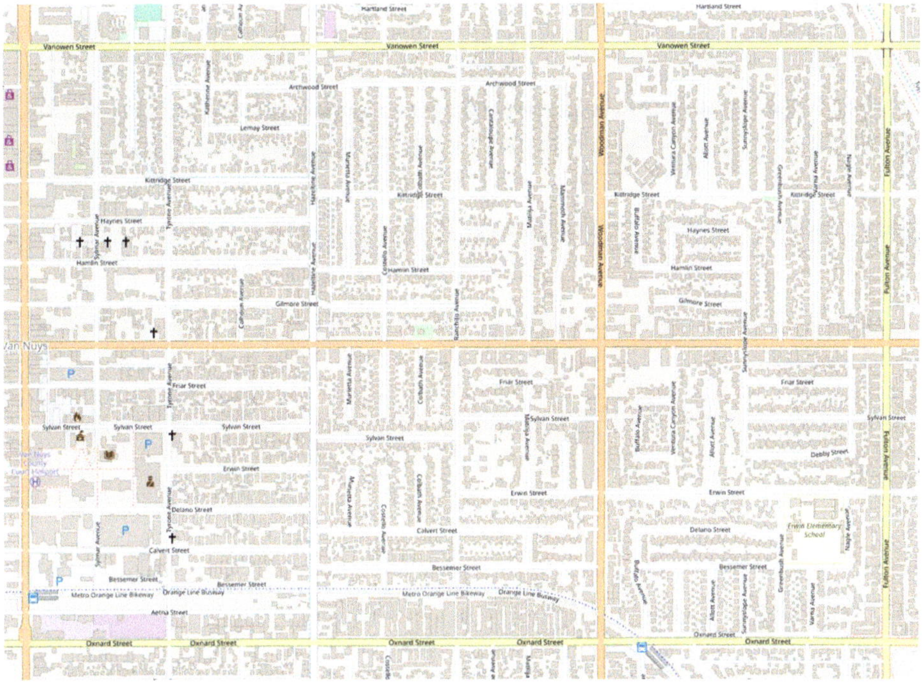

Fig. 3.8 "Optimized" road network in an urban area in Van Nuys, California, USA. (Image rights: OpenStreetMap)

unrestricted operation of today's conventional cars and trucks, EVs—even if the battery technology allows for quick charging—often have to manage with a modest charging capacity of a few kW in order not to locally overload the grid. Even with a 400 V/100 A (40 kW) connection at hand, a battery capacity of 20 kWh, i.e., a range of 100–200 km, would result in charging times of 30 minutes [10] (see also Table 3.2: A circumstance that predicts scenarios with incredibly long waiting times at charging stations).

1. The much-cited battery swapping[3] could indeed reduce the "refueling time" at rest stops but requires that all manufacturers agree on a uniform, standardized battery system. However, this agreement is unlikely to be reached if one follows discussions at the political or even corporate level, such as at the *European Electric Vehicle Congress* 2011 in Brussels. Some lectures and "panel discussions" were dedicated to the topic "Standardization of the EV charging connector," but did not bring any constructive

[3] A further problem of "battery swapping" is that for reasons of optimum center of gravity position, the vehicle battery is usually installed and distributed in the underbody of the vehicle, so that there is no single compact module for easy replacement.

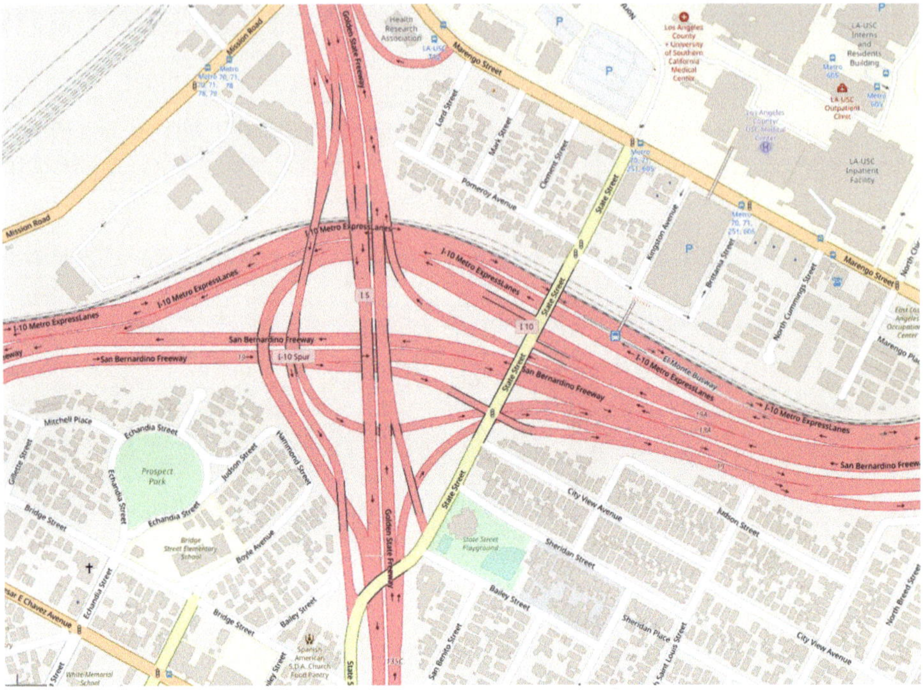

Fig. 3.9 Road network adapted to modern mobility: Highway I10, Los Angeles, California, USA. (Image rights: OpenStreetMap)

Table 3.2 Charging times and corresponding travel ranges of electric vehicles [11]

		Charging time with house connections[a]		Charging time with a fast-charging system[b]
Battery capacity (kWh)	Approximate range (km)	220 V, 15 A (3.3 kW) (h)	220 V, 30 A (6.6 kW) (h)	400 V, 100 A (40 kW) (h)
10	50–100	3.0	1.5	0.25
16	80–160	4.8	2.4	0.4
20	100–200	6.0	3.0	0.5
14	120–240	7.3	3.6	0.6
30	150–300	9.0	4.5h	0.75
100	500–1000	30.0	15.0	2.5

[a]Alternate current (AC) charging was assumed for "normal charging" at the house connection
[b]For the fast-charging system, direct current (DC) was assumed

results [11]. The nonprofit organization *Charging Interface Initiative* (CharIN) e. V. is currently trying to coordinate cross-industry stakeholders to move toward interoperable standardized EV charging, where vehicles, chargers, and software systems work together.

2. Even if hydrogen-powered vehicles have been pushed somewhat into the background in recent years due to advances in battery technology, the establishment of a nationwide hydrogen supply by using the existing gas network could soon make them more attractive. Especially in the commercial vehicle sector and long-distance transport, or in the area of heavy construction machinery, the required energy content is difficult to reach with chemical batteries. This represents an excellent opportunity for the hydrogen fuel cell. Nevertheless, the investment costs for the construction of the necessary hydrogen filling stations are estimated to be at least 300 million euros in Germany alone [12]. Even though the media often talk about the expansion of the hydrogen network, only five public hydrogen filling stations were available in the whole country of Austria in 2016 [13], and the number has not increased until today. For comparison: In 2014, there were already 2622 conventional gas stations in this small country [14]. At present, refueling a hydrogen car is a problem for private customers due to the lack of filling stations.
3. Concepts such as the "air car," which is powered solely by compressed air have also failed to reach a significant market share so far, due to a lack of public filling stations and are therefore of interest at most for commercial vehicle fleets with their own "refueling system" or filling station. Although the vehicle concept, which is currently being further developed and marketed by the company *Motor Development International (MDI)* from Luxembourg, is a good example of a fully functioning and viable *zero-emission vehicle*, the limited availability of compressed air as an energy source is an obstacle to the purchase of this vehicle. Examples of three vehicle concepts that allow locally emission-free operation but are still struggling with the "chicken-and-egg problem" between the vehicle fleet and the refueling or charging station network are shown in Figs. 3.10, 3.11, and 3.12.

3.4.2 Intended Use of the Vehicle

In addition to the dynamics of the driving cycle (see Sect. 4.2.1 for a detailed analysis), it is above all the *predictability of* the cycle or the application profile (use case) that is decisive for an exact design of the energy storage system and the operating strategy of a hybrid vehicle [15]. While commercial vehicles—especially in the public transport sector—offer very good predictability of the underlying driving cycle, this is generally not true in the case of passenger cars. The reason lies not only in the very individual behavior of the driver but often also the "misuse" of the car type. In recent years, this phenomenon has been particularly noticeable in the *sport utility vehicles* (SUV) segment. Although these vehicles were designed for light off-road use and for pulling/transporting heavy loads, they rarely

Fig. 3.10 *AirPod 2.0* compressed air car from *MDI*. (Image rights: Motor Development International)

Fig. 3.11 Hydrogen car *HYCAR 1* at *HyCentA*, Graz University of Technology, Austria. (Image rights: HyCentA Graz)

Fig. 3.12 *nanoFlowcell* vehicle with a redox flow battery, which could (in theory) be quickly refueled by exchanging the electrolyte (There is disagreement about the actual existence and functionality of the *nanoFlowcell* vehicle. However, this does not change the fact that this vehicle concept, similar to the other alternatives mentioned, does not have a charging or refueling infrastructure at the time.). (Image rights: nanoFlowcell IP AG)

leave the urban asphalt roads. Especially in the USA, where the share of these vehicles has already reached almost 30% and is still rising, this type of vehicle is mainly used as a pure city car. Psychological phenomena (see Sections 5.3.1 and 5.3.2) play a particularly important role here. Figures 3.13 and 3.14 give an overview of market shares and trends in the sales of vehicle types.

But even small and not very progressive countries such as Austria have jumped on the SUV trend, as a study published in 2019 by the Verkehrsclub Österreich (VCÖ, a public-benefit organization specializing in mobility and transport) shows. The share of four-wheel-drive and sports utility vehicles increased by a factor of 5 between 2004 and 2018 (see Fig. 3.15).

3.5 Driver and Energy Psychology

Due to the high technical complexity of modern means of transport, it cannot be assumed that the population has the "correct" mobility behavior from an energy perspective. As already mentioned at the beginning, studies by the *Institute of Electricity Economics and*

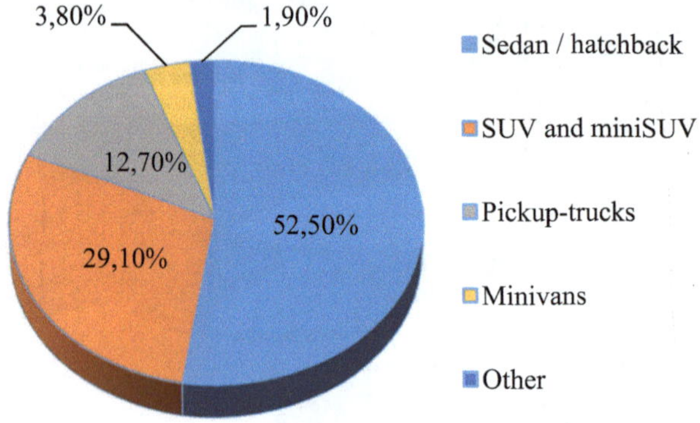

Fig. 3.13 Shares of passenger car types in the USA in 2011 [16]

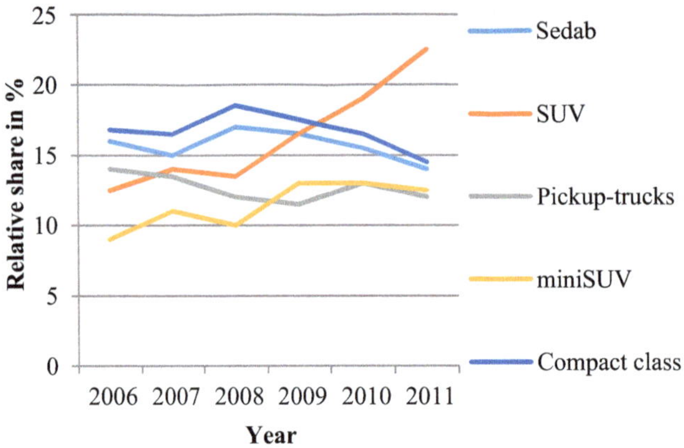

Fig. 3.14 Sales trends of various passenger car classes from 2006 to 2011 [18]

Energy Innovation (Institutes für Elektrizitätswirtschaft und Energieinnovation) at *the Graz University of Technology* have shown that energy is a highly nonintuitive, elusive physical quantity. Among the interviewees who were asked to estimate "what one could do with 1 kWh of energy" were also graduates from technical and other universities (compare Fig. 3.16.).

However, it is not only the misjudgment of the energetic expenditure for the provision of thermal and mechanical power, but also other sociopsychological effects that influence the mobility and driving behavior of the customer and, consequently, the architecture of the powertrain. Mobility and vehicle ownership are associated with the following symbolic attributes [1]:

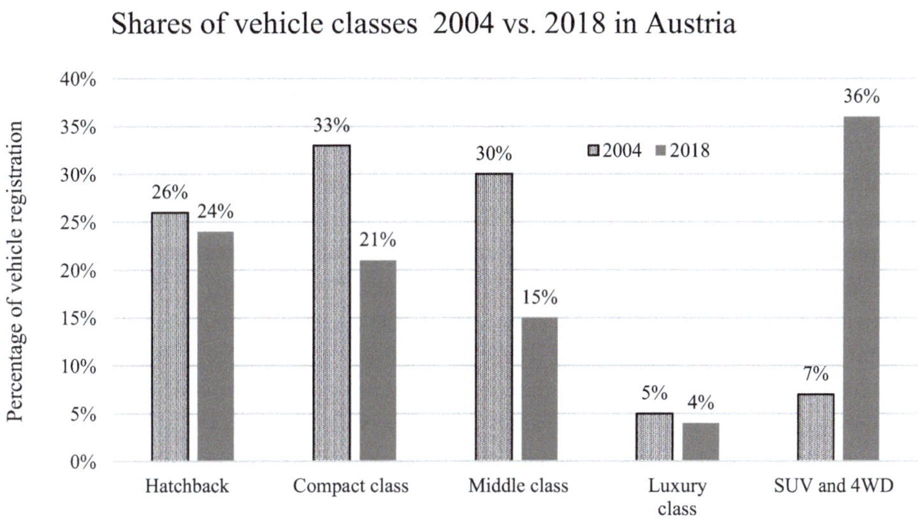

Fig. 3.15 In 2018, Austria's share of SUVs was five times as high as in 2004 [19]

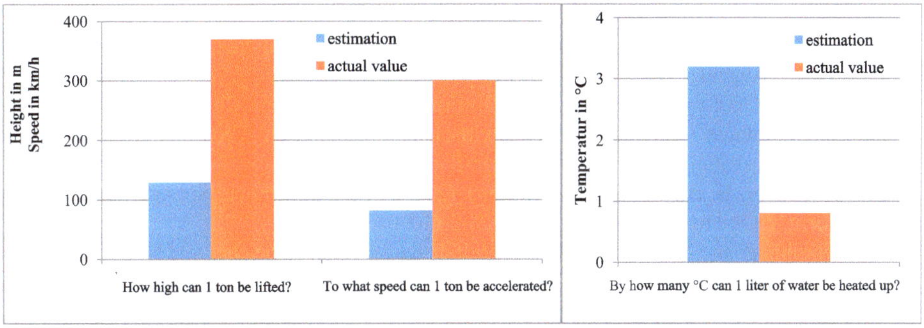

Fig. 3.16 Responses of the Austrian population to the "1 kWh question" [2]. (Image rights: Ludwig Piskernik)

1. *Autonomy:* Self-determination and a high degree of mobility can be considered an almost basic need of the modern citizen in the industrialized Western world. A vehicle conveys freedom to its owner and should in no way restrict it.
2. *Social status:* By choosing a certain vehicle, not only one's social status can be demonstrated but also the willingness to invest in "green" technologies. It is not uncommon for over-motorized, heavy-duty luxury-class vehicles to be offered as hybrid versions (*BMW X6*, *Porsche Cayenne*, *VW Touareg*, etc.), thus merely soothing the buyer's conscience and contributing in no way to sustainable mobility. Likewise, purely electric vehicles are often purchased exclusively as second or pure "city cars" in addition to conventional cars.

3. *Driving experience:* Acceleration, speed, road holding, range, and sound are some of the vehicle characteristics that significantly influence the choice of the vehicle or means of transport in general. This is one of the reasons why the historical development of the automobile has so far been mainly characterized by performance enhancement.

Another aspect that is becoming increasingly important in the course of the electrification of the powertrain is *psychoacoustics*. The so-called overdrive transmission is hardly accepted by customers due to the low engine speed and the resulting low-frequency "humming" noise, despite its efficiency [9] (see also Section 3.2.). A similar fate has befallen the *CVT gearbox*,[4] which allows the VKM to operate mainly at its best point (without discrete shift points) due to its continuously variable transmission ratio [3]. However, the customer expects an increase in vehicle speed directly proportional to the engine speed when the accelerator pedal is pressed. When using a CVT, however, there may even be a drop in engine speed with increasing vehicle speed due to parasitic losses of the control hydraulics [4]. Serial hybrids can exhibit a similar, nonintuitive psychoacoustic behavior, which can lead to rejection among customers.

References

1. H. Stiegler and U. Bachhiesl (2013) Grundlagen der Energieinnovation. Graz University of Technology, Austria
2. L. Piskernik (2008) Erfolgsfaktoren des Infrastrukturanlagenbaus. Chancen und Nutzen der Energiepsychologie für die Energiewirtschaft beim Bau von Kraftwerken und Hochspannungsleitungen. Verlag Dr. Müller, Saarbrücken.
3. H. Naunheimer, B. Bertsche and G. Lechner (2007) Fahrzeuggetriebe. Springer-Verlag Berlin Heidelberg. DOI: https://doi.org/10.1007/978-3-662-07179-3
4. A. Buchroithner and M. Bader (2014) Hybridfahrzeuge, Energiespeicher und Betriebsstrategien in der modernen Mobilität – Eine technologische Bewertung und Hinterfragung der Praxisrelevanz aus Kundensicht im Zuge einer interdisziplinären Systembetrachtung. 13. Symposium Energieinnovation, Graz, Austria.
5. Volvo Cars (2011) Flywheel KERS Component Details – Press Release. Volvo Car Corporation, Göteborg, Schweden. https://www.media.volvocars.com/global/en-gb/media/photos/38469
6. H. Eichelseder and J. Blassnegger (2006) Der zukünftige Ottomotor – Überlegener Wettbewerber zum Dieselmotor? Institut für Verbrennungskraftmaschinen und Thermodynamik, Graz University of Technology, Austria.
7. G. Jürgens (1996) Rekuperation – eine „ewige“ Herausforderung, Graz University of Technology, Austria.
8. Si-Yeon Kim et al (2013) A study on the construction of EV charging infrastructures in highway rest area. 4th International Conference on Power Engineering, Energy and Electrical Drives. Istanbul, Turkey. DOI: https://doi.org/10.1109/PowerEng.2013.6635639

[4] CVT stands for "Continuously Variable Transmission", a continuous transmission without discrete shift points.

9. European Electric Vehicle Conference (2001) Agenda, Forum Europe, Castle House, 1–7 Castle Street, Cardiff, CF10 1BS, UK. https://eu-ems.com/agenda.asp?event_id=72&page_id=523.
10. J. Kohler, M. Wietschel, L. Whitmarsh, D. Keles and W. Schade (2008) Infrastructure investment for a transition to hydrogen road vehicles. 2008 First International Conference on Infrastructure Systems and Services: Building Networks for a Brighter Future (INFRA). Rotterdam, Netherlands. DOI: https://doi.org/10.1109/INFRA.2008.5439664
11. http://www.Netinform.net (2016) Hydrogen / Fuel Cells – Hydrogen Filling Stations Worldwide. http://www.H2-Stations.org, 2016. http://www.netinform.net/h2/H2Stations/Default.aspx. [Accessed September 9th 2016].
12. M. Sackl (2015) Tankstellenstatistik 2014: Immer weniger Major-Branded Tankstellen in Österreich. Wirtschaftskammer, Fachgruppe GTS Wien. http://www.sackl.net/2015/06/26/tankstellenstatistik-2014-immer-weniger-major-branded-tankstellen-in-oesterreich/. [Accessed September 9th 2016].
13. M. Bader, A. Buchroithner, I. Andrasec and A.Brandstätter (2012) Schwungradhybride als mögliche Alternative für den urbanen Individual- und Nahverkehr," 12. Symposium Energieinnovation, Graz, Austria.
14. T. Cain (2011) Auto Sales by Segment in the USA and Canada – May 2011. http://www.GoodCarBadCar.net, Ausgabe 23. 06. 2011. http://www.goodcarbadcar.net/2011/06/auto-sales-by-segment-in-usa-and-canada. html. [Accessed November 3rd 2014].
15. M. Bland (2011) Long Live the SUV – The Light Vehicle Leader. IHS Automotive, Ausgabe 11. 05. 2011. http://blog.polk.com/blog/blog-posts-by-marc-bland/long-live-the-suv-the-light-vehicle-leader. [Accessed Juli 8th 2014].
16. Verkehrsclub Österreich (2018) Verkehrswende braucht Konsumwende. https://www.vcoe.at/themen/verkehrswende-braucht-konsumwende. [Accessed June 5th 2019].

4 Interaction Between *Subsystem* and *Supersystem* of Mobile Flywheel Energy Storage

Even in the case of a modern Li-Io battery with a solid electrolyte and basically position-independent behavior, attributes such as efficiency, self-discharge, maximum power, and, last but not least, service life depend on environmental conditions such as temperature or ambient pressure [1]. Investigations by the *Eindhoven University of Technology (Technische Universität Eindhoven)* in the Netherlands have shown that the nominal range of a battery electric vehicle is drastically reduced in winter when increased auxiliary power is required, for example, for heating, as Fig. 4.1 clearly shows.

The load profile, which is defined by the customer, also has a striking effect on the service life of a battery. For example, the total energy that can be converted over the life of the battery can be maximized by *microcycling*,[1] whereas frequent deep discharge reduces the service life. This influence of environmental conditions on the "cell chemistry," i.e., the *subsystem of* the energy storage device, is even more dramatic in the case of a flywheel system, which contains fast-rotating mechanical components. For this reason, it is necessary to analyze the entire system applying a holistic approach from "coarse" (*supersystem*—environment, vehicle, driver, infrastructure) to "fine" (*subsystem*—details and components within the FESS) and to define a system hierarchy as shown in Fig. 4.2. The potential and importance of a holistic approach for product optimization using the example of flywheel energy storage is confirmed in [2].

Figure 4.2 shows the structure of the flywheel energy storage system in its *sub-* and *supersystem*. For a better overview, the *supersystem* is again divided into vehicle and environment, and the *subsystem* into "entire energy storage" and "component," whereby in this case the ball bearing, which turns out to be a key element in FESS design, was picked

[1] Extracting an amount of energy corresponding to only a small percentage of the battery capacity and then recharging it again

A. Buchroithner, *Flywheel Energy Storage*,
https://doi.org/10.1007/978-3-658-35342-1_4

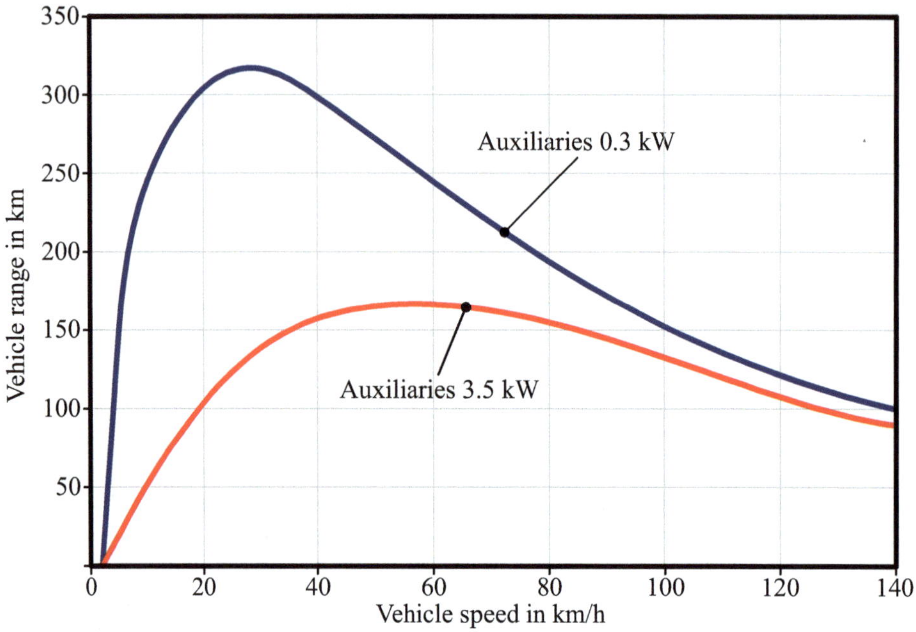

Fig. 4.1 Range of a battery electric vehicle (*ECE VW Golf*) as a function of cruising speed and auxiliary unit power [1]. (Image rights: I.J. Besselink, TU Eindhofen)

out. Concrete examples of the mutual influence and internal interdependencies are given in Sect. 4.1.

4.1 Examples of Direct Influence on the *Super-* and *Subsystem* of FESS

The system hierarchy shown in Fig. 4.2 picks out the ball bearing. Although this machine element will be discussed in detail in Chapter 9, a prominent example of the interaction between this element of the energy storage's *subsystem* and its environment (*supersystem*) is described here:

1. **Road holding and road condition** influence the driving dynamics (pitch, yaw, roll, and maximum acceleration values of the vehicle) and thus define the bearing loads, including gyroscopic reactions due to the gyroscopic moment of the flywheel.
2. **Driving cycle** and required energy storage time define the maximum permissible self-discharge rate and subsequently the tolerable torque loss of the bearing system. The rotational speed collective of the bearings is also defined by the charge state of the FESS.

A summary of the main interactions between aspects of the *supersystem* and components of the flywheel *subsystem* is given in Table 4.1.

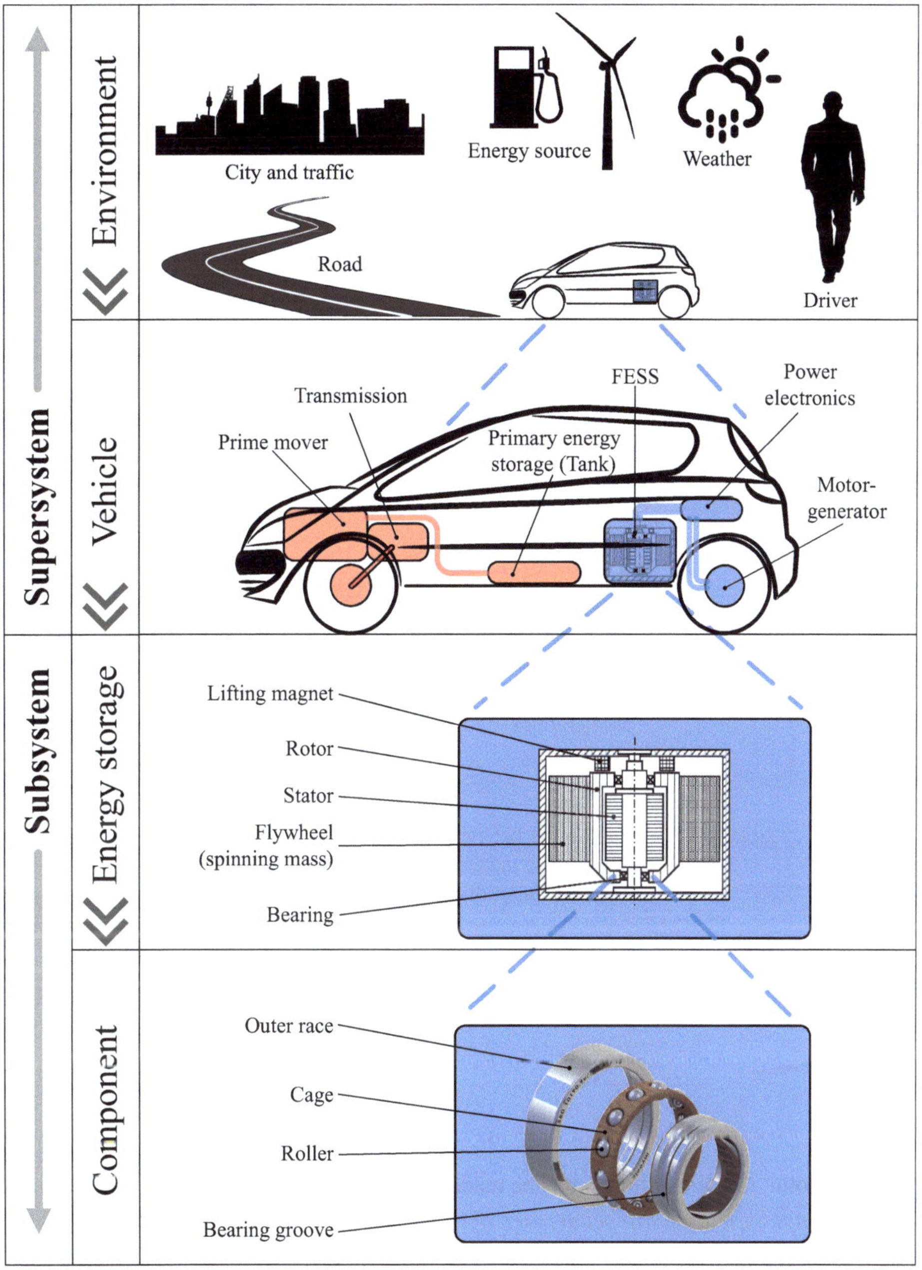

Fig. 4.2 System hierarchy and interaction between the *subsystem* and *supersystem* using the example of ball bearings in a FESS

Table 4.1 Interaction mechanisms between the *subsystem* and *supersystem* of a FESS

Supersystem aspect	Interaction mechanism	Subsystem aspect
Infrastructure and traffic	Dynamics of the driving cycle is influenced by the road network and traffic → recuperation potential and required energy storage duration as well as predictability of the driving cycle/load collective	Permissible loss torque of the bearings Aerodynamic losses and vacuum technology
Energy source (primary energy mix)	Rising prices of fossil fuels and falling electricity prices → savings and profitability of the FESS hybrid system	Increasing numbers of units → manufacturing processes for large series
Weather	Ambient temperature, driving resistance forces, and friction conditions on the road due to precipitation → maximum possible longitudinal and lateral accelerations	Viscosity of lubricants Control system to prevent locking of the wheels during regenerative braking
Driver and psychology	Purchasing behavior or maximum permissible price → public image and perception of the technology	Applicable technologies/ components
Road condition	Maximum accelerations in the vehicle → bearing load and rotational speed collective of the FESS	Design of the bearing system

At this point, reference should also be made to the chapter in Sect. 6.2, which is devoted to the *internal system interdependencies* between the FESS components that have been identified as *critical* and the categorization of these interdependencies.

Special importance must be attributed to the characteristics of the driving cycle, which is why this is examined in more detail in the following section with regard to the recuperation potential using flywheel energy storage.

4.2 Optimization of the *Supersystem*

4.2.1 Influence of the Driving Cycle on the FESS

It is not only the partially irrational purchasing behavior of the customer which is influenced by marketing mechanisms (see Sect. 5.3.1) but also the almost impossible predictability of the driving cycle which makes dimensioning of an energy storage device in a car such a difficult task.

Until now, it has been the standardized *New European Driving Cycle* (NEDC) that has been used for the energetic assessment and optimization of most vehicles in Europe. The powertrain and the operating strategy have thus been adapted to a *synthetic driving cycle* specified by law. Needless to say that using the vehicle in a way that deviates strongly from this standard cycle results in considerably worse fuel consumption values. Figure 4.3 compares different standardized driving cycles and describes their energetic characteristics and implications on the design of the powertrain.

At the latest since the "Volkswagen emissions scandal" (also called "Dieselgate") involving not only the German automobile industry but various large automobile companies, which used modified engine management systems in their vehicles in the event of a standard test cycle in order to reduce CO_2 emissions, even the general public knows that certification on the basis of standardized driving cycles is problematic [3]. It is interesting to note that the "fuel consumption optimization" or automatic detection of the vehicle, whether it is doing test cycle or being driven in real traffic, was an "open secret" with which many vehicle engineers were familiar. Even vehicles such as the *Toyota Prius*,

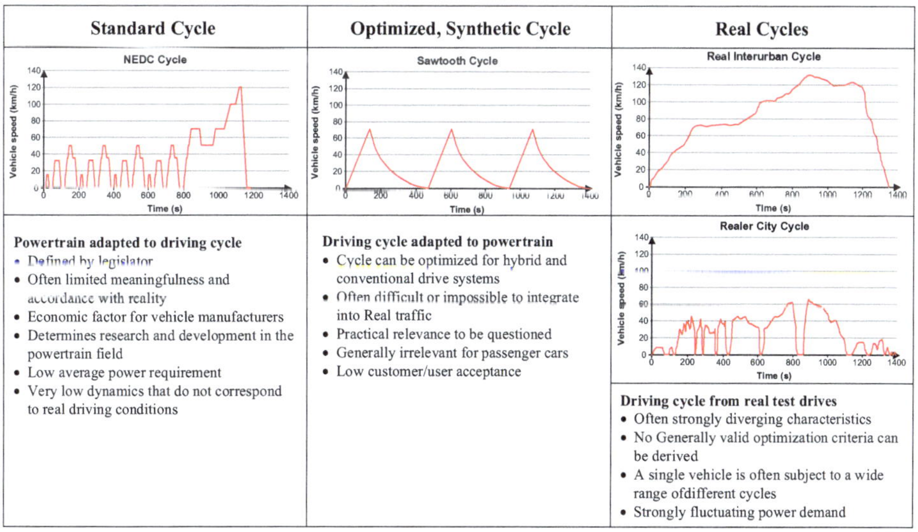

Fig. 4.3 Different driving cycles and their energetic characteristics

which have a reputation of being very environmentally friendly, behave much less efficiently outside the *NEDC*—albeit within the legal limits [4]. It is obvious that the replacement of the *NEDC* by the so-called Worldwide Harmonized Light Vehicles Test Procedure (*WLTP*), a cycle generated from real-world test drives, is a step in the right direction. However, two essential issues remain so far unsolved in this context:

1. It will remain unavoidable that many car users will use their vehicles in a way that deviates significantly even from the *WLTP*. Even though a small, average reduction in CO_2 emissions might be achieved, the problem of the "unpredictability" of the individual user's driving behavior remains.
2. The changeover to a new "stricter" test cycle will probably again bring "only" incremental improvements and a gradual reduction in consumption.[2] The development of completely new vehicle structures or even a new mobility concept will unfortunately not be initiated by this measure.

4.2.2 Energy Requirements of the Vehicle

Looking at the longitudinal dynamics of a vehicle, the power demand of the vehicle during stationary travel consists of the following three components [6]:

1. Rolling resistance F_{roll}
2. Climbing resistance F_{climb}
3. Aerodynamic drag F_{air}

The dependence of the rolling resistance F_{roll} on tire design, tire pressure, road surface, etc. is taken into account in a generalized way in the friction coefficient f_R and thus results:

$$F_{roll} = m * g * f_R \tag{4.1}$$

In theory, F_{roll} would still have to be multiplied by the cosine of the gradient of the incline of the road, $cos(\alpha_s)$, since the wheel contact force decreases with increasing gradient. However, the effect is small for small angles and can thus be neglected here.

Furthermore, since α_s applies to small angles of inclination, it can be assumed that $\sin(\alpha_s) \approx \tan(\alpha_s)$; hence,

[2] Fleet-specific CO_2 emissions of approx. 95 g/km can be expected when using the WLTP, compared to 80–85 g/km in the NEDC [5].

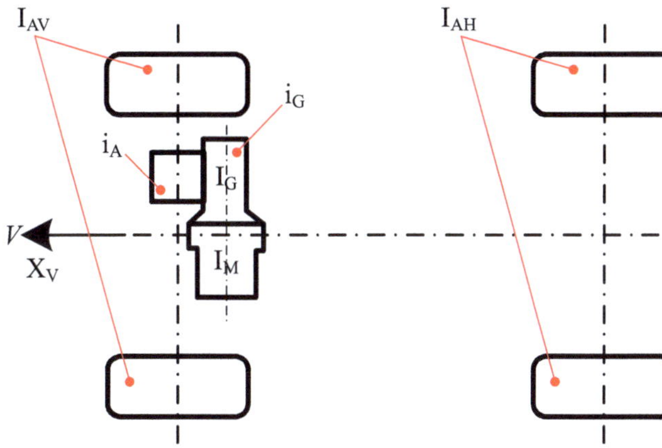

Fig. 4.4 Moments of inertia in the drivetrain of a passenger car with front-wheel drive [6]. (Image rights: Prof. Wolfgang Hirschberg)

$$F_{climb} = m * g * \tan(\alpha_s) \tag{4.2}$$

And according to *Bernoulli's* well-known equation,

$$F_{air} = \frac{1}{2} * c_W * A * \rho_L * v^2 \tag{4.3}$$

The force acting on the vehicle required for stationary travel is the sum of these three forces ($F_{roll} + F_{climb} + F_{air}$). To accelerate the vehicle, the mass forces must be taken into account as well. If the *translational* and *rotational* mass components of the drivetrain are combined, one speaks of the so-called generalized vehicle mass $m*$ [6]:

$$m^* = m_A + m_V + m_H + s_V^* \frac{I_V}{r^2} + s_H^* \frac{I_H}{r^2} \tag{4.4}$$

The *slip* $s*$ at the front and rear wheels plays only a minor role in the approximate energy/force balance and can be assumed to be ~1 in this case. The moments of inertia at the front or rear axle can be calculated according to Eq. 4.5. In the case of a front-wheel-drive vehicle, the following applies:

$$I_{front} = I_{A\_front} + i_A^2 \times \left(I_G + i_G^2 I_M\right) \tag{4.5}$$

$$I_{rear} = I_{H\_rear} \tag{4.6}$$

Figure 4.4 shows the concatenation of the moments of inertia in the drivetrain. As an example, a passenger car with front-wheel drive is shown, which currently has the highest market share.

Dividing the *generalized vehicle mass m** by the actual vehicle mass *m* gives the *rotational mass factor* λ, and the force demand for general, transient driving can be expressed as follows:

$$F_{req} = mg(f_R + \tan(\alpha_s)) + \frac{1}{2}c_W A \rho_L v^2 + m\ddot{x}\lambda \quad (4.7)$$

Applying, $\dot{x} = v$ *general equation of motion of the vehicle* can be formulated:

$$\ddot{x} + \frac{\frac{1}{2}\ c_W\ A\rho_L}{m\lambda}\ \dot{x}^2 + \frac{g}{\lambda}\ (f_R + \tan(\alpha_s)) - \frac{F_{req}}{m\lambda} = 0 \quad (4.8)$$

Integrating the required power over the distance *x* results in obtaining the required mechanical energy at the wheel, which must be provided by the engine to overcome the distance *x*:

$$W_{req} = \int_0^x F_{req} dx \quad (4.9)$$

$$W_{req} = \int_0^x mgf_R dx + \int_0^x mg\ \tan(\alpha_s) dx + \int_0^x \frac{1}{2} c_W A\ \rho_L \dot{x}^2 dx + \int_0^x m\ddot{x}\lambda dx \quad (4.10)$$

Since the energy required is dependent on speed and acceleration, it varies greatly depending on the underlying driving cycle. An empirically determined approximation formula for the *NEDC driving cycle is* proposed by *Prof. Lino Guzzella* of *ETH Zurich* in [7]: (result in kJ/100 km)

$$W_{reqNEDC} \approx A * c_W\ \ 19000 + m\ f_R\ \ 840 + m\ \ 11 \quad (4.11)$$

4.2.2.1 Theoretically Recoverable Energy

This value is essential for the design of a vehicle's flywheel energy storage system, but in no case, it is the same for all vehicle types and should therefore be considered more closely. To determine the recoverable energy, a distinction must be made between areas of positive ($\ddot{x} > 0$) and negative ($\ddot{x} < 0$) acceleration [8]. The following considerations form the basis for the simulation results presented in Sect. 5.2.1.

The geopotential (potential energy due to altitude difference) is not taken into account in the first instance, meaning the case of a horizontal road is considered, which allows the assumption that$\ddot{x} = a$. To give an example, a synthetic cycle is used to explain regenerative

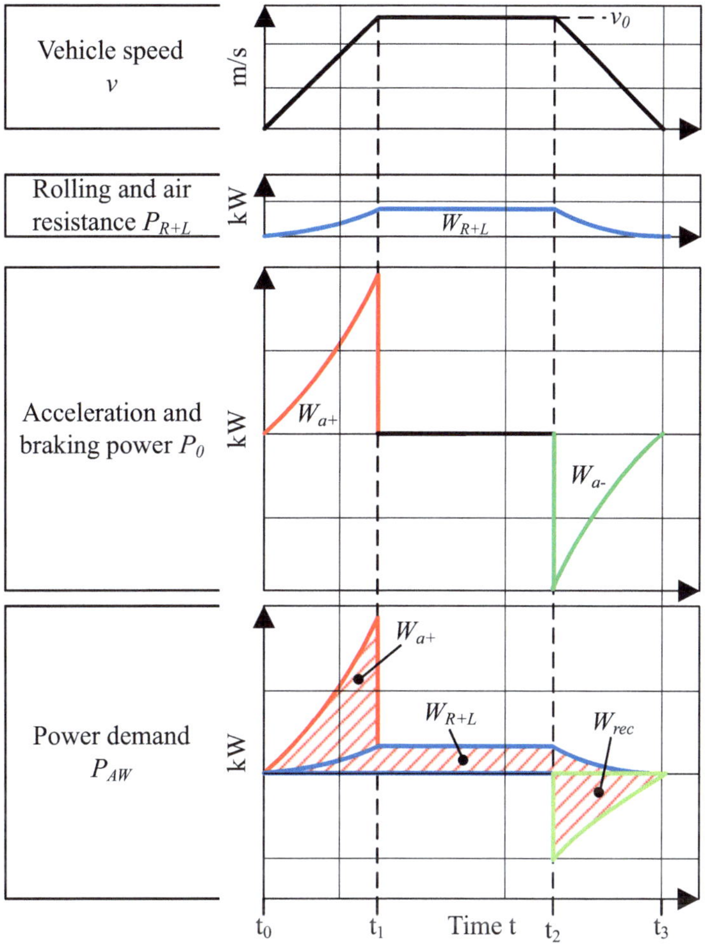

Fig. 4.5 Qualitative progression of speed, power, and energy consumption of a motor vehicle, based on [9]

braking and the corresponding time periods. This cycle is qualitatively illustrated in Fig. 4.5.

It can be seen that the recoverable energy W_{rec}—which is less than that required to accelerate the vehicle—is made up of

$$W_{rec} = \int_{0}^{t_1} m\lambda av\ dt - \int_{t_2}^{t_3} mgf_R v\ dt - \int_{t_2}^{t_3} \frac{1}{2} c_W A\ \rho_L v^3\ dt \tag{4.12}$$

Assuming that

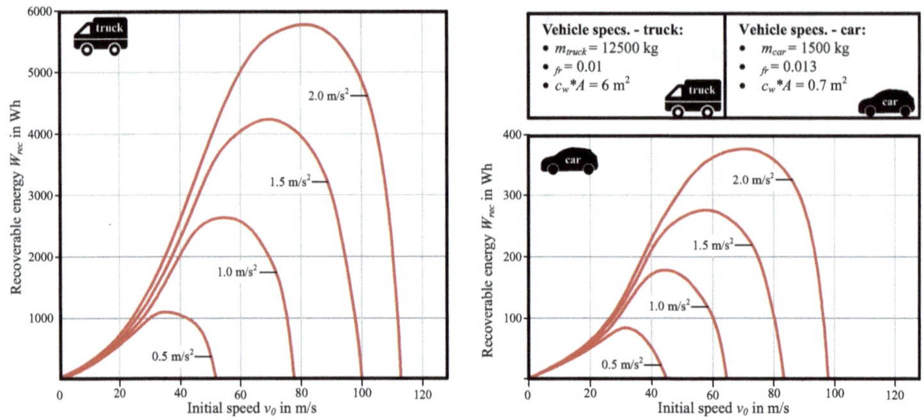

Fig. 4.6 Recoverable braking energy W_{rec} as a function of the initial speed v_O [9]

$$a = constant = \frac{dv}{dt} \tag{4.13}$$

and

$$dt = \frac{dv}{a}, \tag{4.14}$$

the integral can be solved:

$$W_{rec} = m\lambda \frac{v_0^2}{2} - \frac{mgf_R}{a}\frac{v_0^2}{2} - \frac{c_W A\ \rho_L}{2a}\frac{v_0^4}{4} \tag{4.15}$$

Figure 4.6 shows the recoverable energy W_{rec} of two vehicles depending on the initial speed v_O. The left diagram was determined according to [9] based on typical properties of a commercial vehicle (vehicle mass 12,500 kg), and the right diagram used values for a small passenger car (vehicle mass 1100 kg). As can easily be seen, due to smaller air and rolling resistance losses, for constant deceleration a, W_{rec} initially increases with v_O. Because of the increase of air resistance at higher order, from a certain speed—this speed shall be called v_{max}^{rec}—W_{rec} decreases drastically and eventually becomes zero.

If one applies the classical extremum problem to the equation for W_{rec} and sets its first derivative of the velocity to zero, one gets

$$v_{max}^{rec} = \sqrt{\frac{m\lambda a - mgf_R}{\frac{1}{4}c_W A\ \rho_L}} \tag{4.16}$$

It is easy to see that the recoverable kinetic energy of a vehicle also increases with higher deceleration values, but extremely strong decelerations do not lead to a significant increase in efficiency of regenerative braking.

4.2.2.2 Effective Speed Range for Regenerative Braking

Clearly, in the case of coasting to a complete stop, W_{rec} becomes zero and requires a variable deceleration a, since the force demand F_{req} must also be zero. Figure 4.7 shows the share of the energy that can be recovered from the kinetic energy of the vehicle as a function of the deceleration a for a typical transit bus in inner-city traffic. Note that buses like this are most suitable for hybridization and regenerative braking. The figure considers the speed range relevant for regenerative braking.

Up to about 0.2 m/s^2, the assumed vehicle data indicates a *coasting process*, and the entire kinetic energy of the vehicle is converted into irreversible friction (also aerodynamic) losses. Thus, the effectiveness of regenerative braking only begins at deceleration values of 0.4 m/s^2 or higher, in which case between about 50% and 90% of the kinetic energy is available for recovery. In normal comfortable driving operation, decelerations in an order of magnitude of 0.4 to 2 m/s^2 occur.

Since the relative proportion of driving resistances increases with speed, the share of recoverable energy is greater at the constant deceleration assumed here for low initial vehicle speeds.

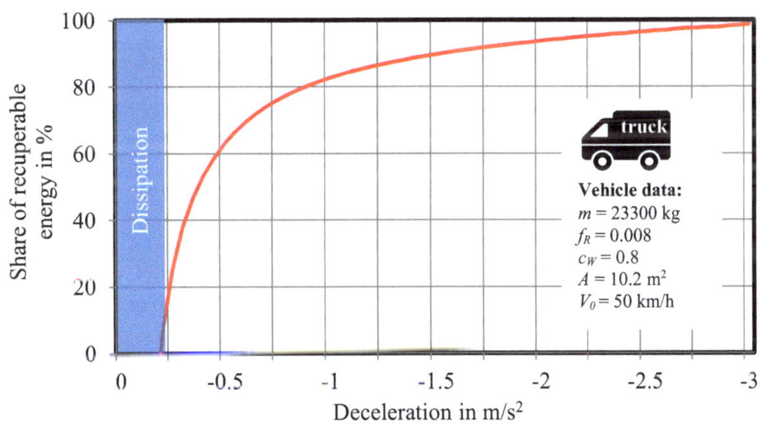

Fig. 4.7 Influence of deceleration values on the share of recoverable energy

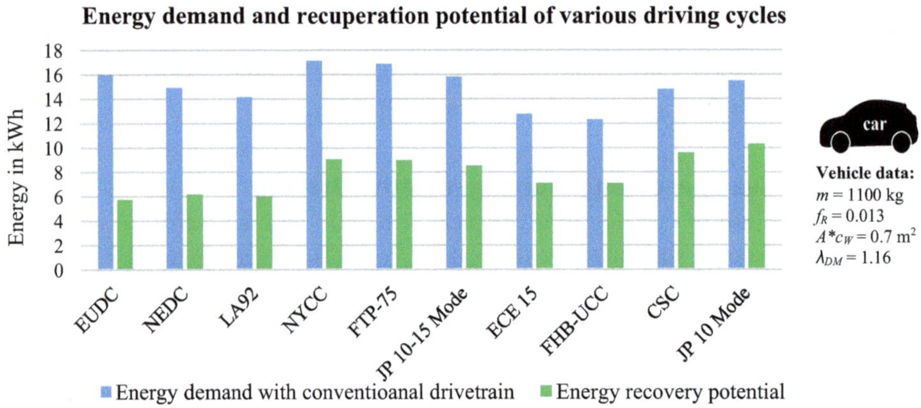

Fig. 4.8 Energy demand and recuperation potential of different driving cycles calculated from the data available in [10]

4.2.3 Profitability of a FESS in a Vehicle

In engineering science, the attempt to evaluate inventions on the basis of *quantifiable specifications* dominates. It is therefore obvious that there is also a desire to assess the profitability of flywheel energy storage systems in vehicles on the basis of one (or few) explicit comparative values. The effectiveness of regenerative braking is primarily subject to the characteristics of the driving cycle. Thus, it makes sense to evaluate the recuperation potential by introducing the so-called *dynamic index* (K_{dyn}), which provides information on whether the use of a FESS in a vehicle is financially or energetically profitable:

$$K_{dyn} = \frac{Theoretical\ recuperation\ potential}{Energy\ consumption} \quad (4.17)$$

The aim of K_{dyn} is to compare the energy recovery potential of different driving cycles. For this purpose, the characteristics of a reference vehicle are defined and the energy recovery potential of different driving cycles is determined on the basis of Eq. 4.10 described in Sect. 4.2.2.

Table 4.2 Selection of standardized driving cycles and their corresponding *dynamic index* K_{dyn} as a measure of the energy recovery potential

Drive cycle	Abbreviation	Figure (from [11])	Duration	Length	Average	K_{dyn}
			Sec.	*Km*	*Km/h*	–
Extra-urban driving cycle	EUDC	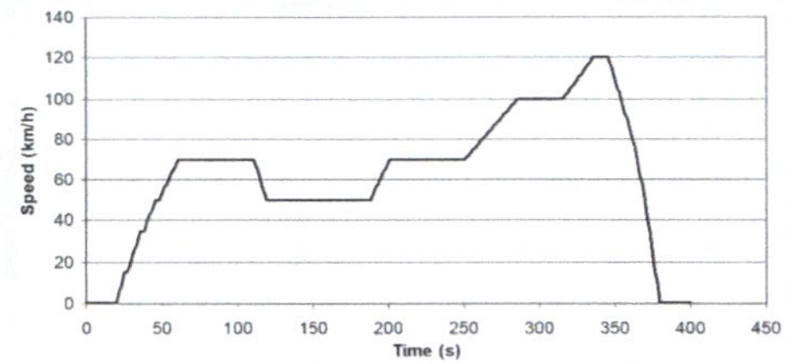	400	6.95	62.6	0.36
New European driving cycle	NEDC	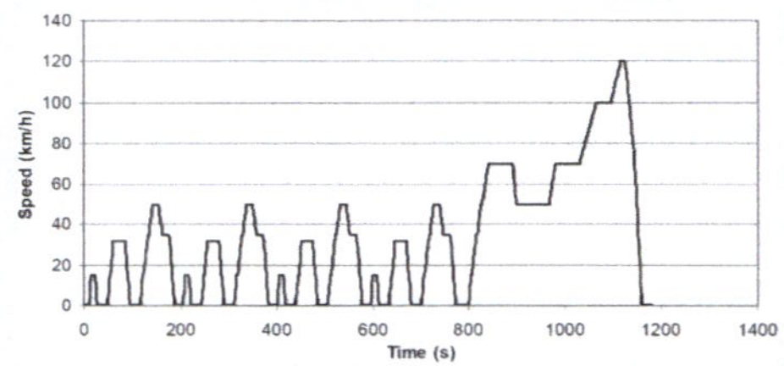	1180	11.0	33.6	0.41
California unified LA92	LA92	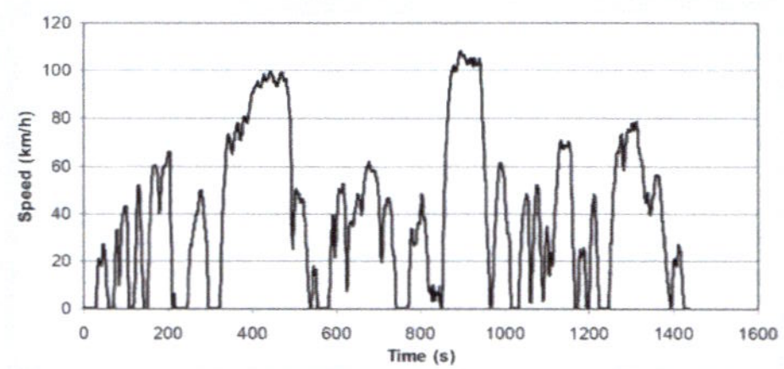	1435	15.8	39.6	0.42

(continued)

Table 4.2 (continued)

Drive cycle	Abbreviation	Figure (from [11])	Duration	Length	Average	K_{dyn}
			Sec.	*Km*	*Km/h*	–
EPA new York City cycle	NYCC	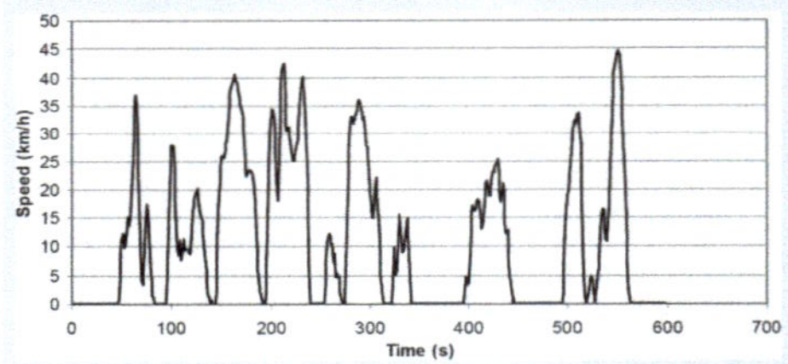	598	1.9	11.5	0.53
ADR 37 Australia	FTP-75	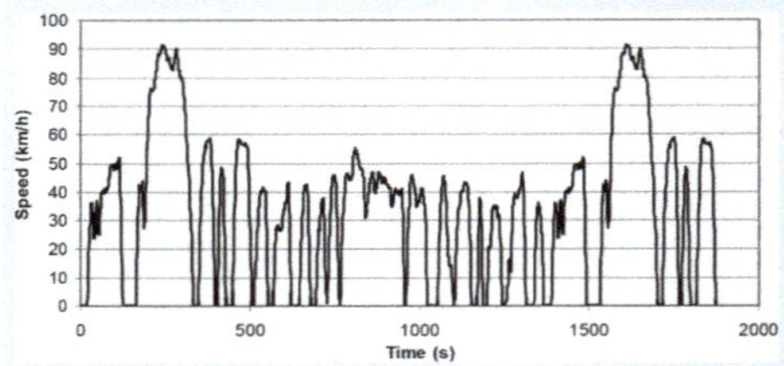	1874	17.8	34.2	0.53
Japanese legislative cycle	JP 10-15 mode	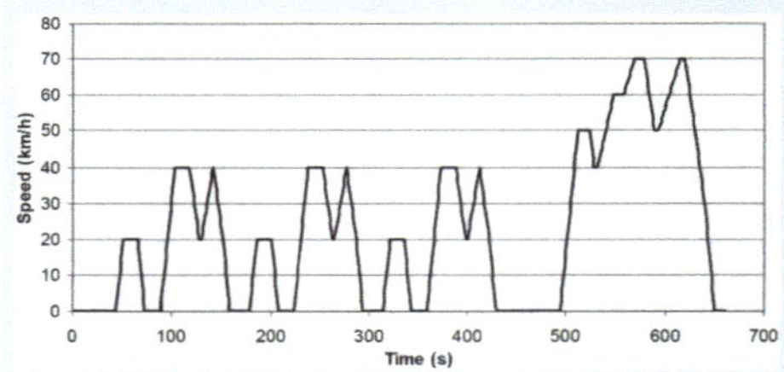	660	4.17	22.7	0.54

New European driving cycle	ECE 15	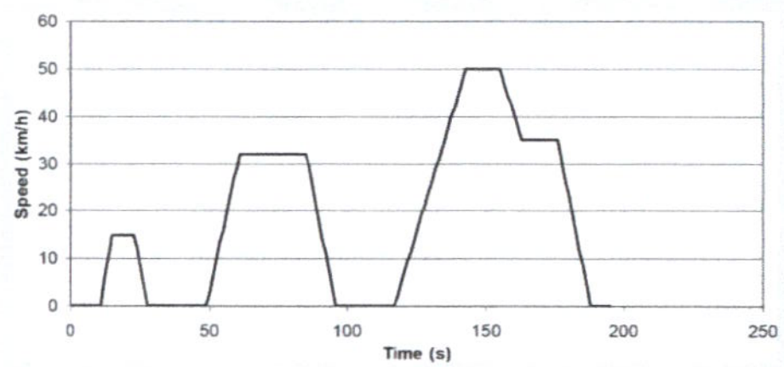	195	0.99	18.4	0.55
University of Applied Sciences Biel Urban City Center	FHB-UCC	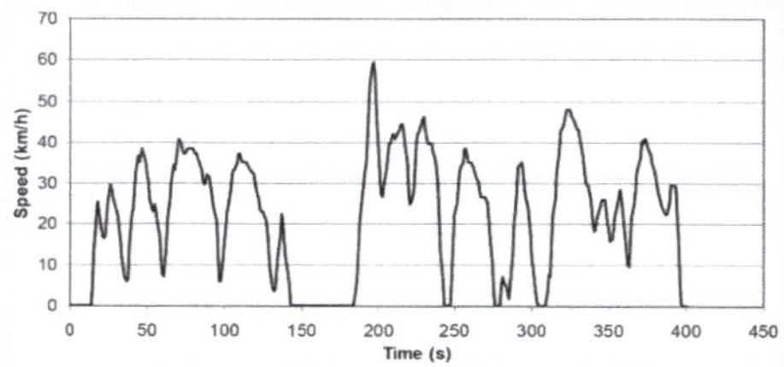	401	2.47	22.2	0.57
City suburban cycle	CSC	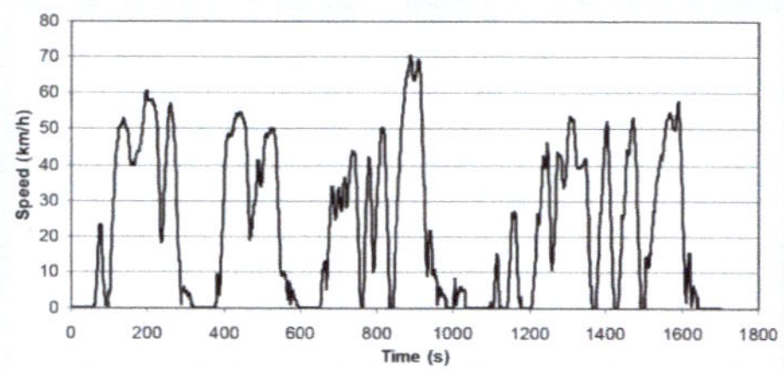	1700	10.75	22.8	0.65
Japanese legislative cycle	JP 10 mode	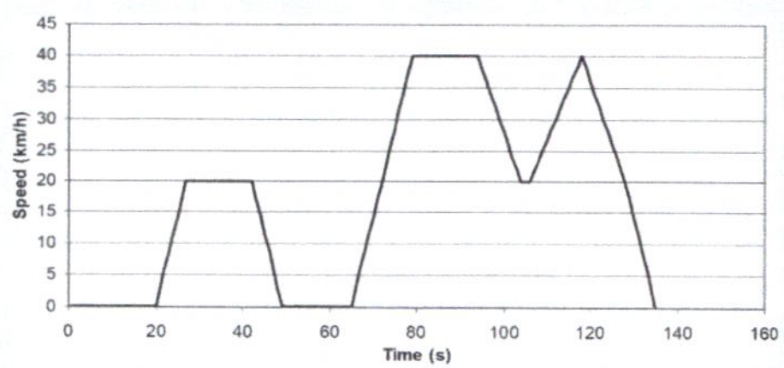	135	663	17.7	0.66

A summary of the different driving cycles listed in [10] is given in Fig. 4.8. As expected, the European *Extra-Urban Driving Cycle* (EUDC), i.e., an overland cycle, has the lowest potential for recuperation. It should be noted that this consideration is a theoretical potential assessment, which assumes the efficiency of the drivetrain to be 100%, i.e., $\eta = 1$.

The *dynamics index* can be seen as the ratio between the energy recovery potential and the total energy required to complete a cycle (see Eq. 4.17). Table 4.2 shows a selection of ten standardized driving cycles and their *dynamic index*. The higher K_{dyn}, the better the suitability for using a FESS in the drivetrain. Logically, an exact quantification of the financial payback period depends on the annual mileage, the current fuel prices, as well as the overall efficiency of the vehicle and the FESS.

References

1. I.J. Besselink, J.A. Hereijgers, P.F. van Oorschot and H. Nijmeijer (2011) Evaluation of 20000 km driven with a battery electric vehicle. Europen Electric Vehicle Congress (EEVC), Brussels, Belgium.
2. A. Buchroithner, H. Lang and M. Bader (2016) A Holistic Approach for Technical Product Optimization. ICDPD 2016 – 18th International Conference on Design and Product Development, Paris, France.
3. Die Presse Online (2015) VW-Skandal: Falsche CO_2-Daten beschäftigen Justiz. http://diepresse.com/home/wirtschaft/international/4860183/VWSkandal_Falsche-CO2Daten-beschaeftigen-Justiz. [Accessed November 19th 2015].
4. M. Burt (2014) MPG and Running Costs. Autocar – First for Car News and Reviews. http://www.autocar.co.uk/car-review/toyota/prius/mpg. [Accessed November 19th 2015].
5. P. Mock, J. Kühlwein, U. Tietge, V. Franco, A. Bandivadekar and J. German (2014) The WLTP: How a new test procedure for cars will affect fuel consumption values in the EU. ICCT – The International Council on Clean Transportation, Washington, D.C., USA.
6. H. Hirschberg (2005) Skriptum Kraftahrzeugtechnik. Institut für Kraftfahrzeugtechnik, Graz University of Technology, Austria.
7. L. Guzzella (2008) Technische Optionen für den Individualverkehr der Zukunft., „Auto der Zukunft – Zukunft des Autos" (Lecture), ETH Zürich, Switzerland. http://docplayer.org/14827192-Auto-der-zukunft-zukunft-des-autos-lino-guzzella-http-www-imrt-ethz-ch.html
8. M. Bader, A. Buchroithner, I. Andrasec and A. Brandstätter (2012) Schwungradhybride als mögliche Alternative für den urbanen Individual- und Nahverkehr. 12. Symposium Energieinnovation, Graz, Austria.
9. H. Schreck (1977) Konzeptuntersuchung, Realisierung und Vergleich eines Hybrid-Antriebes mit Schwungrad mit einem konventionellen Antrieb. Fakultät für Maschinenwesen der Rheinisch-Westfälischen Technischen Hochschule, Aachen, Germany.
10. M. Ackerl (2016) Innovative Fahrzeugantriebe – Simulation von Fahrzeugen mit alternativen Antrieben. FTG – Institut für Fahrzeugtechnik, Graz University of Technology, Austria.
11. T. J. Barlow, S. Latham, I. S. McCrae and P. G. Boulter (2009) A reference book of driving cycles for use in the measurement of road vehicle emissions. IHS, Wolloughby Road, Bracknell, Berkshire, UK.

Optimizing the *Supersystem* of Mobile Energy Storage

5

5.1 Emotion Versus Ratio: Passenger Car Versus Commercial Vehicle

Even if the appearance and technical design of a compact delivery van and a "family van" (or minivan) hardly differ, the respective customer base and use and consequently their suitability for powertrain hybridization are fundamentally different. While in the commercial vehicle sector almost exclusively rational-economic criteria influence the purchase decision, the passenger car sector is highly emotional, as Sect. 5.3 will discuss in more detail.

> In all further considerations, a strict distinction must be made between personal cars and commercial vehicles!

An essential prerequisite for the efficient use of a flywheel energy storage system in a vehicle is a highly dynamic and predictable driving cycle, as it is usually found in commercial vehicles in urban transport. Figure 5.1 shows a qualitative representation of the *suitability* for vehicle hybridization using a flywheel in relation to the *predictability* of the load collective. While buses designed for urban transport guarantee good profitability of a FESS due to their dedicated use, a certain type of passenger car (e.g., SUV) can be predominantly used in city as well as purely interurban traffic, featuring mainly constant speed travel with hardly any potential for energy recuperation.

For the monetary payback period already mentioned, the ratio of operating time to downtime of a vehicle is of great relevance. Here, too, commercial vehicles show a much more favorable behavior, as Fig. 5.2 illustrates.

A. Buchroithner, *Flywheel Energy Storage*,
https://doi.org/10.1007/978-3-658-35342-1_5

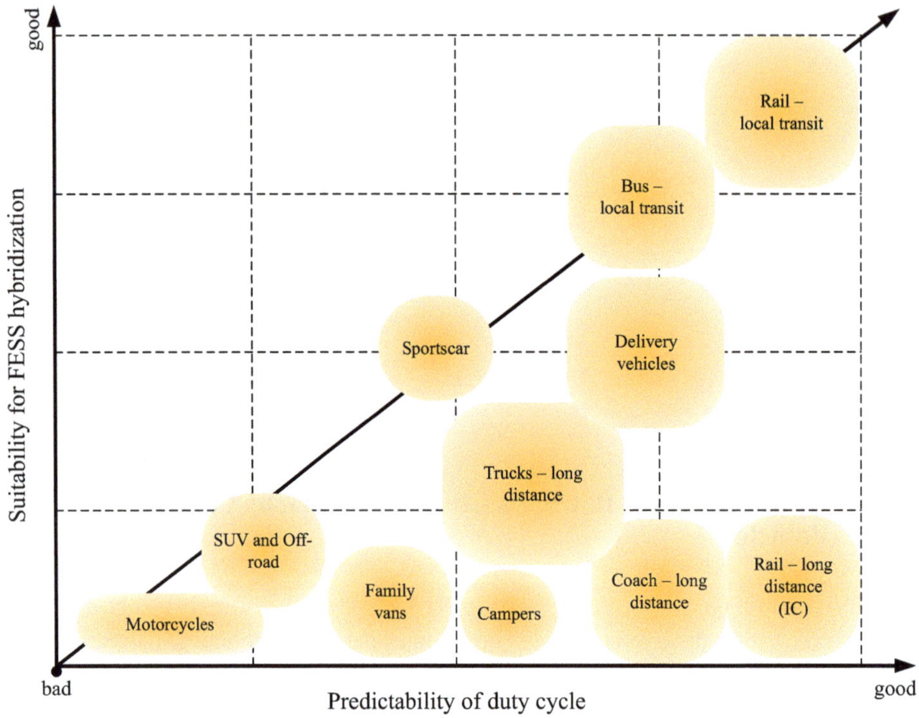

Fig. 5.1 Increase in suitability for flywheel hybridization of vehicles over the predictability of the load collective (approximate qualitative representation)

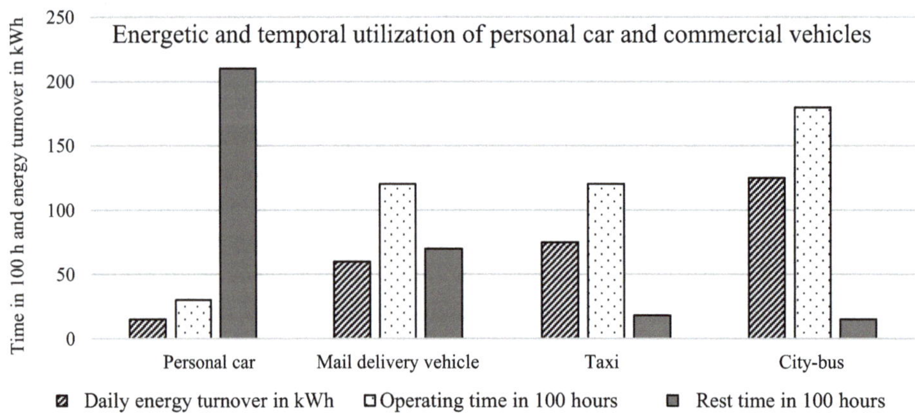

Fig. 5.2 Different use of passenger cars and various commercial vehicles in local transport [1]. (Image rights: Sir John Samuel—redT energy)

5.2 Aspects of the *Supersystem* of Public Transport and Commercial Vehicles

The choice of operating strategy and type of energy storage is primarily determined by the route and economic aspects. If a technology is to establish itself in the commercial vehicle sector, it must enable rapid monetary amortization. Not only the direct reduction of operating costs (e.g., through fuel savings or state subsidies) must be taken into account but also "secondary effects." These sometimes hard to quantify effects include aspects such as image cultivation of the transport operator or the vehicle fleet. A "green image" may help to increase customer numbers, even if no direct reduction in CO_2 emissions is possible.

Table 5.1 summarizes how some characteristics of commercial vehicles for urban transport affect powertrain hybridization (and consequently the suitability for flywheel energy storage integration).

Table 5.1 Characteristics of commercial vehicles in inner-city traffic and their influence on hybridization

Characteristics of commercial vehicles in urban traffic	Positive effect on hybridization
Highly dynamic driving cycle 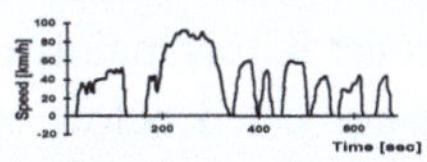	High-energy recuperation potential 
Low downtimes or required energy storage periods 	Low influence of self-discharge 
Good predictability of the load cycle 	Simple energetic dimensioning of the FESS 
High mileage in a short time 	Rapid monetary amortization 
Economic aspects determine purchase decision 	FESS's savings potential convinces customers 
Comprehensive funding opportunities and subsidies and value of a "green image" 	Increased profitability of alternative propulsion systems 
Specific training for professional drivers possible 	Adjusting the deceleration for optimum regenerative braking 

5.2.1 Energy Efficiency of Commercial Vehicles

As mentioned in Table 5.1, the driving behavior, i.e., the implementation of the driving cycle, plays an essential role with regard to the energy recovery potential. In [2], *Prof. Ernst Fiala* proposes a so-called sawtooth cycle for the energetically optimal operation of a conventional drivetrain (see Chap. 4, Fig. 4.3). A commercial vehicle in local public transport is accelerated at the most efficient load point of the combustion engine, i.e., at optimum efficiency, up to a maximum speed and then simply coasts all the way to the next stop without dissipating kinetic energy using the friction brakes. Although this strategy reduces the fuel consumption for the corresponding route to the smallest physically possible possible amount, the following problems arise in practice:

A separate lane is necessary, as this strategy can usually not be integrated into existing traffic.

Passengers would not tolerate extremely low coasting speeds and would get the feeling that they would be better off traveling with another means of transport.

However, the new possibilities gained through hybridization may offer a solution. They enable the operating strategies listed below and facilitate the integration of the vehicle into the existing traffic landscape, even when the ICE is mainly operated at the best load point. Still, it must be noted that the advantages listed below are offset by greater technical complexity and a longer efficiency chain.

1. **Load point shifting:** The vehicle can be accelerated at any speed, even at almost optimum ICE torque and speed.
2. **Regenerative braking:** Instead of letting the vehicle coast to a complete halt, the braking energy can be temporarily stored and reused for later acceleration.

To implement these measures as efficiently as possible, vehicle-specific training, which can be conducted with professional drivers, is advantageous.

In summary, it can be concluded from the above considerations that hybridization can be seen as a sensible, above all short-term solution for vehicles in urban transport or for commercial vehicles with dynamic driving cycles in general.

5.2.1.1 Simulation of Driving Cycles and Operating Strategies in Public Transport

The content of the following section was developed in cooperation with *Michael Bader* and *Ivan Andrasec* within the framework of research projects at the Graz University of Technology [3].

Taking a closer look at motor vehicles in public, inner-city operation, one can find that their usual driving profile consists of a sequence of similar individual cycles, which are

Table 5.2 Input parameters of a typical city bus used for the simulation

Property	Value	Unit
Tare weight	18900	kg
Number of seats	55	–
Average passenger weight	75	kg
Total weight	23025	Kg
Rolling resistance coefficient f_R	0.008	–
Air drag coefficient *cw*	0.5	–
Aerodynamic area *A*	8	m^2
Max. power of the ICE	200	kW

caused by traffic lights and approaches to complete stops. In a generalized and simplified consideration, these micro-cycles consist of three phases:

1. **Acceleration** from standstill to maximum speed with high load
2. **Constant speed**, where only air drag and rolling resistance must be overcome
3. **Braking** with high power to a complete halt

For the simulation described in this section, an analytical approach based on the dynamic equation of driving resistance of the entire vehicle was chosen, which is based on the considerations in Sect. 4.2.2. This approach allows a classical backward simulation and simple, fast parameter variation due to the extremely short computation times. Selected scenarios and operating points were verified using a comprehensive, map-based *MATLAB-Simulink* model of the *Institute of Electrical Measurement and Sensor Systems* (*Institut für Elektrische Messtechnik und Sensorik*) at *the Graz University of Technology.*

The distance between two stops of the synthetic route used as a basis for this simulation is about 400 m. (This is an approximation of the official *Braunschweig driving cycle.*) A constant acceleration of 1 m/s^2 and a constant deceleration of 2 m/s^2 are assumed. This corresponds to the values determined in real-world travel tests [4]. Rolling resistance and air drag, the energy requirements of auxiliary consumers, and the efficiency of the hybrid system are taken into account.

The technical characteristics of this generic, representative city bus, which served as input data for the simulation, are shown in Table 5.2.

From an efficiency point of view, a "sawtooth profile" (see Fig. 5.3) with acceleration in the most fuel-efficient range of the engine's (ICE) BSFC[1] up to maximum speed and subsequent coasting without ICE operation should be aimed for. Although the energy input can be reduced to very low values, this is at the expense of the travel time. To make this approach feasible, a separate lane would be required, as the desired speed profile cannot be integrated into existing traffic. The implementation of the target specification by the driver

[1] Such a "brake-specific fuel consumption" (BSFC) plot is shown and discussed in Sect. 3.4.2 in Fig. 3.3.

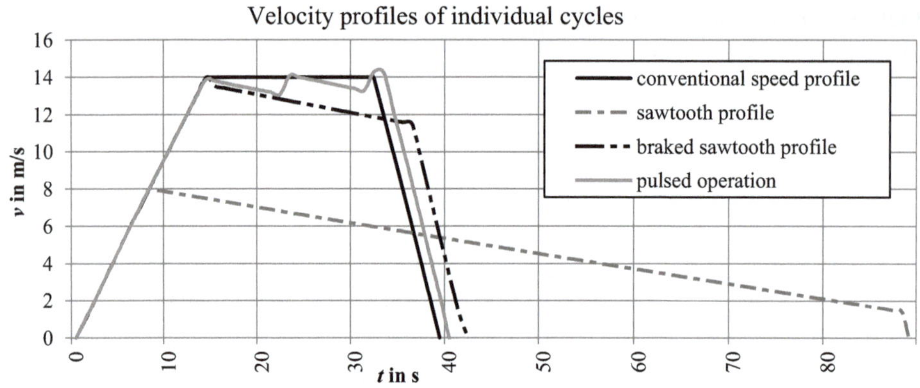

Fig. 5.3 Speed profiles of different operating strategies

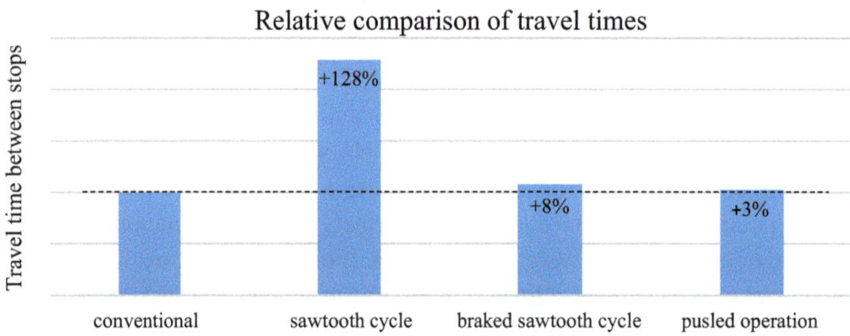

Fig. 5.4 Comparison of travel times for different operating strategies

and the acceptance of low-speed rolling by passengers must also be regarded as critical. Starting from the energy-optimized operation based on the idea of a sawtooth cycle (shown in Fig. 5.3 as an example for an acceleration to approximately 8 m/s and braking below 1.4 m/s), several strategies for increasing the practical suitability are possible. These strategies represent an approximation to the conventional speed profile, with corresponding effects on energy consumption and travel time. If the maximum speed is increased compared to the sawtooth cycle with subsequent coasting with the ICE shutoff, there is an excess of speed when approaching the next stop, which must be reduced using the friction brakes. Another approach is to keep the speed within narrower limits by intermittent operation of the ICE. Figure 5.3 shows the speed profiles in the time domain. Figure 5.4 shows the travel times in relation to conventional vehicle operation. Only the sawtooth cycle shows considerable deviations with regard to travel time and energy consumption.

The braking energy recuperated by means of flywheel energy storage can be reused to only a certain extent due to the limited efficiency of multiple energy conversion processes. Here a relatively conservative 65% (*round-trip efficiency*) is assumed.

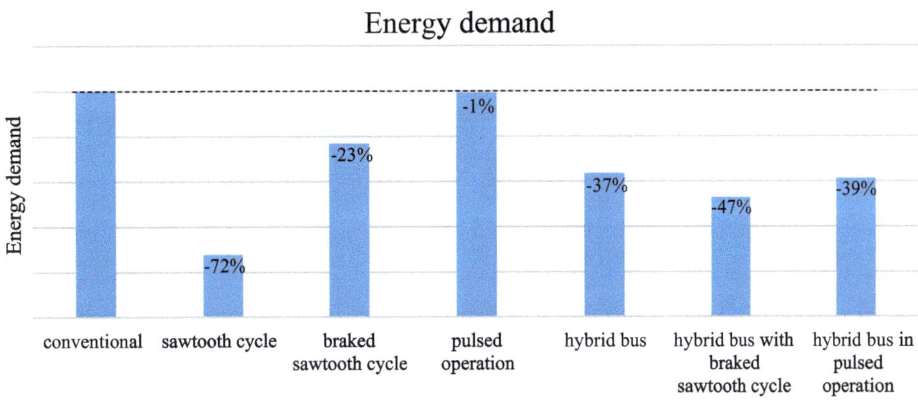

Fig. 5.5 Energetic comparison of the conventional operating strategy with alternative speed profiles without or with hybrid drivetrain (vehicle with flywheel energy storage) and brake energy recuperation

If the vehicle's secondary energy storage system is adequately dimensioned in terms of power and energy content—in this case approximately 330 kW and 0.6 kWh—high-energy savings are already possible compared to the conventional driving cycle due to regenerative braking, despite the pessimistically defined efficiency. In any case, it is necessary to find a compromise between the regenerative braking power and recuperable energy as the technical-economic optimum. Figure 5.5 shows a comparison of the energy-saving potential of the different hybrid system operating strategies with a conventional driving operation.

An example of a transit bus with a hybrid drivetrain and FESS for public transport is given in Fig. 5.6.

5.2.2 Operating Conditions for Hybrid Propulsion Systems and Energy Storage Requirements

The basic prerequisite for the efficient use of hybrid vehicles is a dynamic speed profile, in which case the two aspects of *load point shifting* and *regenerative braking* come into play most strongly. The driving cycle should be characterized by short constant speed sections and relatively high deceleration values in the transient phases. For example, 83% of the kinetic energy of a bus can already be recuperated taking into account the driving resistance forces and a moderate deceleration of 1 m/s^2, and even 94% at 2 m/s^2, which is still acceptable for the passenger from a comfort point of view.

Figure 5.7 shows the speed profile and the corresponding power requirement of the vehicle in the *Braunschweig cycle*. This is a real-world driving cycle in regular urban traffic in Europe. Although not surprising, the high-power peaks (both during acceleration and deceleration), which exceed the moderate average power of 36 kW by about a power of

Fig. 5.6 Bus with flywheel energy storage of the company PUNCH Flybrid, for the recuperation of braking energy. (Image rights: PUNCH Flybrid)

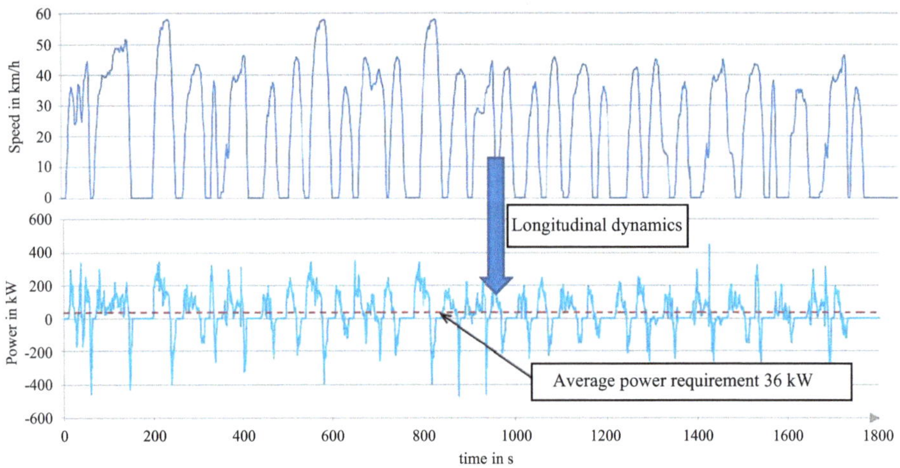

Fig. 5.7 Speed profile and power flow for a city bus in a highly dynamic driving cycle

10, are worth mentioning. This means that an appropriately powerful secondary storage system allows significant downsizing, as well as a more fuel-efficient operation of the internal combustion engine.

Based on these considerations, it can be deduced:

When designing the flywheel energy storage system and its desired performance specifications, it is important to make a compromise between power capabilities and energy content on the one hand and installation space, weight, and system costs on the other.

For the simulated scenario, this means that at a maximum FESS generator output of 165 kW, 95% of the braking energy can already be recuperated. Although the maximum occurring braking power is significantly higher than the maximum available storage power, the unused energy shares are correspondingly low because the power peaks only last a very short duration. This underlines the statement made in Sect. 5.5.1 that the predictability of the speed profile is of crucial importance for a targeted dimensioning of the energy storage system.

From the point of view of the requirements *dynamics of the speed profile* and its *predictability*, bus or rail transport shows the best suitability for hybridization.

Load point shifting and brake energy recuperation in commercial vehicles require a secondary energy storage system with the following characteristics:

1. High cycle life
2. High-power rating at medium energy content
3. High reliability (no aging, temperature independence, etc.)
4. Low maintenance and operating costs

A closer look at Table 3.1, paragraph 3.3, shows that flywheel energy storage devices can meet these criteria and in some cases are a good alternative to the currently popular Li-Ion batteries. However, due to the much smaller production quantities of *flywheels* (usually prototypes or very small series) produced to date, the acquisition costs are currently still relatively high. Possible solutions, which will be presented in detail in Chap. 6 (*Optimization in the Subsystem)*, always pursue the goal of cost reduction in addition to achieving the energetic specifications defined in the *Supersystem Analysis*.

However, the economic profitability of switching from conventional combustion engines to hybrid vehicles for the operator of a transport and traffic company also depends on the current funding opportunities/subsidies and the pricing policy of energy sources. This essential aspect of the *supersystem* requires its own economic-juridical analysis and was examined in detail by Emes et al. in [5] and is therefore not discussed in detail in this book.

It is evident that in addition to all the energy-related criteria, other parameters mentioned in Sect. 5.5.1, such as the larger available installation space, make commercial vehicles in urban public transport a clear favorite for the application of high-performance flywheel energy storage systems, such as the 145 kW/0.75 kWh prototype shown in Fig. 5.8.

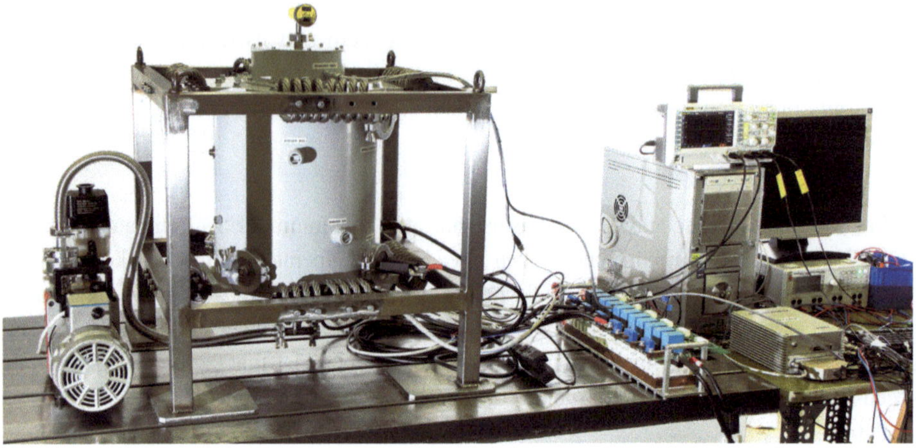

Fig. 5.8 Flywheel energy storage system for a commercial vehicle in inner-city traffic on the test bench of the *Energy Aware Systems* work group, Graz University of Technology

5.3 Individual Transport and Personal Cars

5.3.1 Aspects of the *Supersystem* "Personal Car"

Skeptics of the conventional automobile regard an increase in the efficiency of the drivetrain *alone* as insufficient since the associated reduction in operating costs can lead to a *rebound effect.*[2] What is in fact needed much more urgently is a paradigm shift from a car-oriented attitude in Central Europe or the USA toward a holistic mobility concept, in which especially public transport plays a major role [6]. However, even if the *first mile/last mile* problem[3] can be solved, unrestricted customer acceptance of public transport cannot be expected from today's perspective, as surveys such as shown in Fig. 5.9 have proven.

Although customers have rated the attributes "stress-free" and "environmentally friendly" of public transport better, the personal car wins in all other aspects. A high share of passenger cars can therefore be expected to be part of everyday traffic in the near future. However, the decision criteria for purchasing and operating a passenger car are many times more complex than those for commercial vehicles. The "energy-

[2] *Rebound* or *backfire* means that a certain measure has the opposite effect as originally intended. Reducing fuel consumption, for example, can mean that vehicles are used more often and by larger user groups because of the lower operating costs, which in turn increases the overall CO_2emissions.

[3] *First mile/last mile* refers to the difficulty of getting from the starting point of the journey to a public transport hub and from its end station to the actual final destination.

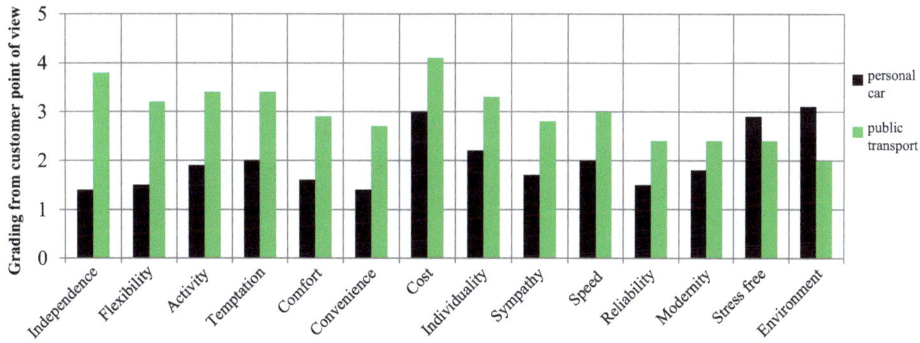

Fig. 5.9 Evaluation (analogous to Austrian high school grades: 1 = best, 5 = worst) of personal cars and public transport from the customer's point of view [7]

psychological" considerations discussed in Sect. 3.5 can essentially be divided into two areas of influence:

1. **Purchase behavior:** Although, from a rational-technical point of view, *purpose and driving cycle* should (!) influence the choice of vehicle, in reality it is mostly the symbolic/psychological attributes discussed in Sect. 3.5 that determine the customer's actual purchase behavior.
2. **Driving performance:** A car is a symbol of autonomy. The driver does not allow himself to be "patronized" and does not want vehicle-specific training, even though this would be reasonable if a regenerative braking option were available. There are two basic ways to influence the driving behavior of passenger cars:
(a) *Create motivation:* The increasing mutation of the road vehicle to an "office on wheels" equipped with Bluetooth, Internet access, and other interfaces bring technologies, which would allow the comparison of driving efficiency on a personal level or even in social networks. Although innovations such as these do not bring about improvement from an "environmental" point of view, they are based on the principle of a so-called market pull innovation [7] and are accordingly successful. The driver experiences personal satisfaction or recognition in the "community," thanks to an efficient driving style. One example is the *next-generation SmartGauge* introduced by *Ford*. The CO_2 savings are displayed in the form of green leaves (or similar) on the dashboard (see Fig. 5.10).
(b) *Technology takes over efficient driving:* It must do so without the customer noticing or feeling restricted (*Technology Push Innovation*).

Intelligent network systems that enable, among many new possibilities, autonomous driving and automatic convoy formation (so-called platooning) can be regarded as the ideal long-term solution. An example is the *EO Smart Connecting Car* of the *German Research Center for Artificial Intelligence* (*Deutsches Forschungszentrums für Künstliche Intelligenz* (*DFKI*)) (see Fig. 5.11).

Fig. 5.10 Next-generation "SmartGauge" from Ford [8]. (Image rights: Ford Motor Company)

Fig. 5.11 Convoy of the *EO Smart Connecting Cars* [9]. (Image rights: German Research Center for Artificial Intelligence—DFKI)

However, as an intermediate step toward this goal, *existing* technologies must be *efficiently linked*! Energy storage systems like the flywheel, which feature precisely determinable energy content, can be optimally operated in conjunction with a GPS-supported energy management. By entering the destination into the vehicles' navigation system, it is not only possible to predict the energy demand by considering influencing variables such as traffic density, road curve radii, and geodetic altitude difference but also to simulate the energy flows to and from the secondary storage in advance and to implement a *predictive control strategy*. In this way, the vehicle takes over efficient "actions" (within the limits of technical possibilities and the predictability of events) without noticeably restricting the driver.

One of the first inventions to pursue this goal—and still in use today—was the automatic transmission invented in 1921 by the Canadian engineer *Alfred Horner Munro*. It allows the definition of fuel-efficient shifting points, thus reducing fuel consumption and increasing comfort through uninterrupted traction.

In order to achieve local zero-emission mobility in certain highly polluted urban regions, virtual boundaries (*geofencing)* can be used to define zones in which combustion engines automatically switch off and the vehicle is propelled only by the (electric) secondary drive system (Fig. 5.12).

While Table 5.2 in Sect. 5.2 shows the characteristics of commercial vehicles in public transport and their effects on vehicle hybridization on a logical-rational level, such a categorization is not possible in the personal car or individual transport sector. The problem

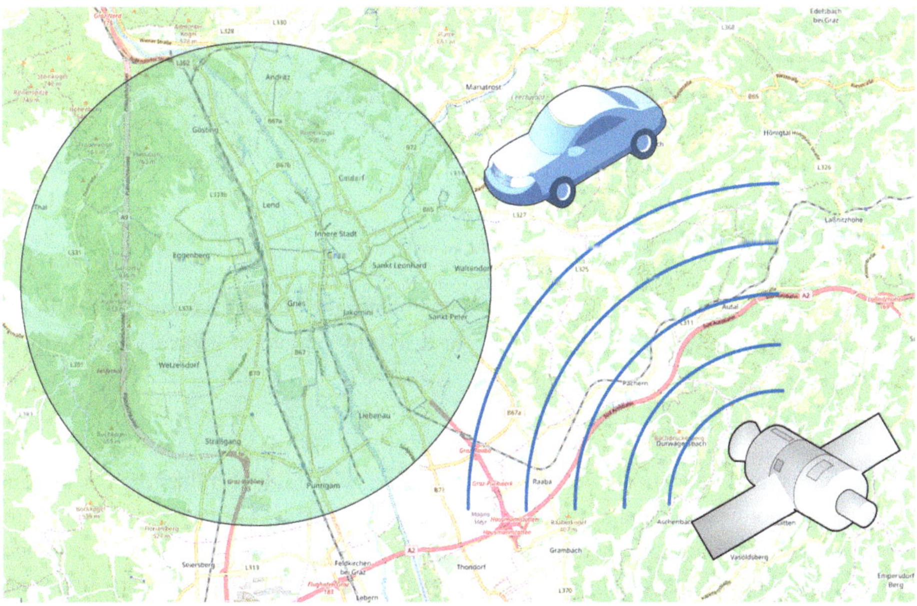

Fig. 5.12 Functional principle of an emission-free zone by "geofencing"

always starts with the driver (who is usually the customer or buyer of the vehicle). Owning a car represents the attribute of *freedom*, which means that the driver would not accept paternalism with regard to her/his driving style or even the purchase decision. Only government legislation can control driving behavior or purchasing behavior by introducing speed limits or financial incentives in the form of tax breaks when buying a hybrid or electric vehicle. It is therefore the driver's behavior based on psychological phenomena that is the main influencing factor, which shapes the interdependencies between the *subsystem* and *supersystem* of the flywheel energy storage system.

The following section focuses on the essential characteristics of private transport by car.

5.3.2 Driver and Psychology

The customer is at the center of the consideration visualized in Fig. 5.13 since his behavior directly and indirectly affects several aspects of the *flywheel energy storage subsystem*. The areas of influence can be divided into the following two categories:

(a) **Driving style:**

The driving behavior of the customer essentially defines the dynamics, service life, and cumulative frequency of certain driving maneuvers and thus results in the driving cycle relevant to FESS design.

The **driving cycle** is the essential basis of the energetic dimensioning of the energy storage unit and defines parameters such as the power of the electric motor generator or the FESS rotor design (energy content).

Driving maneuvers and driving dynamics define, among other things, the bearing loads of the FESS rotor described in detail in Sect. 9.2.1.

(b) **Purchase behavior:**

The purchase behavior of the customer is responsible for the binary decision whether or not a vehicle with flywheel energy storage (or a hybrid vehicle in general) at all will be bought.

(c) **Economy and law:** In this case, we are dealing with mutual influence. On the one hand, economic and legal conditions define the attractiveness of buying a vehicle with flywheel energy storage. On the other hand, it is possible for the collective of customers to shape the free market through the mechanisms of supply and demand. Important influencing factors are:

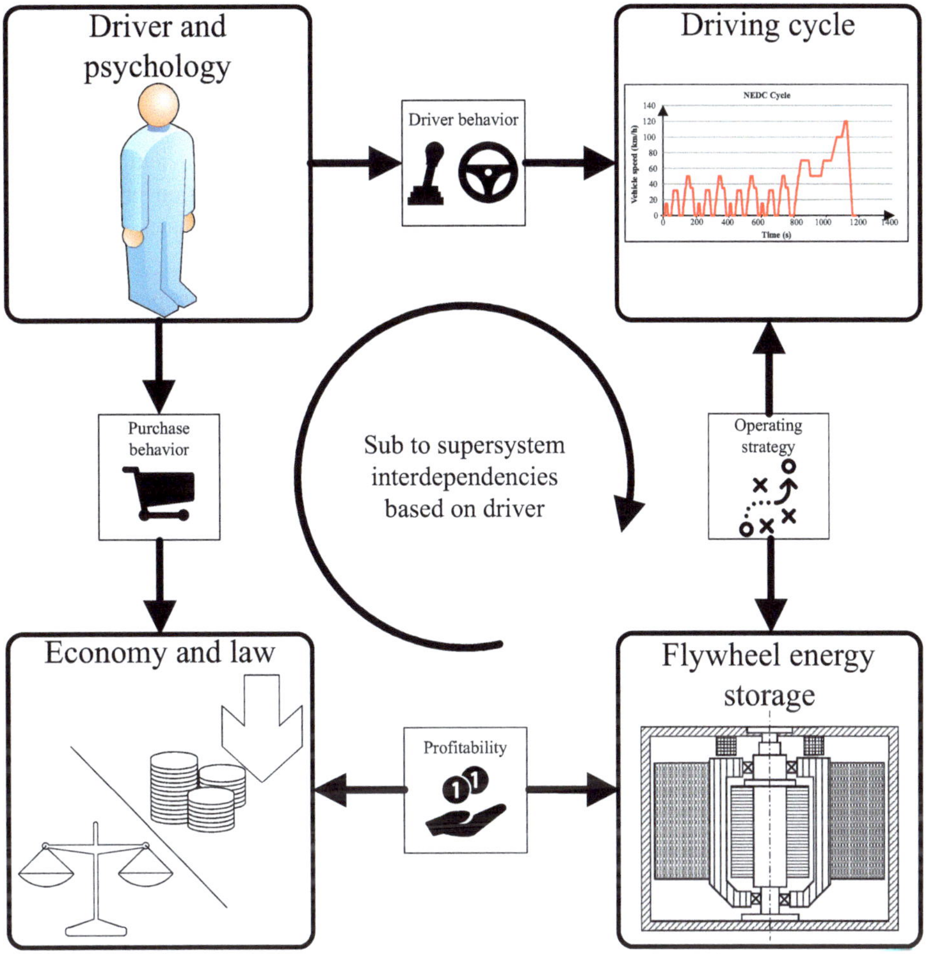

Fig. 5.13 Most important interdependencies between the *subsystem* and *supersystem* of a passenger car in the private transport sector

1. Fuel and energy prices
2. National funding opportunities and subsidies
3. Legal emission and fuel consumption limits
4. Marketing and trends

The main problem in designing a FESS for a passenger car is the poor (or nonexistent) predictability of the load cycle. However, a *predictive control* based on user input of the planned route would be desirable. A fictitious example, which describes a not unlikely scenario, illustrates the problem:

Example

1. Let us assume that during a short journey from the driver's home to a nearby supermarket, the FESS is charged by load point shifting (i.e., by the excess power of the ICE) since the deceleration values in bumper-to-bumper traffic are not sufficient for regenerative braking. The operation strategy anticipates possible acceleration phases (*boost phases*), which, however, never occur due to the heavy traffic and congested streets. So, the energy storage device is not discharged during the journey.
2. When the vehicle is parked, the FESS is 100% charged, but the driver spends too much time shopping so that most of the energy is dissipated by self-discharge. The same applies to the journey home.
3. Ideally, the ICE would have to be switched off a few miles before reaching the final destination (supermarket) in order to cover the last part of the journey (emission-free) using only the flywheel's energy.

In this case, two fundamentally different approaches can be pursued:

1. **Solution in the FESS Supersystem: Predictive Control**

The fact that today almost every vehicle and every smartphone is equipped with GPS has opened up completely new possibilities in terms of predictive control of a hybrid powertrain. After the driver has entered the destination, geo-information data such as altitude differences or slope angles along the route can be determined, thus allowing optimal use of the energy available in the secondary power source. Apps such as the *Google Traffic Estimator* [10], which analyzes the speed of thousands of mobile phones on traffic routes and thus draws conclusions about average travel speeds, possible traffic jams, or potential obstructions, can also be implemented. This has triggered a real boom in the field of predictive control in recent years, although reference is made at this point to technical literature by other authors (e.g., [11–13]). The energy-saving potential of this control strategy is estimated to be a maximum of 10% [14], i.e., significantly less than the almost up to 30% for regenerative braking by flywheel in heavy commercial vehicles and inner-city operation. Predictive control strategy aspects for the application of FESS in a vehicle are graphically illustrated in Fig. 5.14.

2. **Solution in the FESS Subsystem: Reduction of Self-Discharge**

A solution which at first glance appears to be much simpler, since it involves a lower number of *interdependencies in the supersystem* and allows the customer to evade the responsibility of an energy-efficient control strategy, is the *reduction of the self-discharge of the FESS*. If the stored energy content were to be retained even when the vehicle is parked for a long period of time, the need for predictive control would also have a less

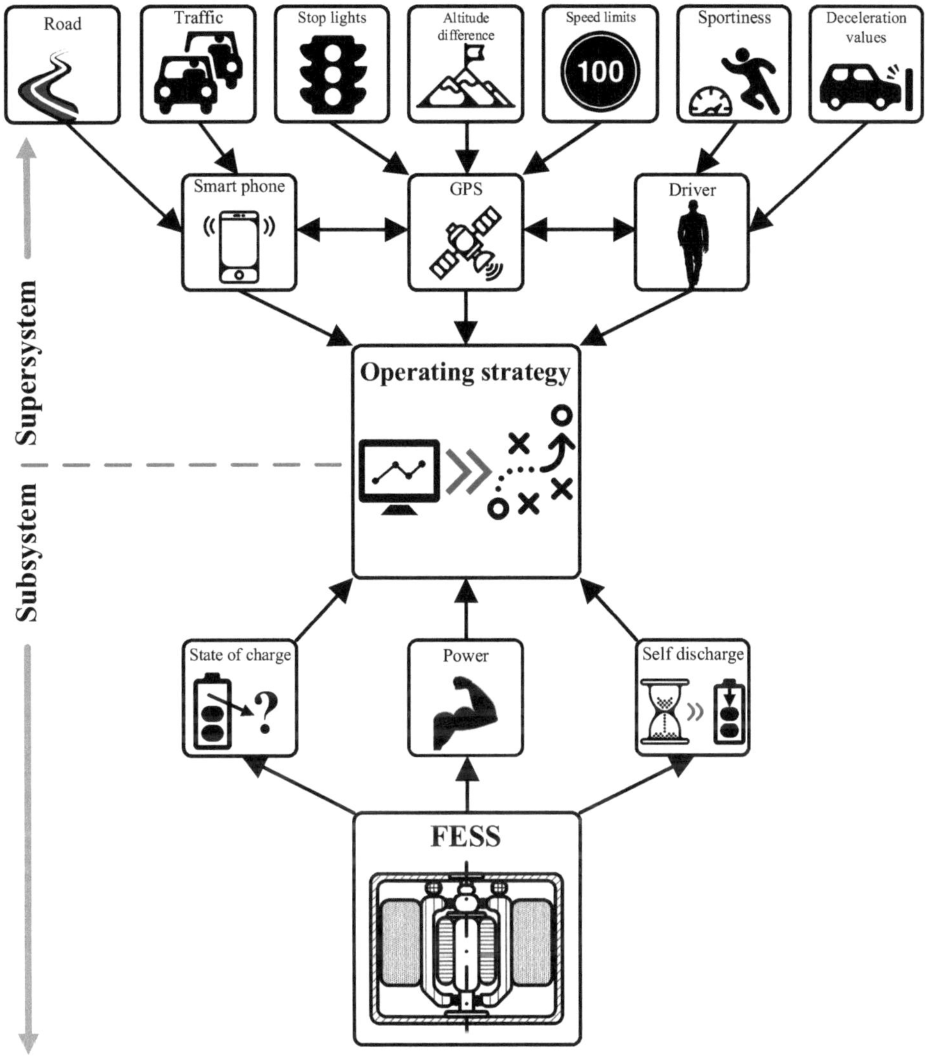

Fig. 5.14 Essential aspects of predictive powertrain control for a flywheel hybrid vehicle

dramatic effect. For a detailed description of technical solutions regarding this problem, please refer to Sect. 10.1 "Reduction of the Torque Loss of FESS Bearings."

5.3.3 Target Characteristics of Mobile Flywheel Energy Storage Devices

5.3.3.1 Economic Consideration

The examples summarized in Sect. 2.3.2, Table 2.4, have proven that in principle a flywheel hybrid vehicle makes sense from a technical-energetic point of view and holds correspondingly high fuel/energy-saving potential. However, increasing economic efficiency of a passenger car alone through hybridization and the addition of a secondary energy storage system is not sufficient for reasonable market penetration. Rather, the energy storage system must provide *added value* that addresses the *psychological attributes* of the buyer listed in Sect. 3.5. In addition to dynamic driving performance (which could be achieved by a *boost function* using the FESS), the car customer attaches particular importance to individualizing the vehicle by purchasing supplementary equipment, so-called extras. The psychologist *Alfred Hermann* emphasizes in [15] that although the customer buys a serial product, he/she wants it to be something "very special." The average *Mini Cooper* buyer invests around 20% of the purchase price in upgrades [15]. Placing sustainable vehicle technology in the segment of individual optional extras offers a way of accelerating the smooth transition to the *zero-emission vehicle*. However, this also defines the price range for the energy storage system and the necessary components in the powertrain. From the *customer's point of view*, the following requirements for a secondary energy storage system in a personal car can be derived:

1. High-power density or specific power (positive influence on vehicle dynamics)
2. Easy, uncomplicated operation
3. Easy integration into the existing vehicle architectures
4. Low or better no maintenance costs
5. Good image/reputation of secondary energy (such as electricity)
6. Easy recycling of the energy storage unit (such as flywheel)
7. Low development and production costs

It may seem surprising that attributes like *high-energy density* and *low self-discharge* are not included in this list. This means that a linear approximation of the FESS to the properties of a virtual reference energy storage device without analysis of the *supersystem* is not appropriate. The customer often buys a solution that is suboptimal from a strictly technical point of view.

The most important requirement for a reasonable market penetration of FESS in the vehicle sector is therefore the reduction of the price and the achievement of certain minimum energy requirements, referred to here as "threshold specifications," which are explained in the following section.

5.4 Energetic Threshold Specifications

The central question behind the term *threshold specification* is

> "When is a technical solution just good enough for the customer to buy it?"

An example should clarify the considerations:

Example
A generic, idealized reference energy store system has infinitely high energy and power density and no self-discharge. Nevertheless, car drivers are satisfied with a range of about 500 to 1000 km between refueling stops in any case, and one can therefore conclude that a tank size of about 30 to 80 liters for a personal car is sufficient based on today's standards. Hardly any customer would demand a larger tank or higher-energy density of the fuel. The situation is entirely different in the case of battery electric mobility. Certain battery technologies, such as the lead battery, for example, achieve only too low-energy densities and are therefore currently not used in electric vehicles. Even a significant reduction in costs cannot compensate for the lack of practicability and performance.

Figure 5.14 shows the specific energies of various mobile energy storage devices. While lithium polymer/lithium-ion batteries are the most widespread energy storage technologies for alternative vehicle propulsion, hydraulic pressure storage and supercaps are not used as primary energy storage in vehicles. All batteries previously used as "main power source" for EVs (including compressed air storage) offer a specific energy of at least 60 Wh/kg (Fig. 5.15).

So how can such *threshold specifications* for flywheel energy storage be determined? What is the least energy density that makes a FESS useful? This question can be answered by analyzing competing technologies and their dissemination on the market.

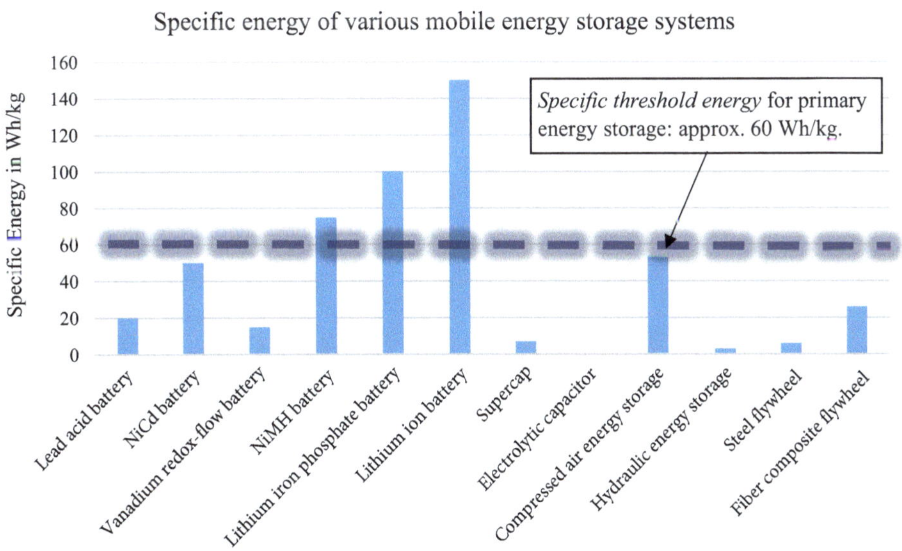

Fig. 5.15 Specific *threshold energy* of energy storage devices for use as primary energy source in vehicles. (created from data based on [16–19])

5.4.1 Determination of Energetic *Threshold Specifications* for FESS

It is a historical fact that flywheels, contrary to various euphoric announcements in the 1970s, have not yet been able to establish on the market [20]. Despite the high theoretical potential (see Table 7.2 in Chap. 7), the achievable specific energies are simply still too low due to the high weight of the surrounding system components (protective housing, vacuum technology, electric motor, control system, cooling circuits). In addition, the high self-discharge rate makes a multiday journey with a purely flywheel-powered vehicle currently unthinkable.

Example

Therefore, the use of FESS in vehicles appears to make sense primarily to enable regenerative braking, for load point shifting of the ICE or short emission-free operation, but not as a primary energy storage system. One of the few exceptions is the *MFO Gyrobus* designed in 1953, which was able to recharge the flywheel at each bus stop (Fig. 5.16).

The requirements for this dynamic *high-power storage system* differ from those of conventional *energy storage systems* used as a primary energy source in alternative propulsion systems. To cover short-term power peaks, devices such as supercaps, Li-Io batteries, hydraulic pressure storage, or flywheels are often used. Table 5.3 gives an

Fig. 5.16 The *MFO Gyrobus* during a test-drive in Yverdon, Switzerland, in 1950. The charging station with pantographs is clearly visible. (Image rights: ABB Schweiz, N.3.1.54627)

Table 5.3 Characteristics of energy storage systems previously used in hybrid vehicles

Energy storage	Spec. energy Wh/kg	Spec. power kW/kg	Energy density Wh/liter	Degree of efficiency %	Self-discharge %/day	Number of cycles	Example vehicle	Ref. [a]
Lead-acid battery	20	0.18	50	50–85	0.1–0.8	500–800	*Audi DUO III (1997)*	[21, 22]
NiCd battery	50	0.2	100	70–90	0.05	2000	*1976 Alfa Romeo EV Conversion*	[23]
Vanadium redox flow battery	15	10	20	70–75	0.001–0.1	2000	*nanoFlowcell Quant E*	[1, 24, 25]
NiMH battery	75	0.7	200	65–70	0.25–3	500–2000	*Toyota Prius II, Honda Insight*	[26]
Lithium iron phosphate battery	100	0.2	200	75–85	1.5	100–2000	*Aptera type-1*	[27–30]
Lithium-ion battery	150	0.6	500	80–90	0.25	400–1200	*Tesla Model S, Nissan Leaf EV*	[31]
Supercap	7	20	10	90–95	1.5	10000	*Toyota Yaris Hybrid-R*	[32]
Electrolytic capacitor	0.05	35	0.5	75–95	3–10	1000–10000	–	[33–35]
Compressed air storage	50	2	50	10–15	0.001	10000	*MDI AirPod, PSA Peugeot Citroën Hybrid Air*	[36, 37]
Hydraulic pressure storage	3	5	2	80–90	0.1	10000	*MAN Hydro Bus*	[18, 38]
Steel flywheel	10	5	5	90–95	20–80	100000	*Oerlikon Gyrobus*	[39]
Composite flywheel	60	6	15	90–95	20–80	100000	*GKN Hybrid Power*	[40]

[a]... These literature references were used in addition to information from [41] and *Wikipedia.org*

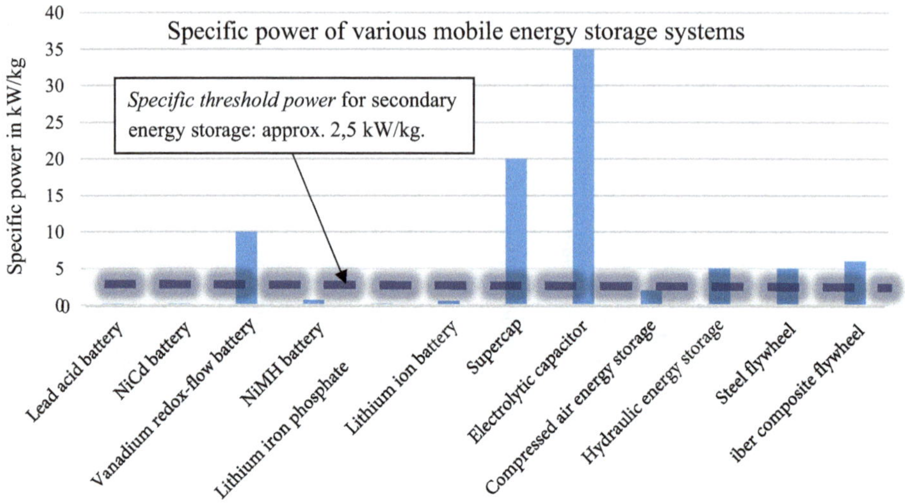

Fig. 5.17 Specific power of different mobile energy storage devices used in vehicles

overview of the properties of energy storage systems that have already been installed in hybrid vehicles and gives examples of these vehicles, which have actually been designed, tested, and documented. It should be noted that the quantifiable specifications per storage technology have a relative bandwidth depending on the specific design and manufacturer. Furthermore, the system boundaries cannot easily be determined for each storage device. However, the values in the table refer to *pack level*, i.e., the necessary "packaging" (housing, power electronics, cooling, etc.) has been taken into account.

In order to objectively compare a property that is highly relevant for the specification of a power storage device, namely, the *specific power*, the representation in the form of a bar chart was chosen. All storage types that were primarily used for brake energy recuperation offer values above or in the range of the *threshold line*, which was defined as approximately 2.5 kW/kg (Fig. 5.17).

Even though all the quantifiable properties listed in Table 5.3, such as the maximum number of charging cycles and self-discharge, play an essential role in the selection of a suitable energy storage technology for a hybrid vehicle, two main energetic objectives for FESS can already be defined:

1. **Specific energy:** When used as *short-term power booster*, this property does not necessarily have to reach the limit of at least 70 Wh/kg required for use as *primary energy storage*, **but should be above 10 Wh/kg** in order to achieve a market advantage over competing "power storage" devices such as supercaps.
2. **Specific power:** This value can be determined by analyzing and comparing the currently available energy storage technologies on the market for hybrid vehicles and was **found to be around 2.5 kW/kg**.

These two "practical" criteria (specific energy and specific power) define the permissible weight (or volume) of the FESS in the vehicle, which provides a first indication of the design or strategic development goals for the flywheel energy storage system.

Other properties such as *self-discharge* and *efficiency* are strongly linked to the driving cycle and the operation strategy and can therefore only be investigated by numerical simulation (as discussed in Sect. 5.2.1.1).

5.5 Relevant Findings of the System Analysis

The decomposition of the hybrid vehicle into a *subsystem* consisting of the powertrain components and a *supersystem* that includes environmental influences and ambient parameters, such as traffic infrastructure and driver, allows a critical *holistic analysis of* modern vehicle topologies. Mobile energy storage systems play a decisive role in this process, whereby they strongly interact with the driving behavior and driving cycle. While hybridization and the transition to alternative (also purely electric) drive concepts in the field of commercial and especially utility vehicles for inner-city use represents a short-term and promising alternative, hybridization measures in the passenger car segment show a significantly lower potential for reducing energy consumption. This is mainly due to sociopsychological aspects. Consequently, it can be concluded that the greatest potential for optimization lies in the influencing parameters of the *supersystem*.

It is therefore indispensable to dare to make the transition from the so far purely technical development of vehicles to an *interdisciplinary system optimization*, which combines different disciplines such as engineering, psychology, politics, legislation, and marketing. In this way, a further step can be taken toward a holistic mobility concept as an alternative to the "status quo" of the conventional passenger car with an internal combustion engine.

The analysis of existing systems and successful applications of competing storage technologies for hybrid vehicles allows an estimation of the *threshold specifications*, i.e., the performance characteristics that must be achieved in order for the flywheel technology to succeed on the market. A specific energy of at least 10.0 Wh/kg and a specific power of more than 2.5 kW/kg are minimum target requirements for the FESS technology.

5.5.1 Summary: Optimization of the *Supersystem* of a FESS

The holistic approach and systematic analysis of the *supersystem* made it possible to identify important influencing variables and parameters with regard to the efficient use of flywheel energy storage systems in vehicles. In summary, it can be said that:

1. Research and development work in the field of energy storage systems for hybrid vehicles in general and flywheels in particular is strongly influenced by political and economic phenomena such as the price of crude oil
2. Considerations of the psychological behavior of the end customer show that the willingness to invest in the individualization of the car is certainly there; however, the misuse of the vehicle type and the individually highly divergent load profile make the design of a FESS as secondary energy storage extremely difficult
3. Commercial vehicles in urban public transport not only offer a dynamic driving cycle and high-energy recuperation potential but are also an optimal platform for the deployment of FESS, which offer high performance at moderate specific energies, primarily because of the good predictability of the load cycle
4. The profitability of the FESS onboard a vehicle is determined (apart from the price, of course) solely by the driving cycle or its *dynamic index* (see Sect. 4.2.3)
5. Especially the comparison of FESS with competing energy storage technologies allows the definition of energetic and economic development goals in the form of so-called threshold specifications, which are summarized in the following section

5.5.2 General, Desirable FESS Improvements

Even though the word “performance enhancement” is omnipresent in the automotive industry, increasing the effective electrical power output of a FESS is not of superior relevance. Rather, consideration of the *interdependencies* of the FESS *subsystem and supersystem* has shown that pursuing the now listed development goals is an indispensable prerequisite for successful market penetration of FESS in the automotive sector:

1. **Increase of the specific energy to over 10.0 Wh/kg**
2. **Achievement of a specific power of 2.5 kW/kg**
3. **Reduction of self-discharge**
4. **Reduction of costs**
5. **Extension of the maintenance intervals**
6. **Improving the inherent safety and image**

The following chapters of the book will present concrete technical solution approaches concerning the FESS *subsystem*. The aim of these solutions is to make it possible to achieve the just mentioned target characteristics or *threshold specifications*.

References

1. J. Samuel (2011) Electric Vehicle Applications of Flow Batteries – Rapid Recharging of EVs by Electrolyte Exchange. RE-fuel Technology Ltd., Wokingham, Berkshire, UK. http://www.paredox.com/foswiki/pub/Luichart/InternationalFlowBatteryForum/Electric_Vehicle_Applications_of_Flow_Batteries_-_Sir_John_Samuel.pdf
2. E. Fiala (2012) Effektive Hybridstrategien. ATZ – Automobiltechnische Zeitschrift, pp. 148–153, Issue February 2012. Springer Fachmedien Wiesbaden, Germany.
3. M. Bader, A. Buchroithner and I. Andrasec (2014) Schwungrad-Hybridantriebe im Vergleich mit konventionellem und alternativen Konzepten. ATZ – Automobiltechnische Zeitschrift, pp. 68–73, October 2014. Springer Fachmedien Wiesbaden, Germany. https://doi.org/https://doi.org/10.1007/s35148-014-0503-2
4. B. Walter, S. Schneider and K. Schimmelpfenning (1999) Stand und Sicherheit im innerstädtischen Verkehr – Eine Untersuchung der tolerierbaren Beschleunigungen. Zeitschrift VKU Nr. 12.
5. M. Emes, A. Smith, N. A. Tyler, R. Bucknall, P. A. Westcott and S. Broatch (2009) Modelling the costs and benefits of hybrid buses from a 'whole-life' perspective. 7th Annual Conference on Systems Engineering Research (CSER 2009), Loughborough University – 20th – 23rd April 2009, UK.
6. P. L. Schiller, E. C. Brunn and J. R. Kenworthy (2010) An Introduction to Sustainable Transportation – Policy, Planning and Implementation. EARTHSCAN, Washington D.C., USA.
7. H. Stiegler (1999) Rahmen, Methoden und Instrumente für die Energieplanung in der neuen Wirtschaftsorganisation der Elektrizitätswirtschaft. Graz University of Technology, Graz, Austria.
8. Ford Motor Company (2013) Fusion Energy Plug-In Hybrid Technology. http://www.ford.com/cars/fusion/trim/titaniumenergi/. [Accessed June 14th 2014].
9. Animation Labs für DFKI GmbH (2012) EO smart connecting car – Innovative Fahrzeugkonzepte. Deutsches Forschungszentrum für Künstliche Intelligenz, Robotics Innovation Center, Robert-Hooke-Straße 1, D-28359 Bremen, Germany http://robotik.dfki-bremen.de/de/forschung/projekte/item.html. [Accessed May 22nd 2014].
10. D. Barth (2009) The bright side of sitting in traffic: Crowdsourcing road congestion data. Official Google Blog. https://googleblog.blogspot.com/2009/08/bright-side-of-sitting-in-traffic.html [Accessed May 9th 2016].
11. B. Gindroz (2014) Optimization of a Predictive Drive Strategy for a Plug-In Hybrid Vehicle (Optimierung der vorausschauenden Antriebssteuerung bei einem Plug-In Hybrid). Royal Institute of Technology, KTH – Department of Vehicle Engineering, Stockholm, Sweden.
12. R. Beck, A. Bollig and D. Adel (2006) Comparison of two Real-Time Predictive Strategies for the Optimal Energy Management of a Hybrid Electric Vehicle. E-COSM – Rencontres Scientifiques de l'IFP.

13. S. Jonesa, A. Huss, E. Kural, A. Massoner, C. Vock and R. Tatschl (2014) Development of Predictive Vehicle & Drivetrain Operating Strategies Based Upon Advanced Information & Communication Technologies (ICT). Transport Research Arena 2014, Paris, France.
14. J. Wang and H. Koch-Groeber (2015) Predictive operation strategy for hybrid vehicles. In: Bargende M., Reuss HC., Wiedemann J. (eds) 15. Internationales Stuttgarter Symposium. Proceedings. Springer Vieweg, Wiesbaden, Germany.
15. K. Kalowitz and S. Anker (2012) Geschäft mit dem gewissen Extra. Die Welt, Welt am Sonntag, Nr. 15./16. December 2012. Axel Springer SE, Berlin, Germany.
16. D. Hackstein and U. Fechter (2008) Seminarvortrag – Regenerative Energietechnik. Fernuniversität in Hagen, Germany.
17. P. Rundel et al (2013) Speicher für die Energiewende. Fraunhofer-Institut für Umwelt-, Sicherheits- und Energietechnik, Sulzbach-Rosenberg, Germany. https://speicherinitiative.at/assets/Uploads/18-Speicher-fuer-die-Energiewende-Fraunhofer-UMSICHT.pdf
18. I. Valentin (2015) Cost Efficient Composite Platform with Integrated Energy Storage for a Hydraulic Hybrid. SPE Automotive Composites Conference & Exhibition, 46100 Grand River Avenue, Novi, MI 48374, USA.
19. C. Fieger (2015) Energiewirtschaftliche und technische Anforderungen an Speicher-Systeme für den stationären und mobilen Einsatz. Forschungsgesellschaft für Energiewirtschaft mbH, München, Germany.
20. A. Buchroithner (2011) Systematische Analyse von Hybridfahrzeugen mit Schwungradspeicher unter Erfassung von Entwicklungstendenzen. Institut für Maschinenelemente und Entwicklungsmethodik, Technische Universität Graz, Austria.
21. Power Sonic Corporation (2009) Sealed Lead Acid Batteries – Technical Manual. Power-Sonic Corporation, San Diego, CA 92154, USA.
22. G. Albright and J. Edie, S. Al-Hallaj (2012) A Comparison of Lead Acid to Lithium-Ion in Stationary Stage Applications. BTH Management, 5942 Edinger Ave #113 - 230, Huntington Beach, CA 92649, USA. http://www.altenergymag.com/content.php?post_type=1884.
23. M. Shoesmith and L. O. Valøen (2007) The Effect of PHEV and HEV Duty Cycles on Battery and Battery Pack Performance. Plug-in Highway Electric Vehicle Conference 2007, Montreal Canada.
24. M. R. Mohamed, S. Sharkh and F. Walsh (2009) Redox Flow Batteries for Hybrid Electric Vehicles: Progress and Challenges. 2009 IEEE Vehicle Power and Propulsion Conference. DOI: https://doi.org/10.1109/VPPC.2009.5289801
25. T. Nguyen and R. F. Savinell (2010) Flow Batteries. The Electrochemical Society Interface. Edition Fall 2010, volume 19, issue 3, pp. 54–56. DOI: https://doi.org/10.1149/2.F06103if
26. Sanyo Twicell (2009) Data Sheet: "eneloop" Cell Type HR-3UTGA. Sanyo Electric Co., Ltd., Osaka Prefecture, Japan.
27. Victron Energy B.V. (2015) Data Sheet: 12,8 Volt Lithium-Iron-Phosphate Batteries. De Paal 35, 1351 JG Almere, Netherlands.
28. Incell International (2010) Comparison – Common Lithium Technologies. Incell Academy, Kistagången 16, 164 40 Kista, Sweden.
29. M. Swierczynski, D. Stroe, A. Stan, R. Teodorescu and S. Kær (2014) Investigation on the Self-discharge of the LiFePO4/C nanophosphate battery chemistry at different conditions. Transportation Electrification Asia-Pacific (ITEC Asia-Pacific), IEEE Conference and Expo, Beijing, China.
30. P. G. Pereirinha, A. Santiago and João P. Trovão (2011) Preparation and characterization of a lithium iron phosphate battery bank for an electric vehicle. XIICLEEE – 12th Portuguese-Spanish Conference on Electrical Engineering, Ponta Delgada, Portugal.

31. G. Nagasubramanian and R. G. Jungst (1999) Energy and Power Characteristics of Lithium-Ion Cells. Lithium Battery Research and Development Department, Sandia National Laboratories, Albuquerque, USA.
32. SKELCAP (2018) High Energy Ultracapacitor Product Information. Skeleton Technologies, Großröhrsdorf, Germany.
33. D. A. Scherson and A. Palencsár (2006) Batteries and Electrochemical Capacitors. The Electrochemical Society Interface, Vol. 15, No. 1. https://www.electrochem.org/dl/interface/spr/spr06/spr06_p17-22.pdf
34. IC – Illinois Capacitor (2017) Aluminum Electrolytic Capacitors – Life Expectancy. Lincolwood, Illinois, USA.
35. C. S. Kulkarni, G. Biswas, and X. Koutsoukos (2009) A prognosis case study for electrolytic capacitor degradation in DC-DC converters. Annual Conference of the Prognostics and Health Management Society, San Diego, USA.
36. P. Fairley (2009) Deflating the Air Car. IEEE Spectrum. https://spectrum.ieee.org/energy/environment/deflating-the-air-car
37. A. Burke (2005) Energy Storage in Advanced Vehicle Systems. GCEP Advanced Transportation Workshop, Stanford University, California, USA.
38. Y. Louvigny, J. Nzisabira and P. Duysinx (2008) Analysis of hybrid hydraulic vehicles and comparison with hybrid electric vehicles using batteries or super capacitors. EET-2008 European Ele-Drive Conference – International Advanced Mobility Forum, Geneva, Switzerland.
39. I. Hadjipaschalis, A. Poullikkas and V. Efthimiou (2009) Overview of current and future energy storage technologies for electric power applications. Renewable and Sustainable Energy Reviews 13(6-7), pp. 1513-1522. DOI: https://doi.org/10.1016/j.rser.2008.09.028
40. H. Wegleiter and G. Brasseur (2009) An Overview of Electrical Energy Storage Systems for Automotive Applications. Alternative Propulsion Systems and Energy Carriers. A3PS Ecomobility Conference, Vienna, Austria.
41. J. G. Patrick and T. Moseley (2014) Electrochemical Energy Storage for Renewable Sources and Grid Balancing, Elsevier Ltd., Amsterdam, Netherlands.

Subsystem Optimization

6

As described in the previous chapters, *the FESS subsystem* comprises all components and parts necessary for the construction and operation of the energy storage unit. By modifying individual components, such as the bearings, rotor, or electric machine, the energetic specifications of the FESS can be influenced and, if necessary, improved. The main goal is to make the FESS properties approach the performance of an idealized reference energy storage system. Of course, this process underlies the scope of current technical possibilities, but at least the *threshold specifications* described in Sect. 5.4 must be reached.

6.1 Deviation of *Desired* from *Actual Characteristics*

To answer the question *which* components require optimization, it must first be clarified which properties of the FESS need to be improved. As summarized in Sect. 5.5.1, the desired properties result from the analysis of the *supersystem* and must be compared with the state of the art in the field of flywheel energy storage systems in order to determine which technical properties must be improved in the first place. Only after a possible discrepancy between desired and actual properties has been identified and precisely quantified, it can be deduced which components of the *subsystem* have a significant influence and therefore need to be modified.

In order to define the deviation between the desired and actual characteristics of flywheel energy storage systems, it is useful to compare the state of the art in the field of FESS with:

A. Buchroithner, *Flywheel Energy Storage*,
https://doi.org/10.1007/978-3-658-35342-1_6

(a) **Reference energy storage devices**
(b) **Competing energy storage technologies**

The definition of these properties is part of the analysis of the *supersystem*. Both of these ideal and real energy storage systems are described in more detail in Sect. 3.3 and Sect. 5.4. However, among the general objectives of improvement are the **increase of the specific energy** (and consequently the reduction of weight) and **the reduction of costs**. In the following section, two prototypes developed by the *Energy Aware Systems* Group at the Graz University of Technology will be examined in more detail with regard to the cost and weight of the components.

6.1.1 Analysis of the Cost and Weight of the System Components of Two FESS Prototypes

In this section, two prototypes are compared and the distribution of weight and costs among the system components is discussed. A short explanation regarding the evaluation of the components as well as a reference to the later following detailed description of the technical solution is also given. Figure 6.1 shows a possible vehicle for which the FIMD flywheel energy storage system presented in Table 6.1 was designed.

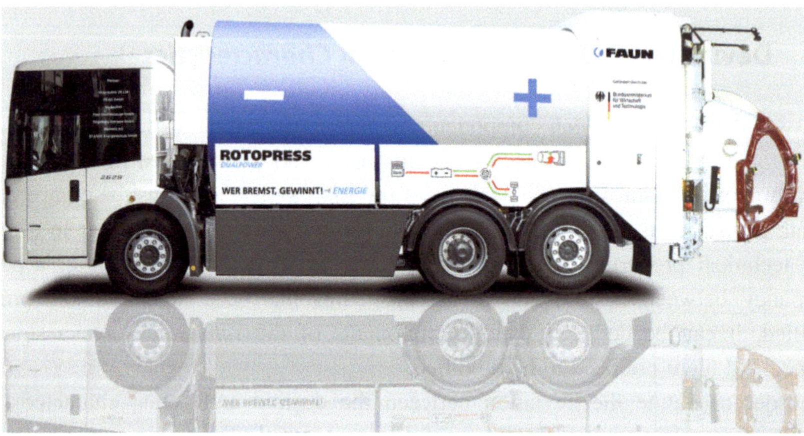

Fig. 6.1 Heavy-duty commercial vehicle for municipal service (FAUN Rotorpress Dual Power [1])—an ideal application for a flywheel energy storage system (Image rights: FAUN Umwelttechnik GmbH & Co. KG)

Table 6.1 Cost and weight shares comparison of the components of two FESS prototypes

<table>
<tr><th>FIMD flywheel</th><th>CMO flywheel</th></tr>
<tr><td>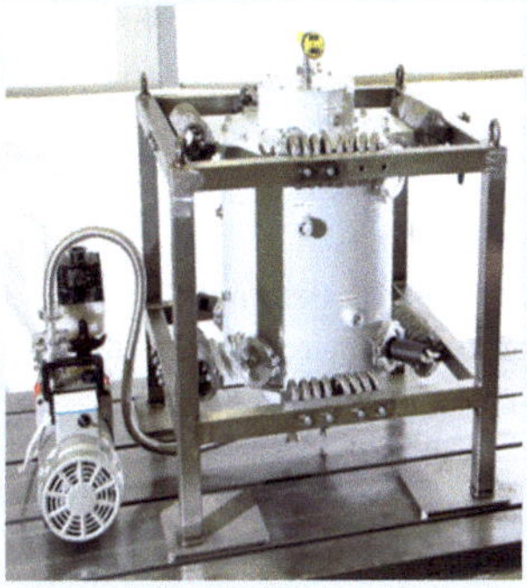</td><td>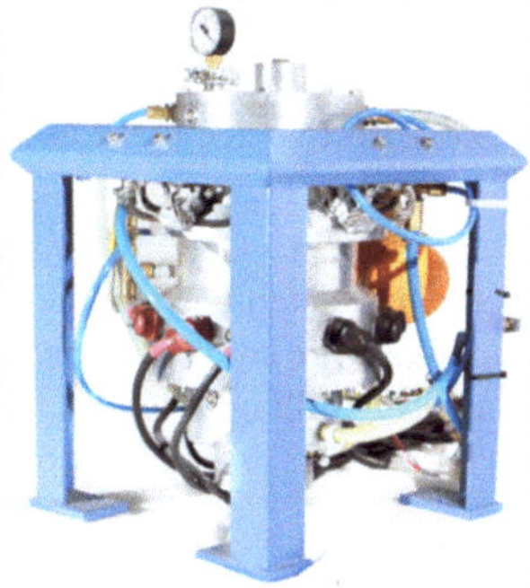</td></tr>
<tr><td colspan="2">Description</td></tr>
<tr><td>FESS for load point shifting installation in a series hybrid (heavy commercial vehicle for inner-city operation)
(see Sect. 7.5 for a more detailed description)</td><td>FESS as secondary storage in a parallel hybrid (demonstrator vehicle for hybrid vehicle technologies developed in Austria)
(see Sect. 7.4 for a more detailed description)</td></tr>
<tr><td colspan="2">Technical data</td></tr>
<tr><td>• Energy content: 0.8 kWh
• Power: 145 kW
• Maximum speed 40,000 rpm
• Rotor weight: 80 kg
• Total weight: 280 kg</td><td>• Energy content: 0.1 kWh
• Power: 40 kW
• Maximum speed 60,000 rpm
• Rotor weight: 11 kg
• Total weight: 35 kg</td></tr>
<tr><td>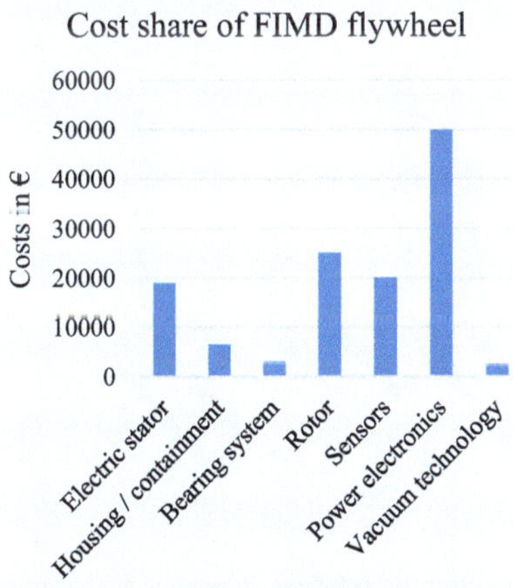
</td><td>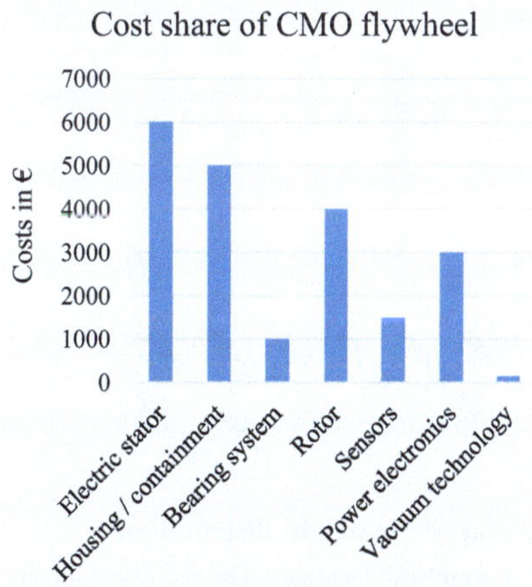
</td></tr>
<tr><td colspan="2">* Considered were the costs for a hand-built laboratory prototype with a quantity of 1! For more detailed description see below.</td></tr>
<tr><td colspan="2">Discussion of cost allocation</td></tr>
<tr><td>Electrical machine (stator): Relatively high share of costs due to the high demand on electrical material properties, and primarily due to the prototype-specific manufacturing process (laser cutting of the metal sheets)</td><td>Electrical machine (stator): Relatively high share of costs due to the high demand on electrical material properties, and primarily due to the prototype-specific manufacturing process (laser cutting of the metal sheets)</td></tr>
</table>

(continued)

Table 6.1 (continued)

FIMD flywheel	*CMO* flywheel
Housing: Cost-effective welded structure made of mild steel	**Housing:** Higher relative costs due to the use of aluminum (turned part from solid material)
Bearing system: Cost-effective due to the use of rolling bearings See Sect. 9.6.1	**Bearing system:** Cost-effective due to the use of rolling bearings See Sect. 9.7.1
Rotor: Costs are primarily determined by the manufacturing processes (wire eroding of the electrical sheets and CNC Milling of the flywheel mass sheets) See Sect. 7.5	**Rotor:** Costs are primarily determined by the manufacturing processes (wire eroding of the electrical sheets) See Sect. 7.4
Sensor technology: Only the sensor technology required for the prototype was considered here. The sensors used in a possible series product (temperature monitoring and speed measurement) are simpler and 10–20 times more cost-effective [2]	**Sensor technology:** Essential sensor technology for condition monitoring (temperature monitoring and speed measurement) plus accelerometers, which dramatically increase costs
Inverters: High costs of power electronics, since a single unit was purchased for the prototype	**Inverters:** Extremely high costs of power electronics, since a custom prototype inverter was developed
Vacuum technology: Low cost due to the use of standard components	**Vacuum technology:** Low cost due to the use of standard components
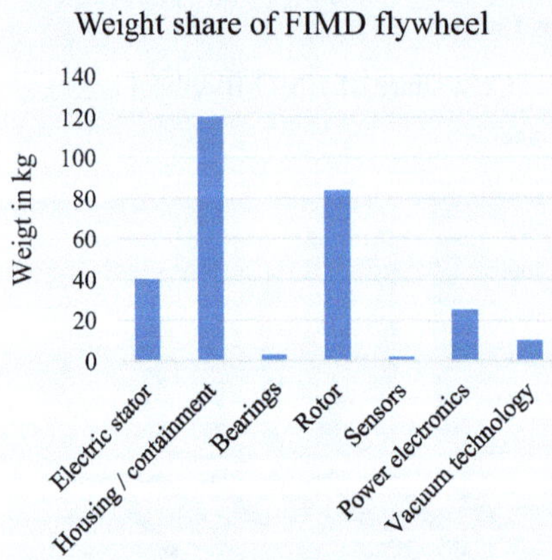	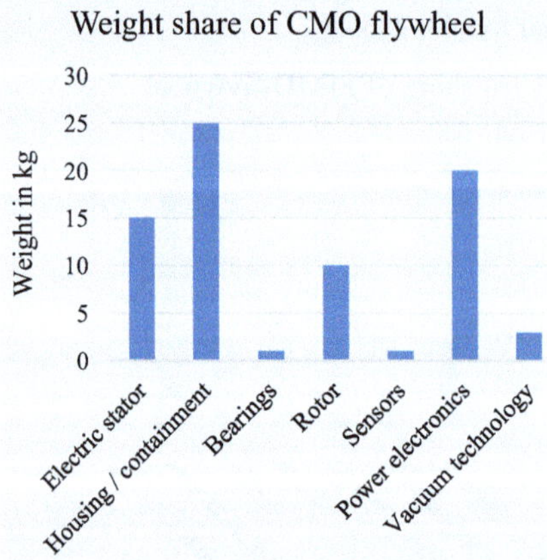
Discussion of the weight distribution	
Electric machine (stator): The weight share of the stator is related to the power rating and technically can hardly be reduced. A high-quality core sheet metal has already been used to achieve high rotational speeds and hence high specific energy. In order to increase the moment of inertia, mass plates made of tempered steel were attached (see Sect. 7.5)	**Electric machine (stator):** The weight share of the stator is related to the power rating and technically can hardly be reduced. A high-quality core sheet metal has already been used to achieve high rotational speeds and hence high specific energy

(continued)

Table 6.1 (continued)

FIMD flywheel	*CMO* flywheel
Housing: Highest weight share due to conservative design and high safety factors. Uncertainty caused by the lack of design guidelines and calculation methods for safety housing See Chap. 8	**Housing:** Highest weight proportion due to conservative design and high safety factors due to the absence of design and calculation specifications See Chap. 8
Bearings: Low cost share by using roller bearings and avoiding expensive magnetic bearings	**Bearings:** Low cost share by using roller bearings and avoiding expensive magnetic bearings
Rotor: High weight due to the required energy content. The reduction can only be achieved by changing the material selection and increasing the diameter (at constant rpm) See Sect. 7.5	**Rotor:** Lower mass share due to low required energy content. A further weight reduction is hardly possible even by changing the choice of material since flywheel mass = mass of the rotor of the electric motor See Sect. 7.4
Sensor technology: Low component volume/ mass compared to the total mass of the FESS	**Sensor technology:** Low component volume/ mass compared to the total mass of the FESS
Inverter: Not a significant proportion due to the high total mass of the FESS	**Inverter:** The inverter was designed as a prototype especially for the CMO-FESS and has a non-optimized cooling system and a heavy copper plate as a base
Vacuum technology: Weight not significant, but reduction possible by using a diaphragm pump instead of a two-stage rotary vane pump	**Vacuum technology:** Weight is not decisive, since a diaphragm pump is used due to the relatively small evacuation volume
Conclusion (valid for both concepts) • A significant cost reduction was achieved in both cases by avoiding magnetic bearings and using roller bearings (for details on bearing system design, see Chap. 9) • The high total costs have their origin in the prototype-specific manufacturing processes • Rotor costs are predominantly determined by the manufacturing process and not by the material • The truly dominant weight share of the (burst) housing is caused by a lack of reliable calculation guidelines and experience with lightweight safety housings (see Sect. 8.9) • The weight of the rotor depends on the required energy content. A weight reduction can be achieved only by changing the rotor material	

6.2 *Internal System Interdependencies*: Interactions Between Critical Components

Although the findings listed in the conclusion of Table 6.1 suggest that, for example, an isolated change in the rotor material would already lead to an increase in energy density, it must be explicitly pointed out that even the change in the properties of *one* component often has far-reaching consequences due to the strong interaction between the highly stressed parts within the FESS. Changing the rotor material from steel to fiberglass, for

Table 6.2 Effects of a change in FESS rotor material on surrounding components

Property of the FESS rotor material			Impact on	
Maximum speed	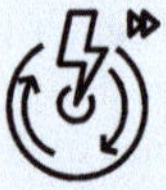	→	Bearing design	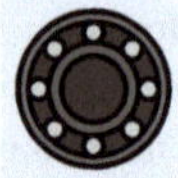
Rotor burst behavior		→	Housing design	
Ageing and damage mechanisms		→	Sensor technology and condition monitoring	
Thermal behavior of the material	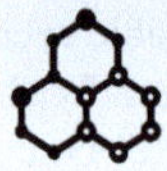	→	Cooling system and max. Operating temperature	

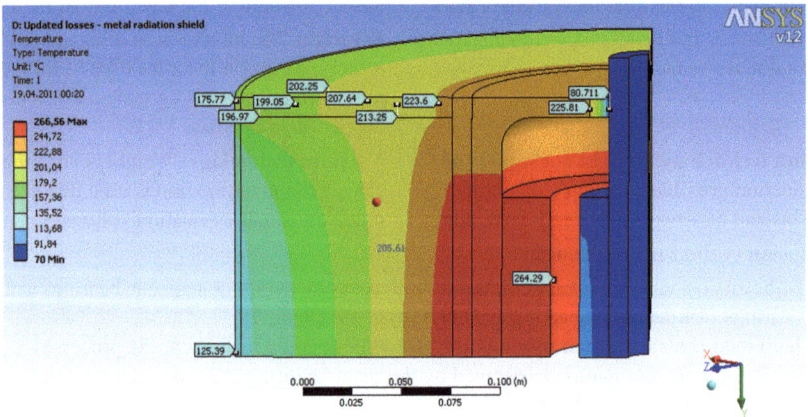

Fig. 6.2 Thermal simulation of an outrunner rotor with carbon fiber bandage (Project E3oN) using the simulation tool ANSYS

instance, brings about a significant change in the properties listed in Table 6.2. (For reasons of better readability, only the essential aspects are summarized here. In the specific case of rotor design, reference is made to Chap. 7).

The choice and design of the components therefore not only influences interactions with the *supersystem* by changing the externally visible properties of the energy storage but also strongly affects *system-internal interdependencies*.

Especially the evacuation of the housing, which is necessary for low windage losses during operation and thus the reduction of self-discharge, causes complex thermal and constructive interdependencies between the components. Figure 6.2 shows the thermal

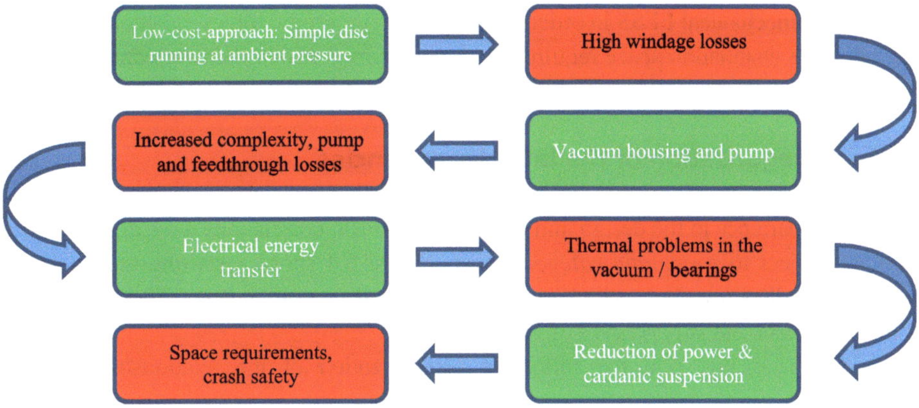

Fig. 6.3 The flywheel energy storage system as a multidimensional optimization problem

simulation of an outrunner rotor with carbon fiber bandage using the simulation tool *ANSYS*. While the stator in the center of the machine is water-cooled, temperatures of up to 265 °C prevail at the inner circumference of the rotor. These temperature levels already exceed the permissible operating temperature of the carbon fiber composite. The heat source is the electrical losses occurring in the rotor, which can hardly be dissipated due to the absence of convective cooling and the additional insulation of the carbon fiber bandage. The only remaining possibility is heat conduction across the ball bearings, which results in high thermal stress on the lubricant and the bearing itself.

It is therefore not possible to improve the properties of the FESS to a satisfactory degree by isolated modification of *only one* component. The apparent improvement of a solution by modifying the topology and morphology of the flywheel energy storage device usually brings about further, different complications. Hence, there is always a trade-off that makes it necessary to make a compromise. An example, which includes a *chain of three conflicting objectives*, is sketched out in Fig. 6.3 and subsequently described in more detail.

- The simplest realization of a flywheel energy storage device, a disk-shaped rotor running freely at ambient pressure (1013 mbar) and with only a mechanical shaft as main connection (= mechanical energy transfer), is subject to considerable windage losses. If one tries to avoid these by evacuation, further losses occur in addition to the blatantly higher design and manufacturing costs of the vacuum housing, namely, those of the vacuum feedthrough and the pump itself.
  - If one now tries to avoid the mechanical vacuum feedthrough losses by using electrical energy transfer and integrating a motor generator, one will encounter thermal problems, since the power loss of the electrical machine is difficult to dissipate in a vacuum.

  Now that the thermal problems have been solved and attempts have been made to fully exploit the advantages of electrical energy transfer by gimballing the electromechanical FESS in order to reduce the gyroscopic forces, it becomes evident that the installation space required for a crash-proof design increases dramatically.

6.2.1 Categorization of the Interdependencies

Based on experience in the design and operation of FESS, correlations were described that must be subject to *systematic categorization.* Figure 6.4 shows—starting at the central component *rolling bearing*—which mechanisms have *direct* and *indirect* effects on aspects of the *subsystem* and *supersystem* of FESS.

According to Fig. 6.4, the following categorization of the relationships can be introduced:

1. ***Horizontal interdependencies:*** Mutual influence of components within the same hierarchical level. The interdependencies are therefore *monosystematic.*

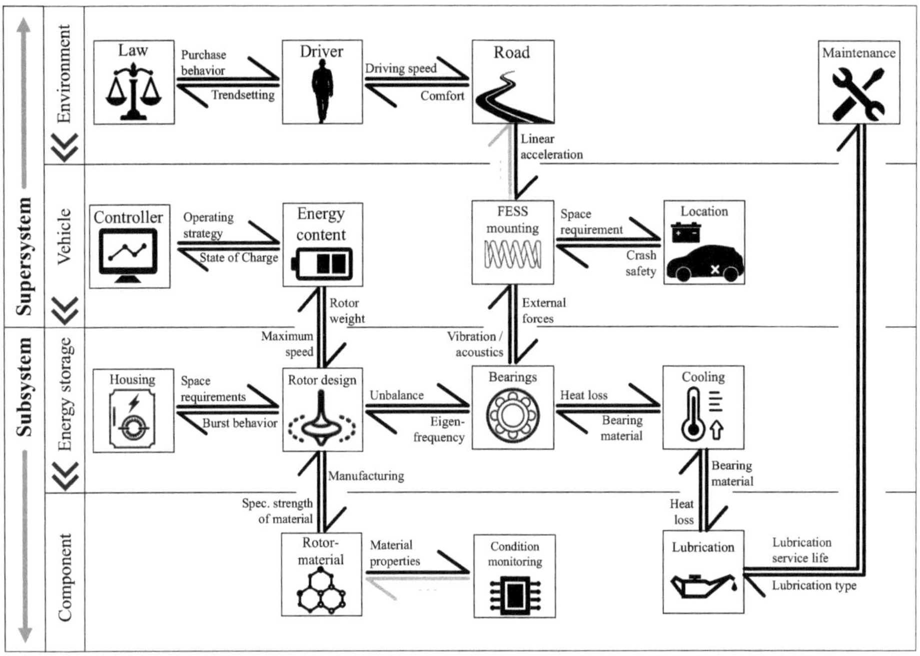

Fig. 6.4 Representation of horizontal and vertical interdependencies in relation to the FESS subsystem and supersystem

Example

- **Example at *energy storage* system level:** The rotor design influences the bearing system via the unbalance forces that occur during operation. Conversely, the bearing system (or its stiffness) influences the natural frequency and dynamics of the rotor.
- **Example at *environment* system level:** The driver specifies the route and consequently the road condition he chooses or the speed at which the car moves on it. Conversely, the road conditions are responsible for the driver's comfort and therefore influence the chosen travel speed.

2. ***Vertical interdependencies:*** Mutual influence of components of different hierarchical levels. The interdependencies are therefore *transsystematic*.

Example

- **Example regarding the *component* and *energy storage* system level:** The general rotor design has a considerable influence on the choice of material via the required manufacturing processes. Conversely, the rotor material determines which geometry makes sense (more about rotor design in Chap. 7).

  Hierarchical levels can also be skipped:

Example
The bearing lubrication (system *component*), for example, has a direct influence on the maintenance intervals over its service life (system *environment*).

3. ***Bidirectional interdependencies:*** There is a *mutual* influence of components, as in the examples above.
4. ***Unidirectional interdependencies:*** One component affects the other, but not the other way around.

Example

- **Example of *unidirectional, vertical interdependence*:** The road influences the design of the compliant FESS-to-vehicle interface (mounting of the housing). Conversely, the condition of the road is in no way dependent on the design aspects of the storage or vehicle system.

6.2.2 Critical Interdependencies in the FESS *Subsystem*

Pursuing the omnipresent goal of cost reduction, computer-aided development tools are increasingly establishing themselves in the automotive sector to replace practical experiments. Assoc. Prof. Michael Bader from the *Institute for Machine Elements and Development Methodology (Institut für Maschinenelemente und Entwicklungsmethodik)* at the *Graz University of Technology* has shown in his habilitation thesis *The Experiment-Based Technical Development Process (Der versuchsgestützte technische Entwicklungsprozess)* that practical testing still represents an indispensable step in the product development cycle and is far from being obsolete. This statement applies in particular to flywheel energy storage systems due to their highly complex *system-internal interdependencies, which* are not obvious in the early development phases and often become apparent only during prototype testing. The reasons for this are not only aspects which are neglected in preliminary designs or first simulations due to lack of experience, but also fundamental uncertainties such as the achievable balance quality due to the rotor manufacturing process (Table 6.3).

The following list is based on experience gained from FESS prototype testing and custom test design: (Admittedly, the number of interdependencies is almost infinite, but the following ones have been *identified as critical* and have been examined in more detail at the *Institute of Electrical Measurement and Sensor Systems* the course of research activities.)

1. **Rotor dynamics:**
   Category: *horizontal, bidirectional interdependence*
   - Involved components/properties: coupling stiffness, natural frequency (eigenfrequencies) shaft/rotor, bearing, bearing plate, housing, suspension, etc.
     - Specific test bench for deeper investigation: LESS (see Sect. 9.7)
2. **Thermal management:**
   Category: horizontal and vertical, bidirectional interdependence
   - Involved components/properties: electric machine, cooling system, vacuum level, bearings, etc.
     - Specific test rig for investigation: Thermal conductivity test bed of the Energy Aware Systems Group, Graz University of Technology (for further information, see [3])
3. **Self-discharge:**
   Category: horizontal and vertical, bidirectional interdependence
   - Involved components/properties: bearings and lubrication, vacuum level, natural frequency of the rotor, rotor material, and lifting magnet/magnetic weight compensation (if applicable).
     - Specific test bench for deeper investigation: self-discharge test bench (see Sect. 10.3)

Table 6.3 Interaction mechanisms between *subsystem components* and aspects of the *supersystem*

Subsystem component		Interaction effect		Supersystem aspect	
Vacuum components	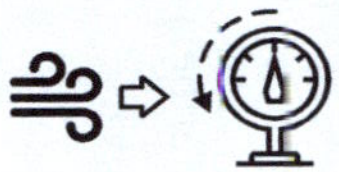	Friction (self-discharge)		Dynamics and load cycle	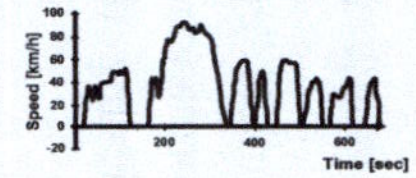
Bearing system				Profitability	
		Wear + service life		Service and maintenance	
Rotor design		Energy density + safety		Suitability for mobile use	
Housing					
Energy transfer (a) Electrical		Max. Power rating		Suitability for boosting	
(b) Mechanical					

6.2.3 Identification of Critical Components

The identification of critical components relates to the finding of technical "enablers," i.e., components that have a critical influence on the service life and performance of the FESS. These are especially components that play a key role in achieving the *threshold specifications* defined in Sect. 5.4.

The conclusion of Table 6.1 already presents some information on critical components, which was obtained based on the cost and weight analysis of two prototypes. However, this comparison also shows that a cost reduction can be achieved by using rolling bearings (instead of active magnetic bearings). In the end, only practical experience can reveal the hidden problems caused by the *system's internal interdependencies*.

6.2.3.1 The Bearing System as Technical Enabler

Experience has indeed shown that when it comes to flywheels, everything literally revolves around the bearings. That is why special attention is paid to this component. All the experience gained related to rolling bearings during the author's first FESS-related research projects had one thing in common: In all prototypes and/or component test rigs, it was difficult or not possible to achieve the planned operating speed as calculated in advance! Table 6.4 compares the planned rotational speeds and the speeds achieved during the initial commissioning of two FESS prototypes and two-component test rigs. In all four examples, the cause for the unforeseen discrepancy in machine dynamics was found to be a problem related to the bearings or an *interdependency within the system*.

6.3 Results: Critical Components in FESS

In addition to the *bearing system* just discussed, systematic analysis and previous project experience have shown that the *safety housing* and the *rotor* are the components that determine the success or failure of flywheel energy storage technology. Table 6.5 describes the current problems of these *three critical* components of the subsystem of a FESS and the related development goals.

Based on this list, the following generally valid measures for the (further) development of FESS for automotive applications can be derived (sorted by priority):

1. **Cost-reducing measures:**
   - Use of roller bearings instead of active magnetic bearings
   - Use of optimized steel rotors instead of wound fiber composite rotors
   - Cost reduction of manufacturing processes for fiber composite rotors
2. **Safety-enhancing measures:**
   - Improvement of the burst containment
   - Improvement of the rotor design
   - Condition monitoring

Table 6.4 Comparison of target and actual rotational speeds achieved during initial commissioning of various FESS and component test benches

Project		Set speed	Reached speed during initial commissioning	Termination criterion	Cause
FIMD		37,500 rpm	24,000 rpm	Bearing damage due to excessive accelerations/ vibrations	Resonance, axial bearing preload changed by thermal expansion of the shaft
CMO	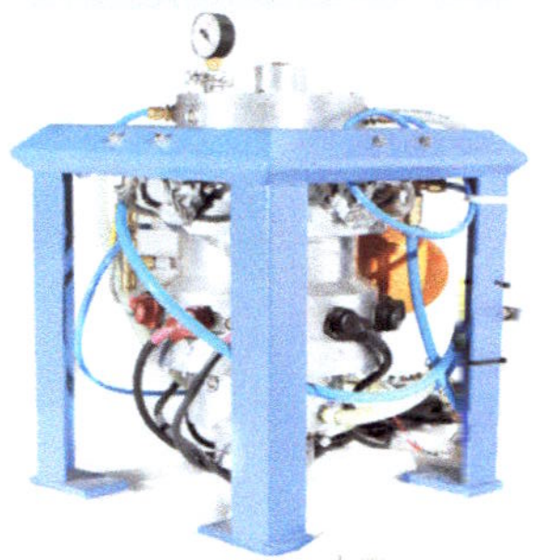	60,000 rpm	10,000 rpm	Excessive power loss and thermal load on the bearings	Probably incorrect setting of the lubricant quantity

(continued)

Table 6.4 (continued)

Project		Set speed	Reached speed during initial commissioning	Termination criterion	Cause
Rotor burst test rig	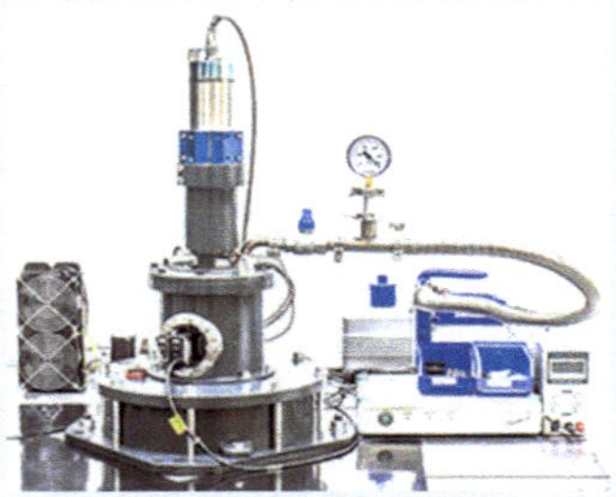	40,000 rpm	24,000 rpm	Bearing damage due to excessive accelerations/ vibrations	Resonance, real bearing stiffness lower than calculated
Thermal conductivity test bench	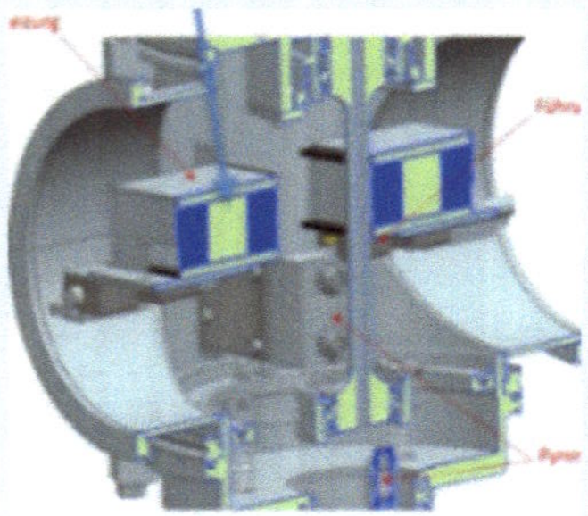	20,000 rpm	10,000 rpm	Bearing damage due to excessive accelerations/ vibrations	Resonance, real bearing stiffness lower than calculated

3. **Service life-enhancing measures:**
   - Reduction of bearing loads
   - Increase of balancing quality
   - Predictive maintenance
4. **Performance-enhancing measures:**
   - Reduction of the housing weight
   - Increase of rotor speeds

Subsequently, an in-depth examination and optimization of the three subsystem components identified as critical, rotor, bearings, and housing, is presented.

Table 6.5 *FESS subsystem components* identified as critical and description of current problems and development goals

Component	Current problems	Development goals
Rotor 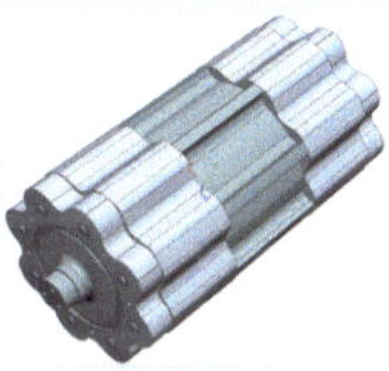	In order to reach the threshold energy density, an increase in rotor speed is necessary. High costs due to complex manufacturing processes and choice of material (fiber composite). Bursting of the rotor represents a safety risk	Reduction of costs, Achievement of high safety levels or favorable crash and burst behavior Increase in energy density
Bearing system 	Magnetic bearings are too costly and require a complex control system. Rolling bearings are a low-cost alternative, but they are the only components subject to wear, which dominate self-discharge and service life among other properties	Reduction of the loss torque with a simultaneous increase of the bearing and lubricant service life
Housing 	The housing dominates the weight distribution among the FESS components and thus reduces the specific energy of the entire system. It is a safety-critical component; hence, approval and certification must be considered very carefully, especially in the automotive sector	Reducing weight while maximizing safety in the event of rotor failure or vehicle crash

References

1. FAUN Umwelttechnik (2018) Rotopress. FAUN Umwelttechnik GmbH & Co. KG Feldhorst 4, 27711 Osterholz-Scharmbeck, Germany.
2. A. Buchroithner et al (2016) Decentralized Low-Cost Flywheel Energy Storage for Photovoltaic Systems. International Conference on Sustainable Energy Engineering and Application (ICSEEA 2016), Jakarta, Indonesia. DOI: https://doi.org/10.1109/ICSEEA.2016.7873565
3. A. Buchroithner, P. Haidl, H. Wegleiter, M. Simonyi and T. Murauer (2019) Design, operation and results of a low-cost test rig for investigation of thermal properties of rolling element bearings in vacuum, 18th ESMATS – European Space Mechanisms and Tribology Symposium, Munich, Germany.

7 Rotors for Mobile Flywheel Energy Storage

In order to better understand the motivation and design process of the rotors, which are described in detail in Sects. 7.4 and 7.5, the following paragraphs summarize the essential mechanical principles. In Sect. 7.2, the state of the art is surveyed and analyzed. It should be noted that at this point, the chapter *Rotor* is only considered with a focus on energy density and machine dynamics behavior. Rotor-specific burst and failure scenarios are considered in the chapter *Housing* (see Chap. 8).

7.1 Essential Physical Relationships of FESS Rotor Design

Considering the aspects discussed in Sect. 2.2.1, it becomes clear that the maximum energy content of a flywheel energy storage device is defined by the permissible rotor speed. This speed in turn is limited by design factors and material properties. If conventional roller bearings are used, these often limit the speed, as do the heat losses of the electrical machine, if it forms one unit with the flywheel in a so-called integrated design. Imbalance forces—as explained in Sect. 9.6.1 describing a case study—can also limit the speed and lead to insurmountable resonances or at least significantly reduce the bearing life.

In most cases, however, it is the centrifugal forces that drive the rotor material to its limits and thus determine the FESS's maximum energy content or energy density. In order to ensure sufficient operational safety and to avoid critical operating states, the stresses in the material must not exceed the maximum permissible values. Further information on the following derivations can be found in literature, i.e., [1] and [2].

Solid bodies, which rotate at a given angular velocity ω, are subject to centrifugal forces according to the *d'Alembert principle*. These inertial forces of negative mass acceleration can be assumed to be

A. Buchroithner, *Flywheel Energy Storage*,
https://doi.org/10.1007/978-3-658-35342-1_7

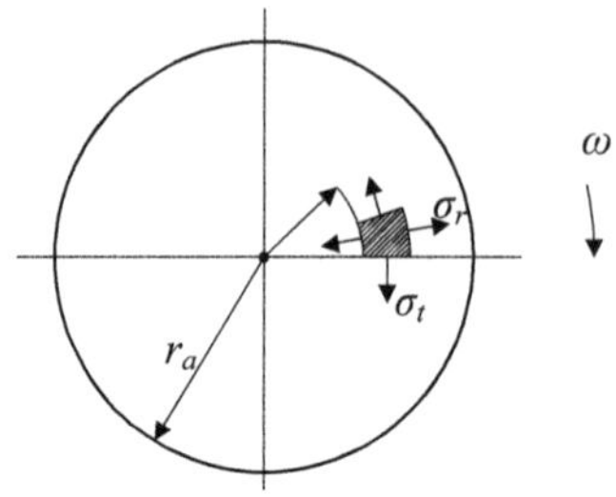

Fig. 7.1 Stresses occurring in a rotating solid disk

$$\omega^2 r dm = \omega^2 r \ \rho \ dA \ dr \tag{7.1}$$

It is now necessary to trace the centrifugal forces back to stresses and determine their maximum. The first case to be considered is a rotating solid disk (Fig. 7.1).

The greatest stresses are the tangential stresses σ_t at the center of the disk since the supporting cross section is smallest here, but the centrifugal forces are greatest due to the accumulation of masses located at the outer radius [1]. The following applies:

$$\sigma_{t \ \max} = \rho\omega^2 r_a^2 \frac{3+\mu}{8}\left[2 + \left(\frac{r_i}{r_a}\right)^2 \left(1 - \frac{1+3\mu}{3+\mu}\right)\right] \tag{7.2}$$

where μ describes the transverse contraction coefficient (Poisson's ratio) and can be assumed to be approximately *0.3* for steel. After simplifying the mathematical expression and introducing the radius ratio $r_i/r_a \approx 0$ in the case of a *small inner bore*, one obtains (Fig. 7.2)

$$\sigma_{t \max 0} = \rho\omega^2 r_a^2 \frac{3+\mu}{4} \tag{7.3}$$

Based on these equations, the relations for a solid disk without central bore can be derived. If $r_i = 0$, then one obtains

$$\sigma_{t \ full} = \rho\omega^2 r_a^2 \frac{3+\mu}{8}\left[1 - \frac{1+3\mu}{3+\mu}\left(\frac{r}{r_a}\right)^2\right] \tag{7.4}$$

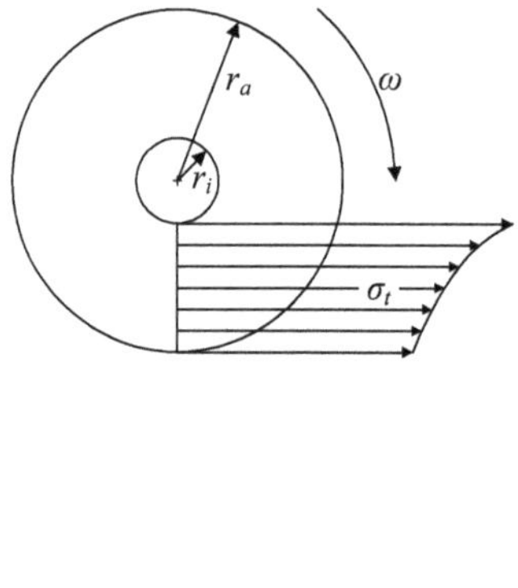

Fig. 7.2 Stress curve in a circumferential disk with central bore

Or

$$\sigma_{r\ full} = \rho\omega^2 r_a^2 \frac{3+\mu}{8}\left[1 - \left(\frac{r}{r_a}\right)^2\right] \tag{7.5}$$

Regarding the center of the solid disk, r = 0 and subsequently $\sigma_{t\ full} = \sigma_{r\ full}$ with the mathematical term:

$$\sigma_{\max\ full} = \rho\omega^2 r_a^2 \frac{3+\mu}{8} \tag{7.6}$$

It can be seen that the maximum stresses occurring in a disk with central bore are twice as high as the ones in a solid disk. Highly stressed, fast-rotating components should therefore be designed without bores if possible. A solution for this difficult shaft-disk connection is a forged shaft end or a screwed-on shaft flange according to *Zwerenz and Schauberger* [3] (see Fig. 7.3). The further away the axial bores are from the center of the disk (axis of rotation), the less significant is the reduction of the maximum speed, as the discussion of the *FIMD* rotor in Sect. 7.5 shows.

Looking only at the strength of the flywheel alone and ignoring aspects of manufacturing or system integration, it does not always make sense to consider a disk of equal thickness and calculate its stress distribution. Instead, it would be beneficial to specify an optimum stress distribution and derive the thickness of the disk from this. A solution to this problem has been known for a long time and is called *de Laval disk*, or *disk of constant stress*.

The material can be used optimally for the case that

$$\sigma_r = \sigma_t = \sigma \tag{7.7}$$

If one determines the height of the disk based on this criterion, the result is

$$h(r) = h_0 e^{-\rho(\omega r)^2/(2\sigma)} \tag{7.8}$$

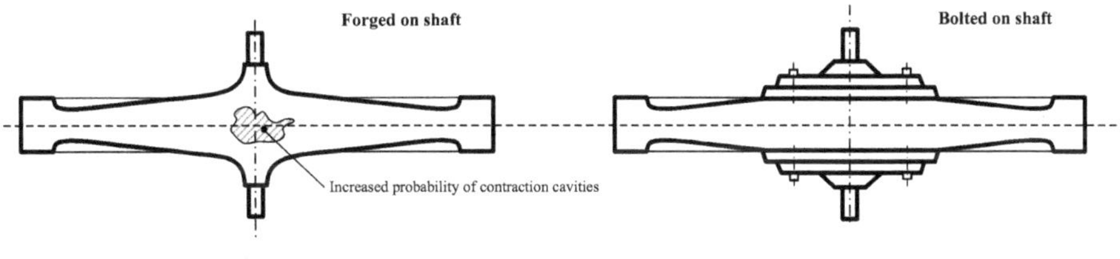

Fig. 7.3 Possible shaft-rotor connections for reducing the centrifugal stresses in the center according to [3]

Table 7.1 Shape factors K_{shape} of different rotor shapes [4]

Description	Sketch	Shape factor K_{shape}
Ideal disk of constant stress	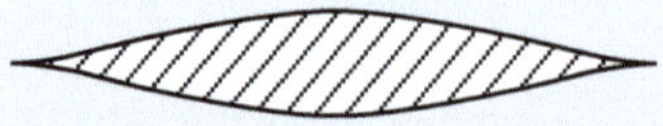	*1.0*
Real disk of approximately constant stress		*~0.7–0.9*
Conical disk	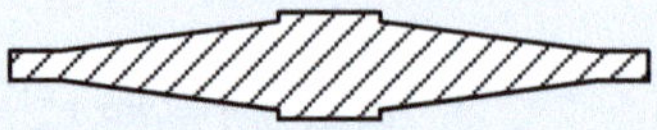	*~0.7–0.85*
Cylindrical disk		*~0.6*
Thin-walled cylindrical ring	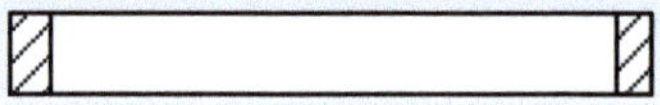	*~0.5*
Cylindrical disk with central bore	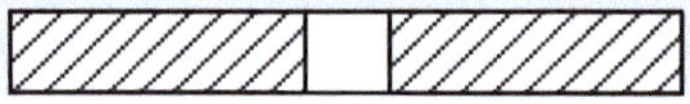	*~0.3*

The determined height h in Eq. 7.8 corresponds to a shape factor of $K_{shape} = 1$ in Table 7.1.

In any case—with or without central bore—the density of the rotor material is included linearly in the calculation of the tangential stresses of the flywheel. Sect. 2.2.1 has shown that the energy content increases linearly with the mass moment of inertia of the rotor, but is proportional to the square of its speed.

It can be concluded that light, high-strength materials are best suited for flywheel rotors of high-energy density, as they cause lower centrifugal forces and thus allow higher speeds. The maximum storable kinetic energy of a disk thus depends on the ratio of the tensile strength σ_{max} to the density ρ of the material and can be expressed as follows:

$$E_k = K_{shape} \frac{\sigma_{max}}{\rho} \tag{7.9}$$

K_{shape} is the so-called form factor of the flywheel, which essentially takes into account the decrease in the theoretical specific energy of the rotor material through practical shaping and is described in Table 7.1. The form factors were empirically determined by the Italian flywheel pioneer *Giancarlo Genta*.

The maximum permissible speed depending on the rotor material and shape is therefore

$$n_{max} = \frac{30}{\pi} \sqrt{4K_{shape} \cdot \frac{\sigma_{max}}{r^2 \cdot \rho}} \tag{7.10}$$

Based on this equation, the theoretically achievable gravimetric energy densities for flywheels made of different materials can be calculated, as shown in Table 7.2. It must be noted that these values cannot be achieved in practice, as no safety margins and real design influences such as notches, etc. are taken into account.

Table 7.2 Material properties and related theoretical energy densities of different FESS rotor materials

Material	Tensile strength σ_{max} *N/mm²*	Density ρ *Kg/dm³*	Specific energy σ_{max}/ρ *Wh/kg*
Structural steel	340	7.8	12.1
Standard electrical steel	400	8	13.9
Tempered steel (*42CrMo4*)	1100	7.8	36.6
Birch wood	137	0.65	58.5
Aluminum (*Ergal 65*)	600	2.72	61.3
Titanium (*ZK 60*)	1150	5.1	62.6
High-strength steel (*AlSi 4340*)	1790	7.83	63.5
Glass fiber (E-glass/EP 60%)	960	2.2	132
Kevlar ("*Aramid 49EP*"/60%)	1120	1.33	234
Carbon fiber (*T1000G*)	3040	1.5	563

It is therefore clear that the specific energies of flywheel energy storage systems will continue to rise as long as there are advances in materials science. If it were possible to build a rotor from carbon nanotubes, it could achieve specific energies of almost 15,000 Wh/kg. Although this scenario must still be regarded as a dream of the future, it shows the high theoretical potential of this technology.

7.2 Analysis of Existing Systems/State of the Art

7.2.1 Composite Flywheels

Table 7.2 illustrates the great potential of composite rotors and explains why research in the field of FESS has relied on this rotor design for several decades and still does. At 563 Wh/kg, the theoretically achievable energy density of the carbon fiber *TG1000* even exceeds by far today's commercially available Li-Io batteries, which usually achieve 100 to 200 Wh/kg at the cell level.[1]

However, the mass share of the matrix material (resin), an inherent safety factor, and above all the weight of surrounding system components, such as the housing, electric machine, frequency inverter, cooling system, etc., significantly reduce this theoretical energy density, as shown in Fig. 7.4. This substantial divergence between real and theoretical specific energies can also be observed in the example of the *NASA G2 Flywheel*, which is shown in Fig. 7.5. This circumstance also applies to steel rotors. Although the first

[1] At the " pack level", which includes the housing, balancing board, and cooling, the specific energy is further reduced

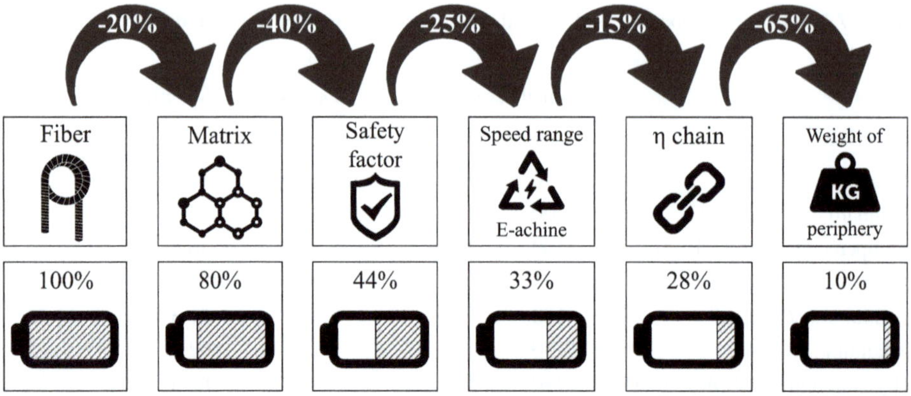

Fig. 7.4 Reduction of the theoretical energy content of the fiber composite rotor material due to design influences

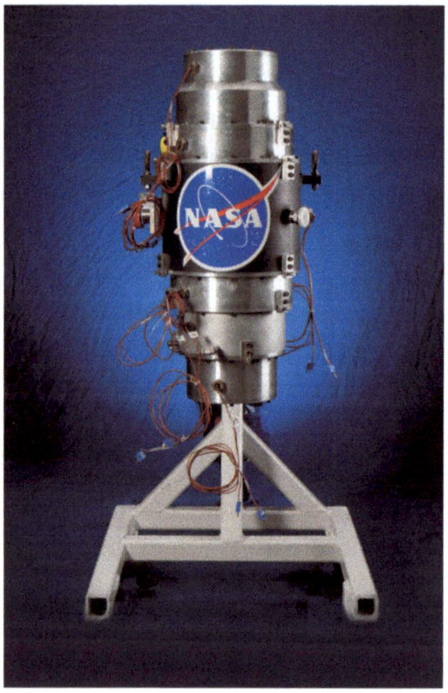

Fig. 7.5 *NASA G2* Flywheel for space applications. (Image rights: NASA)

step, the reduction from the pure fiber strength to the composite material using a matrix, is omitted, steel rotors already offer significantly lower specific energy anyway. Nonetheless, composite rotors have represented the lion's share of FESS applications since the 1970s. Some of these composite rotors have been intensively investigated with regard to the maximum achievable energy density, and results have been published, as shown in [5] and the references in Table 7.3. Below, the reader finds a selection of representative flywheel rotors made of fiber composites (Figs. 7.6 and 7.7).

Table 7.3 Overview: composite flywheel rotors

Designation	Manufacturer	Year	Design structure	Speed	Spec. Energy[a]	Ref.
APL Filament Flywheel	Johns Hopkins Applied Physics Lab	1980	Braided Aramid fiber ring	30,000 RPM	~ 100 Wh/kg	[6]
Curved Kevlar Spokes Flywheel	Johns Hopkins Applied Physics Lab	1980	Kevlar cylinder with flexible spokes	36,000 RPM	~ 80 Wh/kg	[6]
PowerRing	LaunchPoint Technologies, Inc.	2005	Shaftless carbon fiber ring with permanent magnets	8400 RPM	~ 80 Wh/kg	[7]
ComFess	NEDO—New Energy and Industrial Technology Development Organization, Japan	2003	Wound carbon fiber rotor	24,000 RPM	~ 22 Wh/kg	[8]
GKN HP MK4 Flywheel	GKN Hybrid Power	2014	Carbon fiber wound on steel hub	36,000 RPM	~ 45 Wh/kg	[9]
BeaconPower	Beacon Power LLC	2010	Carbon fiber wound on steel hub	15,500 RPM	~ 25 Wh/kg	[10]
NASA G3 Rotor	NASA/Glenn Research Center	2006	Carbon fiber wound on aluminum hub	52,500 RPM	~ 120 Wh/kg	[11]
Flybrid F1 KERS	Flybrid Systems Inc. Torotrac	2009	Carbon fiber wound on aluminum hub	64,500 RPM	~ 30 Wh/kg	[12]
UT-CEM composite flywheel	University of Texas, Center for Electro-mechanics	1998	Multi-layer carbon fiber on steel hub	40,000 RPM	~ 42 Wh/kg	[13]

[a] The theoretical and maximum *specific energy of the rotor* itself is considered, not that of the entire energy storage system

7.2.1.1 Advantages of Composite Flywheel Rotors

In the following, the main advantages of fiber composite rotors compared to steel rotors are described. The findings come from extensive literature analyses, which include publications between 1960 and 2019.

Fig. 7.6 Exploded view of the *Ricardo* "Kinergy High-Speed Flywheel" prototype with magnetic drive. The carbon fiber rotor can be seen on the right side of the picture. (Image rights: Ricardo PLC)

Fig. 7.7 Flywheel by PUNCH Flybrid, designed for the KERS used in Formula 1. (Image rights: PUNCH Flybrid)

1. **High-energy density through exploitation of the specific strength of the fiber**
   (a) The high specific strength of the fiber absorbs the dominant tangential stresses in wound rotors. The anisotropy of the material is thus optimally exploited.
2. **Favorable burst behavior**
   (a) In most cases, centripetal force-induced breakage of composite rotors is indicated by the occurrence of mass imbalance forces due to delamination. In contrast to a spontaneous brittle fracture, the system can be shut down safely if this unbalance is detected in time.

Fig. 7.8 Comparison of the burst behavior of a steel and fiberglass rotor tested at ETH Zurich, Switzerland [5]. (Image rights: Peter von Burg, ETH Zurich)

(b) Good-natured fracture behavior through "pulverization" of the composite material, resulting in a more homogeneous pressure load on the burst containment. Large fragments occur only rarely (Fig. 7.8).

7.2.1.2 Disadvantages of Composite Rotors

3. **Production and costs**
   (a) The winding process is subject to the highest accuracy requirements and is mastered by only a few, specialized manufacturers. Fiber pretension, resin content, curing time, and similar manufacturing parameters must be precisely determined and monitored.
   (b) Due to the thermal energy released during the curing of the resin, solid structures cannot be produced with any desired wall thickness. Therefore, concentric rings are usually manufactured, which are conically ground and axially press-fitted (Figs. 7.9 and 7.10).

Fig. 7.9 Left: wet-winding of a carbon fiber composite hoop; right: finished 5 kWh flywheel rotor consisting of a steel shaft and axially press-fitted concentric hoops. (Image rights: FWT Composites & Rolls GmbH, Austria)

Fig. 7.10 Winding of a 160 kWh rotor for the *Advanced Locomotive Propulsion System* at the *Center for Electromechanics* at the *University of Texas*, Austin [14]. (Image rights: Center for Electromechanics, University of Texas)

4. **Reproducibility of rotor properties**
   (a) Investigations and burst tests of composite flywheels have shown that even test objects of identical design exhibit strongly varying bursting speeds and a quite different failure behavior [15]. This deviating behavior possibly also negates the previously mentioned advantage of the usually more benign burst behavior (see point 7, "Unforeseeable spontaneous rotor burst also possible").
5. **Balancing and maintaining the balancing quality**
   (a) Phenomena such as setting and creep—especially in the matrix material—which can be caused by temperature influences may change the balance quality over time. Also, the installation of a balancing groove (see Sect. 9.6.1) or balancing holes is possible in the only metallic shaft of the composite flywheel (and therefore with a small effective radius).
6. **Aging and high cycle fatigue**
   (a) Under correspondingly low loads, steel shows a pronounced *high cycle fatigue (HCF) behavior*.[2] Fiber composites, on the other hand, offer characteristics rather similar to aluminum, which for the most part cannot be described as high cycle fatigue resistant. One of the main arguments in favor of FESS compared to chemical batteries is their high number of charge/discharge cycles and long service life, which can only be achieved if the rotor can guarantee a correspondingly high number of load cycles. Figure 7.11 shows the fatigue strength behavior of various fiber

[2]The term *high cycle fatigue* (HCF) was coined by August Wöhler (∗ 22 June 1819 in Soltau; † 21 March 1914 in Hannover, Germany). He researched the materials steel and iron. The "Wöhler diagram" named after him represents the relationship between the number of cycles to failure and the stress amplitude for a material under oscillating load [21].

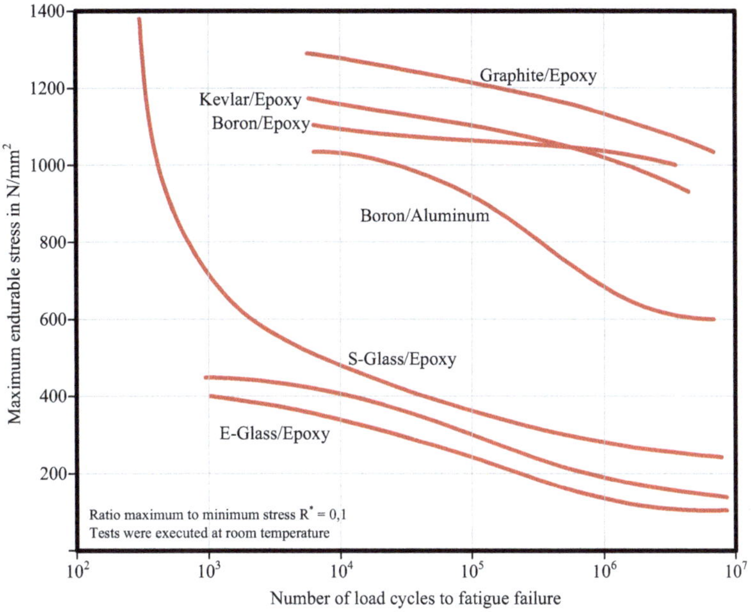

Fig. 7.11 Wöhler diagram (HCF diagram) of different fiber composite materials. (Image rights: McGraw-Hill Education)

composites, and it is obvious that *S-glass/epoxy*[3] shows a particularly striking drop. It should be noted that *E-glass/epoxy*[4] has an absolute tensile strength that is around 25% lower than that of *S-glass/epoxy*.

Since flywheel rotors are safety-relevant components, the strength of the rotor must be proven to endure the whole projected service life. *Anthony Colozza* from the *Glenn Research Center* of *NASA* describes in [17] that empirical strength proofs of a fiber composite rotor typically take more than 6 months since the speed spectrum has to be run through several tens of thousands of times. The rotor is accelerated to maximum speed approximately every 5 minutes and decelerated again to reach a cyclic positive stress load. Since it is a scientifically proven fact that steel can be assumed to be high cycle fatigue resistant under certain conditions, this effort is not required for steel rotors.

---

[3] The “E” in the designation has the historical background that these optical fibers were originally developed for *electrical* applications.

[4] The “S” in the name comes from the English word “stiff” and already indicates an increased modulus of elasticity and tensile strength.

(b) Aging of plastics, in general, poses another problem regarding the reliability and safety of composite rotors. The following aspects are given in [16] as the main reasons for the degradation of the mechanical properties of fiber composite materials:
- Aging of the polymeric matrix
- Humidity
- Temperature cycles
- Ultraviolet radiation
- Chemical products (acids, lyes, etc.)
- Deformation
- Fatigue due to load cycles
- Biological influences (fungal infestation)

In the case of flywheel rotors, of course, *thermal cycles*, cyclic positive stress, and *aging* of the polymer matrix are the most important factors. (Fungal infestation of rotors running in a vacuum is considered rather unlikely).

Aging of the fiber composite material affects primarily the matrix material and reduces the permissible stress of the rotor in radial direction [18, 19]. For example, if the matrix material has maximum permissible stress of approximately 120 MPa, then it is assumed that the entire fiber composite material can transfer about two-thirds of this stress in the radial direction. This assumption is based on the *stress concentration factor*, described in more detail in [20, 21]. If the material still has its full load-bearing capacity at the beginning, it can be assumed that it will only have half of this initial value after 10,000 hours according to an exponential function (Fig. 7.12).

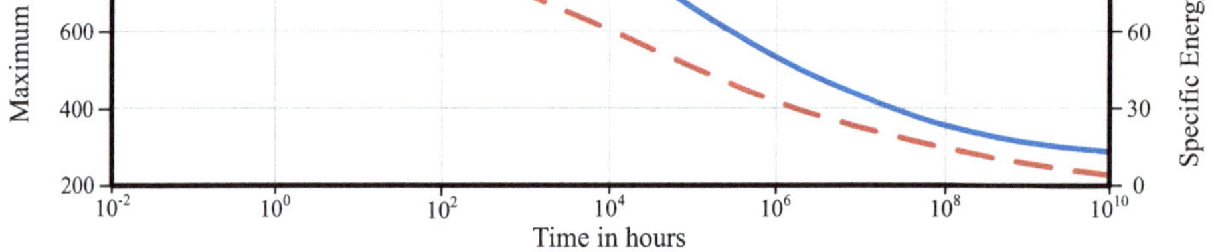

Fig. 7.12 Maximum tolerable radial stresses and specific energy of a composite flywheel, calculated based on the investigations of *Koyanagi* [21]. (Image rights: Springer Fachmedien Wiesbaden GmbH)

Fig. 7.13 Test rig for investigating the burst behavior of composite flywheels. The optical image for the high-speed camera is deflected by a mirror system. (Image rights: Ricardo PLC)

7. **Unforeseeable spontaneous rotor burst also possible**
   - Although the early detectability of rotor failure by delamination is an often mentioned advantage of wound fiber rotors [15], accidents are also known in which spontaneous, brittle-like fracture failure occurred without notice in advance. Systematic investigations of the real failure behavior of flywheels made of fiber-reinforced composites require complex, high-performance imaging measurement technologies and inherently safe test rig design. Figure 7.13 shows a skidding overspeed test stand with a high-speed camera system developed by the *School of Computing, Engineering and Mathematics* (University of Brighton, UK).
8. **Formation of harmful dust in case of rotor failure**
   - The "pulverization" of a fiber composite rotor in the event of failure has a positive effect on the design of the burst containment since the design can be based on a homogeneous pressure distribution, that is, the fiber composite particles behave like a fluid. Usually, no large, sharp-edged fragments can penetrate the casing, but the fine pulverization of the rotor poses further dangers. Particularly in the case of carbon fibers, fine carbon dust is produced during rotor failure, which is not only extremely harmful to the human respiratory tract but also highly flammable. The US space agency *NASA* published a report as early as 1979, in which the potential dangers of carbon fiber dust due to its high electrical conductivity were pointed out [22].

- An example is the explosion of a carbon fiber flywheel of a UPS system of the company *Beacon Power* in Stephentown in July 2011 [21]. Although the actual impact of the rotor fragments could not penetrate the burst containment, the subsequent coal dust explosion destroyed the steel housing and the surrounding concrete foundation.

9. **Poor thermal conductivity**
   - The rejection of heat losses of the electric machine of a flywheel energy storage system is a fundamental problem caused by the absence of convection due to operation in a vacuum. In a *fully integrated design* (see Sect. 2.2.2), the active part of the electric machine also serves as flywheel rotor mass. Since most electrical sheets usually have relatively low mechanical strength, their diameter is limited at a given rotational speed and—in the case of an outrunner topology—the laminated rotor can be wrapped in a thick bandage of fiber composite material.[5] In this case, an increase in the maximum rotor speed is achieved on the one hand by introducing compressive stresses into the electrical sheet stack, while on the other hand, the moment of inertia and thus the energy storage capacity are further increased. In the course of a research project funded by the Austrian Research Promotion Agency (Austriaische FFG) entitled "Efficient Electrical Energy Storage for Urban Public Transportation" (*E3oN*), which was carried out at the *Graz University of Technology*, simulation results showed that this fiber composite bandage has a strong thermally insulating effect that may cause problems regarding heat rejection. Figure 7.14 shows a thermal simulation of the outrunner rotor with carbon fiber bandage. The stator (in the right part of the picture) is water-cooled and therefore experiences much lower operating temperatures than the rotor. The highest temperature is reached in the active part of the rotor and rises above 200 °C due to the insulating effect of the carbon fiber bandage. Even inside the bandage, temperatures of far more than 100 °C occur. These high temperatures not only represent a considerable thermal load for other system components such as rolling bearings, lubricants, and even electrical steel but above all reduce the strength and durability of the fiber composite material, as described in Fig. 7.15.
   - Due to the anisotropy of composite materials, the thermal conductivity also shows a strong directional dependence, whereby the heat conduction in fiber direction is usually higher by a factor of about 10, than perpendicular to the fibers. The problem here lies in the insulating layers of the matrix. While glass and carbon fibers conduct relatively well, epoxy resin only achieves values between 0.2 and 0.3 W/mK [24]. Mild steel, on the other hand, conducts heat significantly better with values up to 60 W/mK [25]. Table 7.4 summarizes the thermal conductivity of various fiber composites.

[5] In the case of an integrated rotor topology, a carbon fiber bandage wound around the electrically active rotor would increase the air gap of the machine and thus reduce its efficiency.

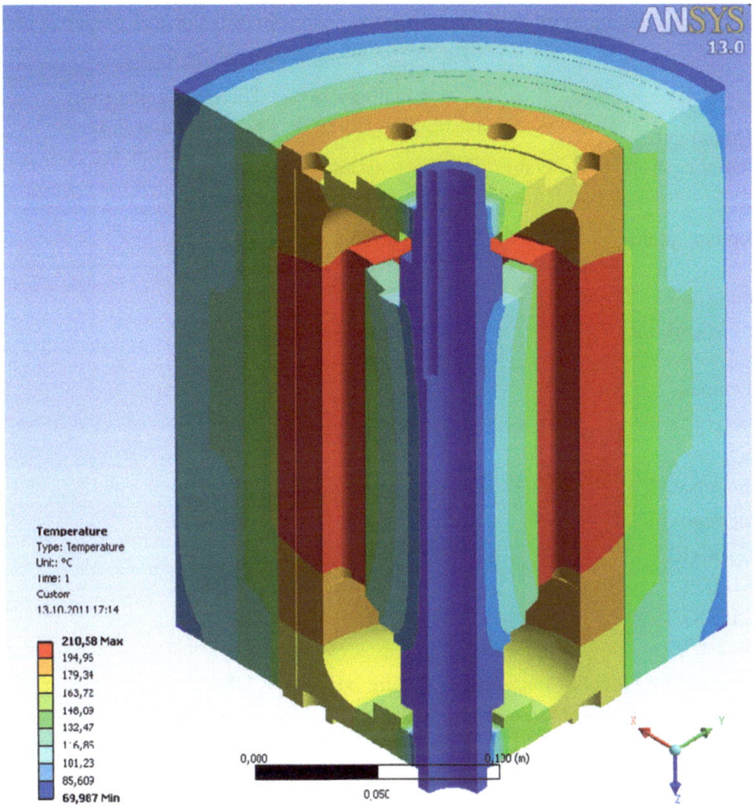

Fig. 7.14 Thermal simulation of the E3oN outrunner rotor concept

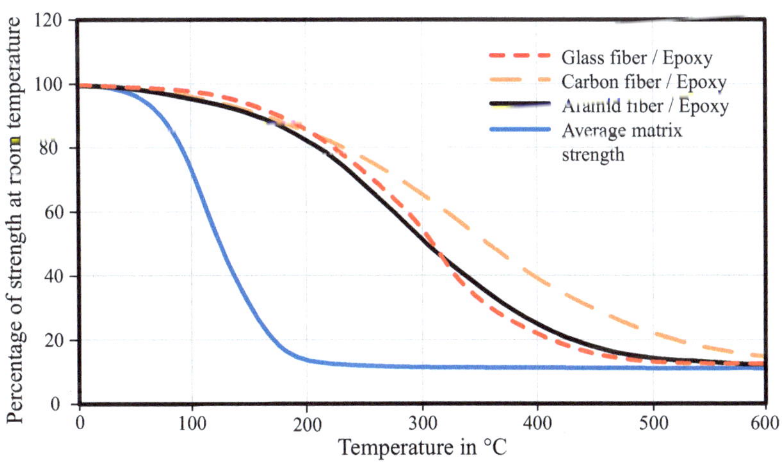

Fig. 7.15 Tensile strength of fiber composites as a function of temperature [31]. (Picture right: Luke A. Bisby)

Table 7.4 Thermal conductivity of various fiber composites in direction and perpendicular to it [26]

Material (fiber/matrix) (material number/short name)	Literature reference	Thermal conductivity in fiber direction, k_x (W/m∗K)	Thermal conductivity perpendicular to the fiber direction, k_y (W/m∗K)	Average value k_x/k_y
Graphite/epoxy "plain-weave fabric composite"	[27]	5.36	0.43–0.50	11.5
Hexcel F593 "carbon/epoxy plain-weave pre-preq laminate"	[28]	2.0–3.5	0.50–0.80	4.2
Graphite/epoxy "matrix lamina"	[29]	3.8–8.0	0.40–0.80	9.8
Carbon fiber/epoxy[a]	[26]	–	0.30–0.80	–
Carbon fiber/epoxy[b]	[26]	5.0–7.0	0.50–0.80	10.4
1.2080-X210Cr12 "Böhler K100"	[25]	20.0	20.0	1
1.0570/S355J2G3 "Böhler St52–3"	[25]	35.0–45.0	35.0–45.0	1

[a]Values measured by *Lincoln Composites*, Lincoln, NE 68507, USA
[b]Values measured by Tian Tian, *University of Nebraska*, Lincoln, USA [26]

At this point, it must be noted that, in addition to thermal conductivity, the thermal expansion coefficient of fiber composites also plays an essential role with regard to design details of the rotor. While metallic materials have thermal expansion coefficients (α) of around 10 to $15*10^{-6}\,K^{-1}$, carbon fiber composites can even show negative values! This is an aspect that must be taken into account when selecting the material for the shaft and hub of the rotor, especially when designing interference fits.

10. **Poor temperature resistance**
    - The high-strength fibers themselves usually offer significantly higher temperature resistance than the matrix material. While carbon fibers lose their strength at about 300 °C, glass fibers can withstand temperatures of up to 850 °C before a significant loss of mechanical strength occurs. The problem, however, is the much lower temperature resistance of the matrix, which is usually designed as some kind of epoxy resin. The strength of this epoxy resin already suffers a significant drop at about 120 °C [30]. Figure 7.15 shows the degradation of the strength of aramid, carbon, and glass fiber composites as a function of their temperature. Since energy storage flywheels are operated in a vacuum, the absence of convection for the rejection of the heat losses of the electric machine can lead to operating temperatures above 200 °C, as the simulation results shown in Fig. 7.14 indicate.

11. **Ferromagnetic properties**
    - At the beginning of this book (and again in Sect. 5.5), it was pointed out that FESS technology can only become established if consistent *low-cost systems* achieve corresponding market penetration and broad application. This, however, can only be achieved by using low-cost bearing systems, such as spindle bearings. In order to reduce the basic (axial) mechanical load and subsequently the torque loss, partial or complete magnetic lifting (i.e., weight compensation) of the rotor can be useful (also see Sect. 10.3). For this purpose, however, a large part of the rotor face must be rotationally symmetrical and ferromagnetic. Composite materials can a priori not be lifted magnetically. There are certain polymer-based materials, so-called magnetically loaded composites (MLCs), which have ferromagnetic properties [31], but the inhomogeneity of the magnetic field would most likely cause extensive eddy current losses in this case. An example using such a material is a carbon fiber rotor with MLC developed by *Williams Hybrid Power*, which has also been used in the *GKN* hybrid bus concept *Gyrodrive* since 2014 [33].

7.2.2 Steel Flywheels

Table 7.5 gives an overview of energy storage flywheel rotors made of steel. It should be noted that almost all historical concepts used a solid, isotropic rotor, and the achieved specific energies are significantly lower than those of composite rotors. Some examples are shown in Figs. 7.16 and 7.17.

7.2.2.1 Development Goals for Steel Rotors

Table 7.6 shows a comparative summary of the essential properties of steel and fiber composite FESS rotors, which can be derived from Sect. 7.2.1.

Tables 7.2 and 7.6 indicate clearly that an increase in the specific energy (i.e., permissible maximum speed) of steel rotors is required to be able to compete with composite rotors. Due to the complex interaction of the rotor failure mechanisms with the containment, the topic safety housing design cannot be considered in isolation and assessed in a general way (see also Chap. 8). In any case, the goal is a rotor design, which shows the following characteristics:

- Safe operation at the rotor's strength limit with maximum exploitation of the rotor material
- Early detection of overload or rotor damage
- Favorable failure mechanisms matching the specific safety containment design

Compared to a large number of research articles in the field of composite rotors, there seem to be hardly any publications on "low-cost/high-performance" steel rotor concepts.

Table 7.5 Overview: Steel flywheel rotors from selected FESS

Designation	Manufacturer	Year	Topology	Speed	Spec. Energy[a]	Ref.
Lockheed Flywheel Transaxle	Lockheed Martin Corp., USA	1973	Solid, shaft bore	24,000 RPM	~25 Wh/kg	[38]
NYC Subway Flywheel	Garrett AiResearch Manufacturing, USA	1974	Solid, shaftless rotor	14,000 RPM	~8.6 Wh/kg	[39]
MAN Gyrobus Flywheel	M.A.N/Mercedes-Benz, Germany	1980	n.a.	12,000 RPM	~13.6 Wh/kg	[40]
Gyreacta	Robert Clerk, England	1961	n.a.	15,000 RPM	~4 Wh/kg	[41]
Gyrobus	Oerlikon Werke, Switzerland	1953	Solid	3000 RPM	~6 Wh/kg	[42]
Dynastore	Compact Dynamics, Germany	2006	Layered (high-strength electrical sheet)	80,000 RPM	~6 Wh/kg	[43]
Hybrid III Flywheel	ETH Zurich	1999	Solid, heat-treatable steel (42CrMo4)	6000 RPM	~2.5 Wh/kg	[44]
PPM 60 Flywheel	Parry People Movers, England	1992	Solid, with shaft bore	2500 RPM	~11.8 Wh/kg	[45]
VW T2 FESS Hybrid	RWTH Aachen, Germany	1975	Solid (42CrMo4)	13,400 RPM	~3.6 Wh/kg	[46]

[a]The theoretical and maximum specific energy of only the rotor itself is considered, not that of the entire energy storage unit

Although some literature, such as [4] and [5], points out that solid steel rotors usually break into two to three large (and therefore high-energy) chunks, the consequence of this statement seems to be the transition from steel to other materials. In the course of extensive literature research, only two steel rotor concepts could be found, which due to their topology show a good-natured burst behavior:

1. *N.V. Gulia*, **rotor made of wound steel strip** [47]
2. *Compact Dynamics System*, **rotor designed as a stack of dynamo sheets** [43]

The second of the two rotor concepts is shown in Fig. 7.18.

Fig. 7.16 Flywheel energy storage system *Piller Powerbridge* with steel rotor (3D sectional view). (Image rights: Piller Group GmbH)

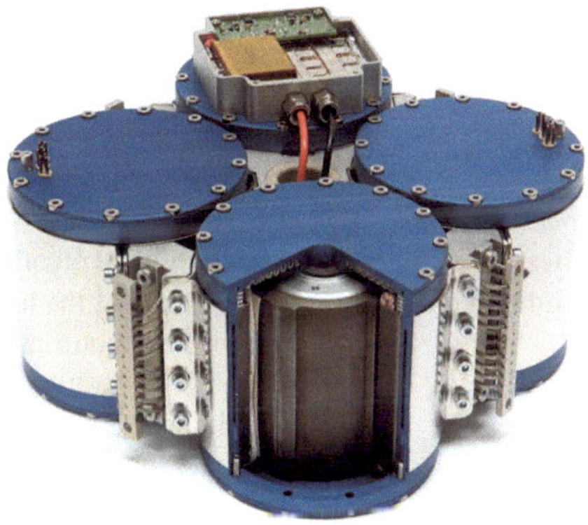

Fig. 7.17 Half-section through a model of the *Compact Dynamics* Flywheel module. The steel rotor can be seen on the inside. (Image rights: Compact Dynamics GmbH)

7.3 Requirements Derived from the *Supersystem Analysis*

Hardly any other component of the flywheel energy storage system shows such extensive and complex interrelationships as the rotor itself. The most essential of these interdependencies can be described as the *eight paradigms of FESS rotor design* and are shown in Fig. 7.19.

Taking all the properties mentioned in Fig. 7.19 into account, it is now necessary either to avoid the specific disadvantages of current fiber composite rotors by using new types of

Table 7.6 Summary of essential properties of steel and fiber composite rotors for flywheel energy storage

	Steel rotor		Composite rotor	
Specific energy	–	Low specific energy due to high material density → High rotor stresses induced by centrifugal forces	+	High specific energy due to low material density and good exploitation of the high specific strength in the fiber direction
Burst behavior	–	Unfavorable, because the rotor fractures usually into few large fragments with high mass and high kinetic energy → Sharp-edged rotor pieces require robust and heavy burst containment	+	Often advantageous, as rotor failure usually begins with delamination and can be detected early. The disintegration of fiber material absorbs energy and produces fine fragments → Homogeneous pressure load on the containment walls
Maintenance of the balance quality	+	Good, because of good dimensional stability, no aging, easy mounting of balancing weights in bores or dovetail groove	–	Poor, because aging of matrix material can release residual stresses. Mounting of balancing weights difficult
Temperature resistance	+	Good. Significant reduction in strength occurs only at temperatures above the operating temperature of bearings and electrical steel	–	Bad. Significant decrease in strength already at low temperatures around 120–200 °C depending on thermoset matrix material
Thermal conductivity	+	High, thus favorable for dissipating the heat loss of the electric motor in the case of FESS with fully integrated design	–	Low, therefore thermal insulation of the heat losses of the electric machine and rolling bearings
Production costs	+	Low material costs, but more material necessary to achieve a certain absolute energy content. Low manufacturing costs	–	Higher material costs, but less material is required for a certain energy content. Medium to high manufacturing costs depending on rotor size

materials in combination with smart design[6] or to develop a steel rotor concept that meets two essential design targets:

Increase of the energy density: Only if a *threshold energy density* of about 10 Wh/kg (see Sect. 5.4) is reached, FESS will be competitive.

[6] Attempts have been made to design matrix-less rotors, using only flexible bundles of fibers, but so far this approach has not been realized successfully.

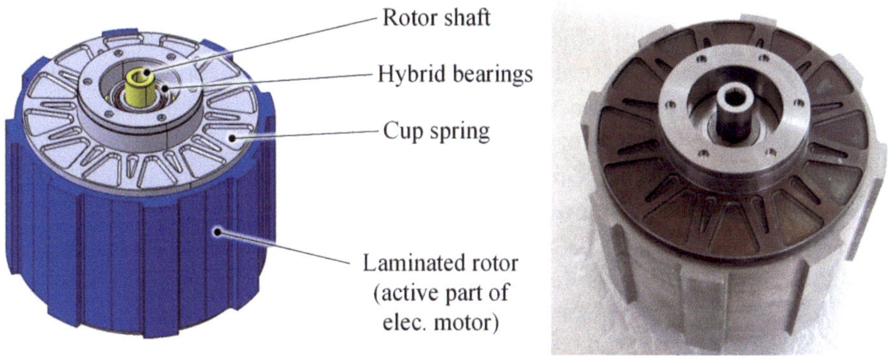

Fig. 7.18 Laminated electrical sheet rotor of the *Dynastore* FESS by *Compact Dynamics*, Germany [48]. (Image rights: Compact Dynamics GmbH)

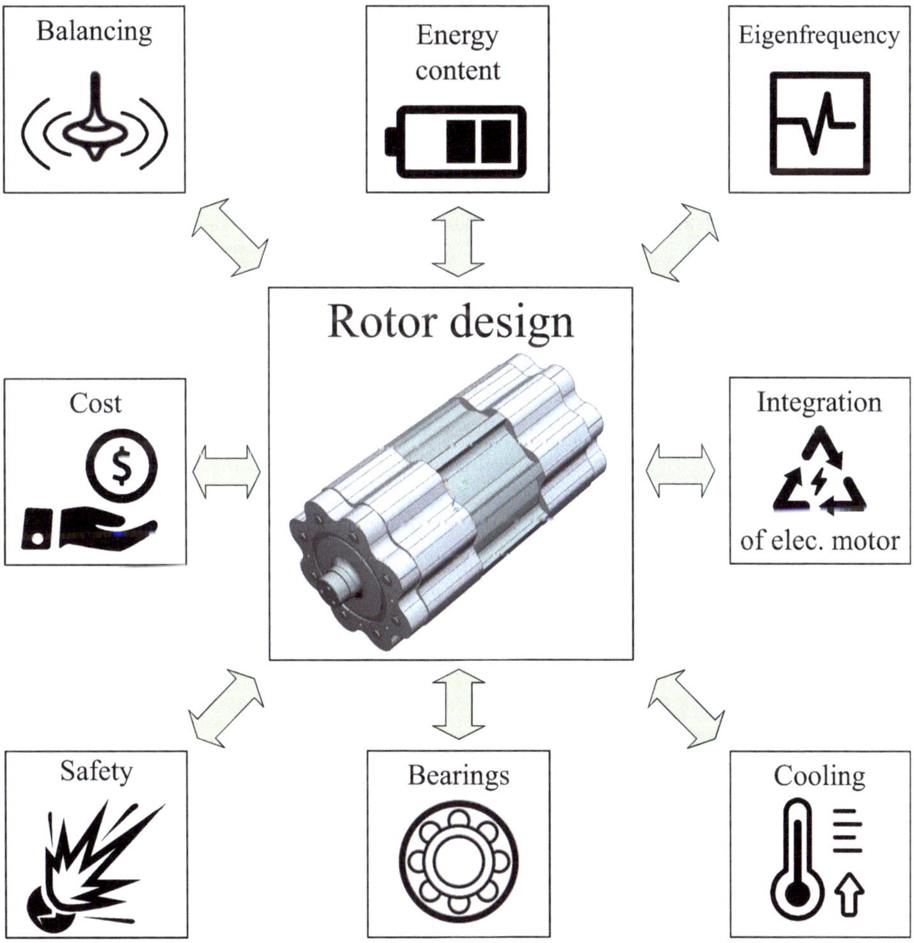

Fig. 7.19 The eight paradigms of FESS rotor design—major influences on the design of the flywheel

Control of the burst behavior[7]**:** Only if a FESS is considered a *safe technology* by customers, it has a chance of sufficient market penetration.

The many advantages of steel rotors over fiber composite rotors—as described in Table 7.6—can only be realized when these two prerequisites are fulfilled. A narrow-minded or "blind" optimization of the energy density to a level beyond the *threshold* and relying solely on fiber composite rotors can be an obstacle to the market success of FESS due to the high costs. In the following, two rotor concepts co-developed by the *Institute of Electrical Measurement and Sensor Systems* (*Institut für Elektrische Messtechnik und Sensorik*) at Graz University of Technology will be presented, which rely on steel as an inexpensive and easily controllable material.

7.4 Solution Approach/Case Study: *CMO Rotor*

7.4.1 System Description *Clean Motion Offensive* Flywheel

The *Clean Motion Offensive* (CMO) is an initiative funded by the *Climate and Energy Fund* (*Klima- und Energiefond*) of the Republic of Austria, which aims to give companies and research institutions a key position in the nationwide market launch of electric mobility [32]. The inability of the "big" players in the electromobility sector to solve certain problems, such as insufficient ranges, service life and costs of vehicle batteries, insufficiently developed infrastructure, etc., is seen as an opportunity for smaller tech companies. The following project contents have been defined by the funding body, the Austrian *Climate and Energy Fund* [35]:

- Increased EV range through innovative range extenders
- Cost reduction in energy storage through technical and commercial innovation
- Testing of possible variants of "intelligence" distribution in the system (intelligent battery system, intelligent vehicle, intelligent charging station, intelligent load distribution, intelligent network, *smart metering*)
- Integration of the energy storage into the Smart Grid and billing systems via telematics
- Development of electrically propelled vehicles for special niche applications
- Infrastructure (load distribution and satellite systems, "Easy2use application")

One of the main tasks of the CMO project was to build a demonstrator vehicle that would showcase the capabilities of the technologies, which were newly developed in the course of the initiative. The relatively large consortium, which included several Austrian automotive companies, was complemented by the *Institute of Electrical Measurement and Sensor Systems* of the *Graz University of Technology*, whose task was to design a mobile flywheel

[7] A more "good-natured" bursting behavior of the rotor allows the use of a lighter bursting housing and thus also increases the specific energy (Wh/kg) of the system.

Fig. 7.20 Concept of the CMO demonstrator vehicle [36]. (Image rights: Clean Motion Offensive/ Business Upper Austria)

energy storage device for load point shifting in the vehicle. (An overview sketch of the demonstrator vehicle is shown in Fig. 7.20.)

The energetic key data of the flywheel energy storage module were defined as 300 kJ (approx. 83 Wh) of energy content and 20 kW continuous power (with 40 kW peak). An internal rotor concept of a five-phase switched reluctance machine was chosen. The rotor was supported by two precision ball bearings with circulatory oil lubrication, which is described in more detail in this book (see Sect. 9.7.1). The machine is housed in a burst containment made of aluminum, which is attached to a steel frame by means of vibration damping wire rope isolators. These wire rope isolators show a highly progressive characteristic curve, which not only has a positive effect on the acoustic behavior of the FESS but is also an essential safety feature, as it prevents the burst housing from being torn away from the base frame in the event of rotor damage. A 12 V diaphragm pump evacuates the housing to reduce windage losses. The manufacturing of the flywheel energy storage was taken over by the company *rosseta Technik GmbH* (*Dr. Frank Täubner*). The company, formerly located in Derenburg, Germany, was specialized in the development and production of fast-rotating electric machines and flywheel energy storage systems. The concept of the *CMO* flywheel is shown in Fig. 7.21 and a photo of the system is shown in Fig. 7.22.

7.4.2 The CMO Rotor Concept

The motivation for a laminated rotor design can be found in the fact that the active part of the electric machine also simultaneously acts as the flywheel spinning mass. This concept of a “fully integrated design” allows compact dimensions, i.e., a higher volumetric energy density, which is a decisive advantage in mobile applications. However, depending on the electric machine type, the rotor must have appropriate electrical and magnetic (material) properties. The choice of machine type is essentially influenced by two considerations:

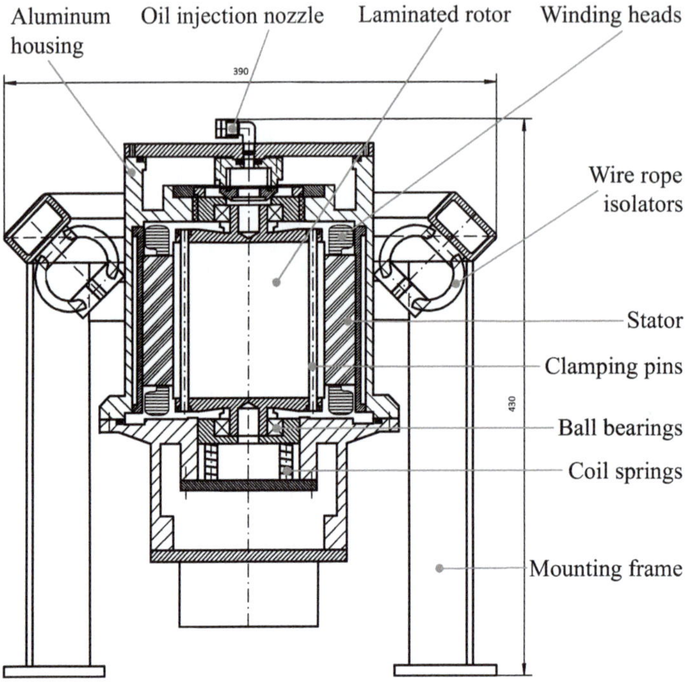

Fig. 7.21 Sectional drawing of the CMO flywheel

Fig. 7.22 Photograph of the entire CMO flywheel system. The housing is suspended in a steel frame (blue) on wire rope isolators

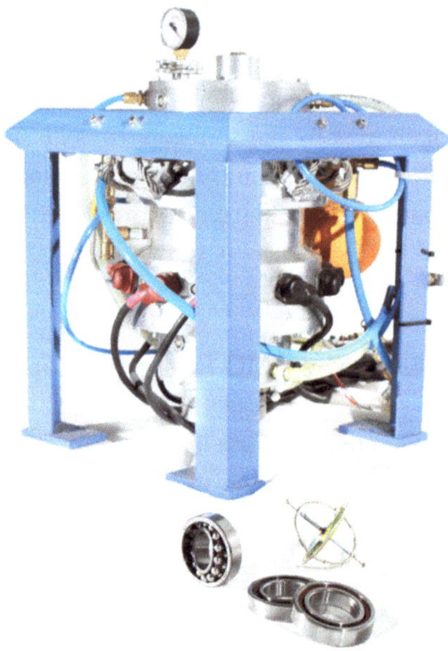

1. **Centrifugal forces:** Due to the high rotational speeds, the use of copper windings and/or glued on permanent magnets in the rotor can be problematic. (Balance quality and strength.)
2. **Idle losses:** In order to keep self-discharge low, only machines with low idle losses (i.e., not permanently excited ones) are considered.

Hence, it can be deduced that *switched* or *synchronous reluctance machines* (and in some cases also asynchronous machines) are well suited for FESS. Since the cooling of the rotor in the evacuated atmosphere of the flywheel housing is problematic, it is not only desirable to keep the electrical losses as low as possible but also to allow these losses to occur predominantly in the stator (which in the case of the *CMO flywheel* is water-cooled). This can (to a certain extent) be achieved by appropriate design of the converter switching characteristics [49].

Eddy current losses are mainly caused by induction due to the temporal change of the magnetic field in the rotor. In order to keep these losses low, the rotor is made up of stacked, individual 0.35-mm-thick metal sheets, which have an electrically insulating full-face bonding layer (*backlack*) in the separation interstices. The stack of sheets is additionally held together by eight clamping pins running axially along the rotor teeth. Due to its advantages regarding optimal stress design, a shaftless rotor, which is shown in Fig. 7.23, was chosen, which has the following key properties:

- Rotor diameter (air gap): $d_{rotor} = 120$ mm
- Length approximately 110 mm (magnetically active 100 mm)
- Outer stator diameter: $d_{outer} = 170$ mm
- Rotor material: *Vacodur S Plus*, mechanically optimized (tensile strength $R_m = 800$ N/mm^2)
- Rotational speed range: 13,000–60,000 rpm

Fig. 7.23 Photo of the CMO flywheel rotor

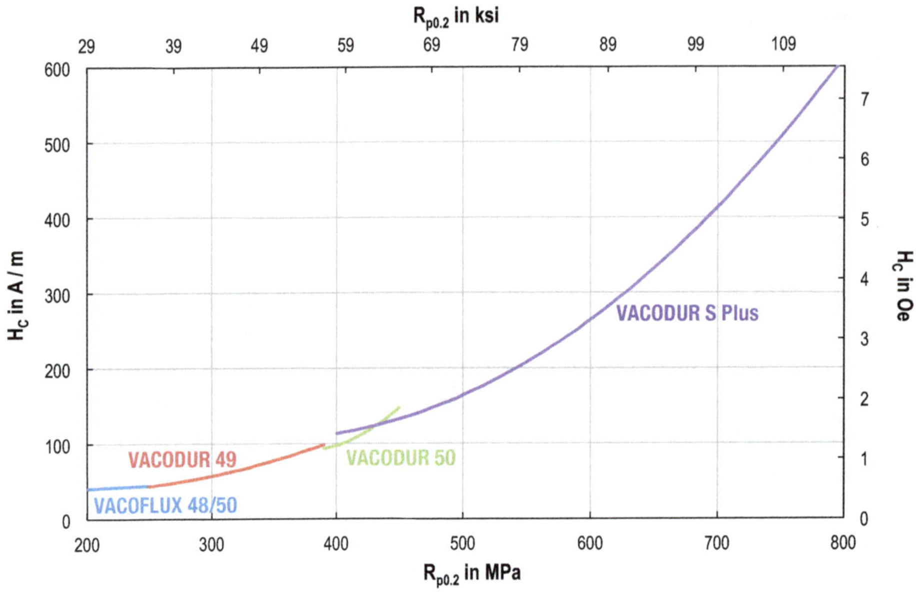

Fig. 7.24 Yield strength $Rp_{0.2}$ and coercive field strength Hc adjustable by annealing for VACOFLUX and VACODUR (strip material 0.35 mm), Vacuumschmelze, Germany [37]. (Image rights: Vacuumschmelze GmbH and Co. KG)

Since no "external" flywheel is flanged to the electrically active part of the machine's rotor, the stack of electrical sheet metal acts as the sole kinetic energy storage device. The energy density is a function of the ratio σ/ρ (see Table 7.2), which shows that there is a conflict of objectives between the electrical and mechanical properties of the rotor.

Soft magnetic cobalt-iron alloys, which are preferably used for rotors of reluctance machines, allow high flux densities and magnetic saturation of up to 2.35 T. However, with a yield strength ($R_{p0.2}$) of 190–250 N/mm^2 and ultimate tensile strength (R_m) of 220–550 N/mm^2, the mechanical properties of these materials are conceivably unsuitable for enduring high centrifugal forces. However, the mechanical strength of cobalt-iron alloys can be modified by adjusting the temperature during the final annealing. Annealing at high temperatures can achieve optimized magnetics, while low temperatures result in improved mechanical properties [37] (Fig. 7.24).

The material selected for the *CMO* rotor is *Vacodur S Plus* with a tensile strength of 800 N/mm^2 and a density of about 8.12 kg/dm^3. Assuming an ideal flywheel with shape factor of $K_{shape} = 1$ and inserted in the formula

$$E_k = K_{shape} \frac{\sigma_{max}}{\rho},$$

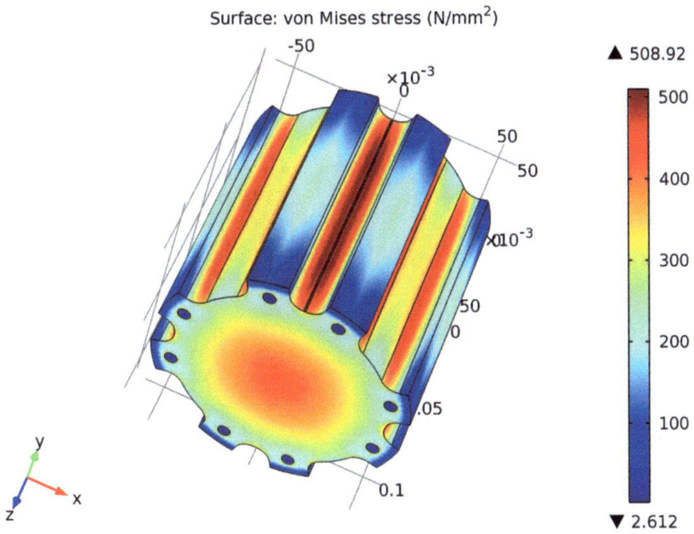

Fig. 7.25 FEM simulation of the stresses (von Mises stress) in the CMO rotor at maximum speed of 60,000 rpm. (*Energy Aware Systems Group*, TU Graz). (Image rights: Bernhard Schweighofer)

a theoretical, maximum specific energy of 27.375 Wh/kg can be calculated. However, the real stress distribution in the rotor and a required safety margin (maximum stresses of approx. 500 N/mm^2 are permitted) reduce this theoretical maximum to about 50%. The results of a FEM simulation of the rotor are shown in Fig. 7.25.

7.4.2.1 Balancing of the *CMO Rotor*

Due to the short, compact rotor geometry (diameter to length ratio is about 1), the rotor can be assumed to be a rigid body, and the first natural frequency of the system is primarily determined by the stiffness of the bearing system. In order to keep the excitation by a rotating force vector low, the rotor must be finely balanced. (This process is explained in more detail in Sect. 9.6.) As there is a thin film of full-face bonding (*Backlack*)[8] between the electrical sheet metal layers of the rotor, which glues the parts together, it is not necessarily to assume that setting phenomena will cause a significant loss of pretensioning force of the clamping pins and change the balancing quality over time. The rotor is therefore designed for one-time balancing by removing material at the endplates. Figure 7.26 shows a photo of the lower endplate of the rotor, on which a blind hole, which was drilled during the balancing process, is clearly visible.

[8] *MAGNETBONDER HT-01* by the company *Vakuumschmelze* with a density of 1.1 g/cm^3 and a maximum shear stress of 7100 N/cm^2.

Fig. 7.26 Front view of the CMO rotor with bearing shield, rotor endplate, and the sheet stack

7.4.2.2 Burst Behavior of the CMO Rotor

Due to the high costs and the tight project schedule of the *CMO* project, no spin test/ overspeed burst test was carried out with the *CMO* rotor. However, there is a strong similarity to the only other flywheel system with the same rotor design currently available—the *Compact Dynamics Dynastore* rotor—which also offers a fully integrated design. (This rotor is shown in Fig. 7.18 as well as in Fig. 7.27 before and after a burst test, respectively). It can be seen that the electric machine has been completely "pulverized" and thus a relatively even pressure load was applied to the burst containment without penetrating it. The largest fragments represent the disk spring for the axial preload and the rotor shaft itself. In the case of the *CMO* flywheel, even these two components are not present due to the shaftless architecture, but two endplates with a larger mass (approx. 350 g each) could form larger, high-energy fragments.

7.5 Solution Approach/Case Study: *FIMD Flywheel*

Based on the findings of the FFG project *E3ON*, the prototype of a FESS with 1.0 kWh energy content and 145 kW peak power was designed and built. The designation *FIMD* stands for "fully integrated multi-disk rotor."

One of the decisive factors in the design of the concept was the request of a manufacturer of special vehicles to be able to use a flywheel energy storage system as an alternative to supercaps in the hybrid drivetrain of a heavy commercial vehicle. The vehicle concept is shown in Fig. 7.28.

Fig. 7.27 Burst tests of laminated rotors: Prototype of the Compact Dynamics F1 module. (Image rights: Compact Dynamics GmbH)

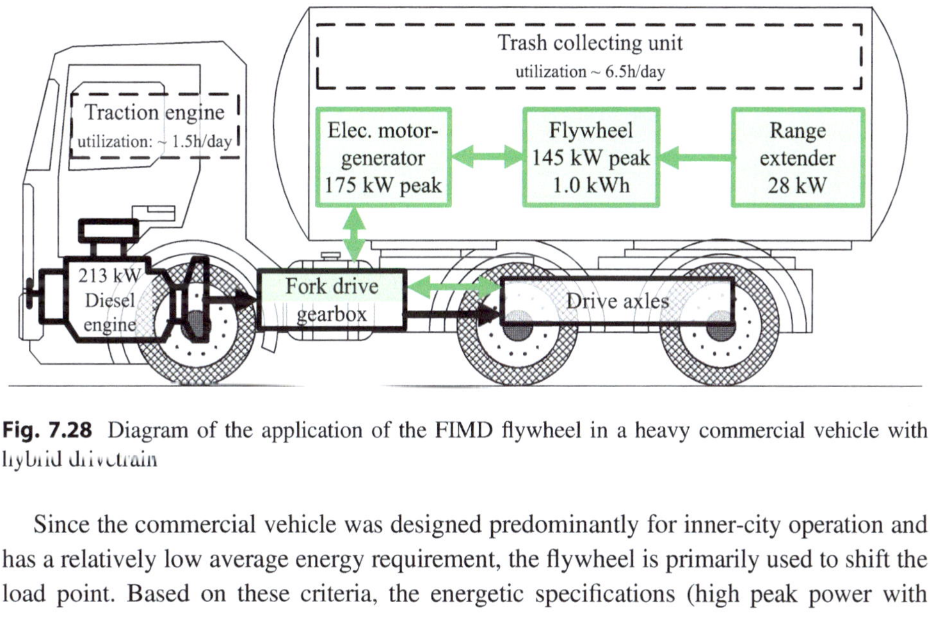

Fig. 7.28 Diagram of the application of the FIMD flywheel in a heavy commercial vehicle with hybrid drivetrain

Since the commercial vehicle was designed predominantly for inner-city operation and has a relatively low average energy requirement, the flywheel is primarily used to shift the load point. Based on these criteria, the energetic specifications (high peak power with relatively low-energy content) can be derived. The development goals of the project are shown in Table 7.7.

The concept follows a fully integrated topology for reasons of compactness. The active part of the electric machine and the flywheel masses are combined in a compact, cylindrical rotor and supported by spindle bearings. (For more detailed information on the bearing system, see Sect. 9.6.1.) In contrast to the preceding project *E3oN*, during which an external rotor concept (outrunner) was designed, motivated by the higher moment of inertia, the *FIMD flywheel* has a classic internal rotor. The advantage of this arrangement

Table 7.7 Profile of requirement of the *FIMD* flywheel

Energy content	1 kWh
Max. power	145 kW
Efficiency (*round trip*)	~70%
Spin-down Time (self-discharge)	>1 h
System weight	<300 kg
Service life	50,000 h

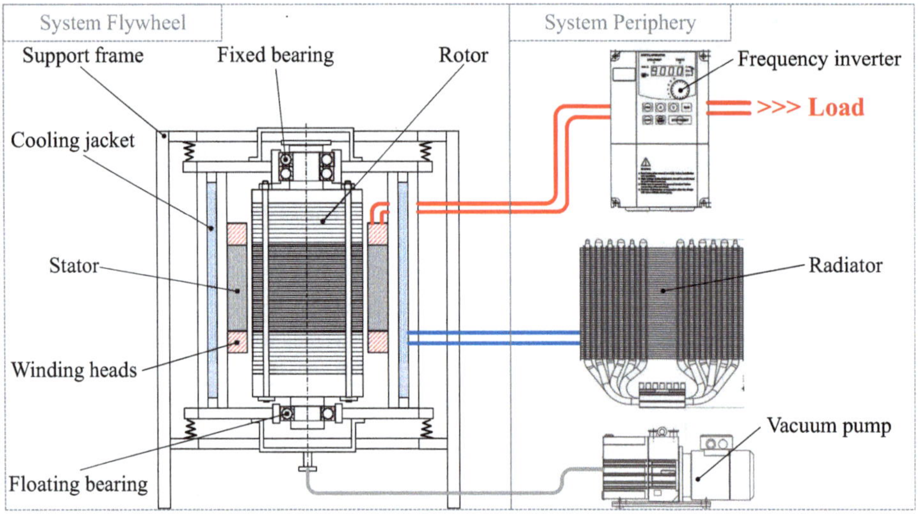

Fig. 7.29 Schematic sketch and system overview of the *FIMD flywheel*

is that the stator core and windings serve as additional burst protection in the unlikely event of rotor failure or crash.

A three-pole-pair *synchronous reluctance machine* with a nominal power output of 75 kW and a peak power of 145 kW was selected as the motor generator. The speed range is 13,000 to 40,000 rpm, with the V/f inflexion point at 21,000. The three phases are connected in a star configuration. Two *SKAI* frequency converters (*Type SKAI45A2GD12-W24DI*), which were purchased from *Compact Dynamics*, are required for the operation of the machine. The peak value of the phase current is 3 × 320 A at 680–750 V. The stator and winding heads (where most of the heat losses occur) are installed directly adjacent to the water-cooled housing wall, thus ensuring good heat rejection. The entire system is shown schematically in Fig. 7.29.

The sensors intended for condition monitoring of the FESS are shown in Fig. 7.30:

- **Acceleration/vibration:** One single-axis piezo accelerometer (acting in radial direction) at each bearing position

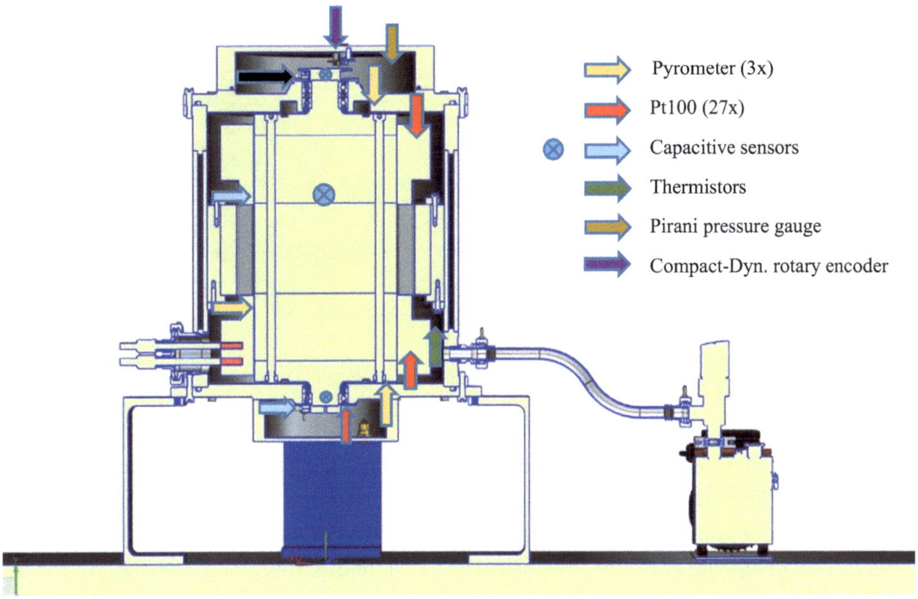

Fig. 7.30 Overview of the sensors in the *FIMD* flywheel

- **Speed and rotor position angle:** Inductive encoder at the end face of the rotor, one signal every 90°
- **Concentricity and balance quality:** Two eddy current sensors offset by 90°, which are directed radially onto a circular rotor disk plate
- **Temperature:** In total 30 temperature sensors (Type *PT100* and non-contact *pyrometers*) for measuring the temperature of outer bearing races, winding heads, stator, and cooling water

Detailed information on the bearing system is given in Sect. 9.6.1.

7.5.1 Structure of the *FIMD* Rotor

In order to minimize material stresses caused by centrifugal forces, the rotor has a shaftless design (see Sect. 7.1). However, once the flywheel and electric machine are integrated into *one* rotor, it is not possible to completely avoid axial bores. As shown in Fig. 7.31, the multilayer (laminated) design of the rotor is held together by so-called clamping bolts. The boreholes are located relatively close to the outer circumference of the rotor, and the material has been removed between the respective bolts by "recesses." This so-called petaloid contour of the flywheel rotor is a result of the numeric FE optimization of the centrifugal stresses present in the rotor material. However, the contour of the electric

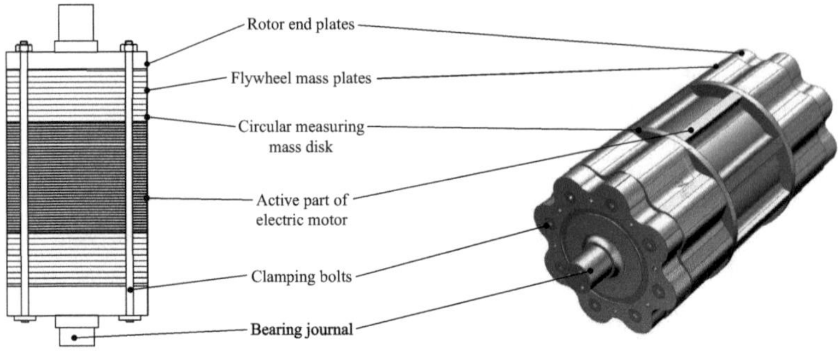

Fig. 7.31 Structure of the FIMD rotor

machine (or its rotor teeth) is determined exclusively by electrical design criteria. The task of the circular "measurement mass rotor disk" is described in more detail in Sect. 9.6.1.

At this point, it must be mentioned that in total two rotors were manufactured during the project. The *prototype rotor* is shown in Fig. 7.31 and a shortened *burst rotor* for the *overspeed test* described in Sect. 7.5.2.

The special design of the rotor allows the following development goals to be achieved:

1. Cost reduction: The change from wound carbon fiber structures to steel rotors allows a theoretical price reduction by a factor of 20[9] [50]. The material 42CrMo4 used for the flywheel plates is available in Austria for about 1.79 €/kg (as of 2016) [51].
2. Increase of energy density: The following properties of the FIMD flywheel allow an increase of the specific energy compared to conventional steel rotors:
   - The use of high-strength steels or electrical steel sheets optimized for mechanical properties
   - A shaftless design that nearly doubles the shape factor K_{shape} (see Table 7.1 and Eq. 7.9 in Sect. 7.1) and thus the energy content
   - Utilization of high specific material strength by avoiding massive, thick-walled components (compare Table 7.8) due to the layered rotor design
   - The optimization of the centrifugal force stresses in the rotor by introducing the "petaloid contour"
3. Improvement of the thermal behavior: The flywheel mass plates directly adjacent to the active part of the electric machine act as a heat sink or "heat storage" and, thanks to their high mass, can "absorb" some of the electrical heat loss of the rotor and, due to their good thermal conductivity, transport it via the bearings to the water-cooled stator.
4. Increased safety: The inherent safety of the FIMD rotor has been increased as follows:

[9] The exploitation of this reduction potential does not only depend on the material price, but also requires optimized, cost-effective production processes!

Table 7.8 Reduction of the specific strength of *42CrMo4* with increasing component size [53]

Component thickness (mm)	to 8	8–20	20–60	60–100	100–160	160–250
Tensile strength *Rm* (N/mm²)	900	750	650	550–500	460–500	390

- Selection of steel as a material with high cycle fatigue-resistant characteristics in contrast to fiber composites (see Fig. 7.11).
- Reduction of the maximum fragment size/mass and energy through layered design (laminated rotor). (The burst behavior of rotors or the implications for the housing design are discussed in Chap. 8.)
- The choice of high-strength (but brittle) electrical steel sheets, high-strength endplates, and ductile flywheel plates allows early detection of overload due to centrifugal forces by measuring the relative geometric expansion of the rotor, as described in Sect. 7.5.2.

7.5.1.1 Choice of Material

1. **Electric motor/generator:** *Vacodur S Plus* from the company *Vacuumschmelze* is used [34]. It is a soft magnetic cobalt-iron alloy and its mechanical properties can be adjusted by heat treatment (final annealing). Yield strengths beyond 800 N/mm^2 are possible, as the stress-strain test performed in Fig. 7.49 shows.
2. **Endplates with bearing journals:** Since the endplates are the largest and most solid components and the force is transmitted via the integrated bearing journals, the highest quality steel was selected for these parts. This is the hot-worktool steel *W400 VMR* from *Böhler*, which can also be found under the material number 1.2343 and the designation X37CrMoV5–1. The designation *VMR* means that it is hot-working steel cast in a vacuum, with a particularly pure microstructure that can be adapted by soft annealing, enabling tensile strengths of more than 1300 N/mm^2 [23].
3. **Flywheel mass plates:** The low-cost heat-treatable steel 42CrMo4 (material number 1.7225) was selected for this purpose. It is mainly used for parts with high toughness requirements in aircraft construction. Its tensile strength as a function of component size is shown in Table 7.8.
4. **Clamping bolts:** Due to electromagnetic requirements of electric machines, clamping bolts needed to be made from non-ferromagnetic material. *Austenitic steels* are suitable for this purpose, whereby the high-strength material 1.4573 (or X6CrNiMoTi1812 according to DIN 17006) was selected. Its tensile strength reaches up to 740 N/mm^2 [52].

7.5.1.2 Assembly and Conditioning of the Rotor

Due to several factors, the assembly of a rotor in the described design is a process that influences the performance, service life, and safety of the entire system. The following problems had to be solved:

- **Static overdetermination of the clamping bolts (hyperstatic bore pattern)**: Due to production conditions, the bore pattern for the tensioning bolts must be created individually in each rotor plate (see Fig. 7.31). In the active part of the electric machine, no machining is possible due to the low sheet thickness of 0.1 mm.
- **Different coefficients of thermal expansion:** Operating temperatures of 150 °C can be expected. The coefficient of thermal expansion α of the clamping screws made of *X6CrNiMoTi1812* is about 17.5 10^{-6} K^{-1}, while the flywheel mass plates only have 11.9 10^{-6} K^{-1} [54, 55].
- **Setting phenomena in the sheet metal stack:** Settlement causes a loss of pretensioning force of a bolt by leveling out production-related roughness peaks between the clamped components or bolt head and base on a microscopic scale.
- **Unknown mechanical properties of the electric machine laminates:** The mechanical (or dynamic) properties of the full-face bonding layer (*VAC HT-01* high-temperature adhesive), which electrically insulates the individual 0.1 mm electrical sheets from each other, and the filling degree were unknown. It was necessary hence to measure the elastic modulus of the entire stack of electrical sheets. For this purpose, specimens were fitted with strain gauges as shown in Fig. 7.32 and loaded with a hydraulic press.

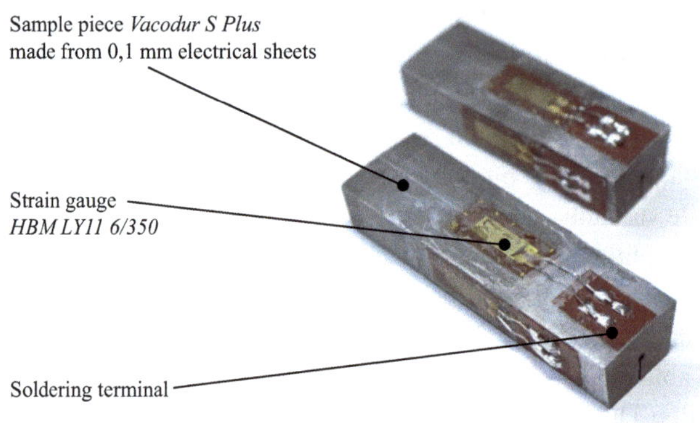

Fig. 7.32 Samples of the electrical sheet stack of *Vacodur S Plus* prepared for determining the elastic modulus

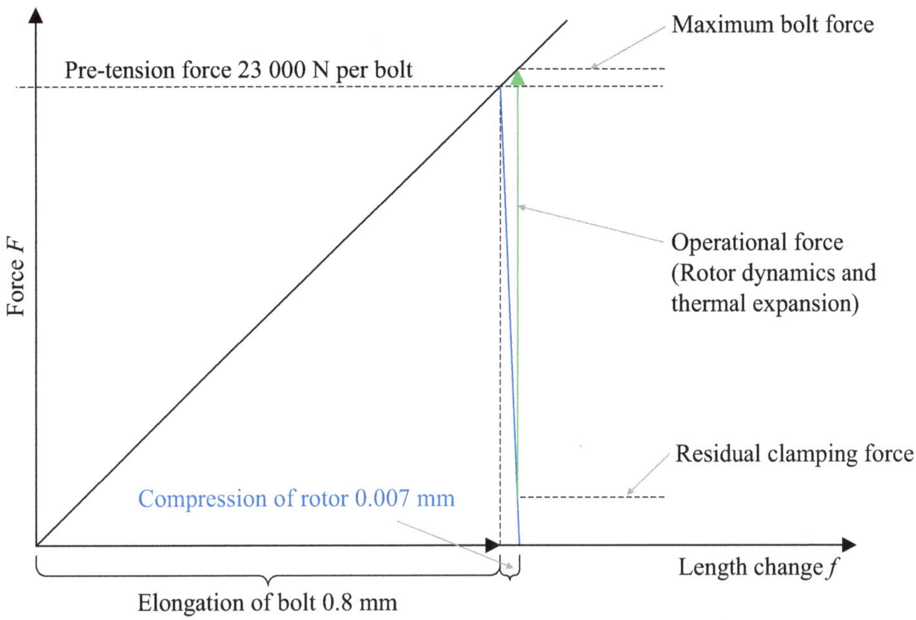

Fig. 7.33 Force-deflection diagram of the clamping bolts of the FIMD rotor (real installation situation; values were determined by force/displacement measurements)

Solutions

Static overdetermination can only be countered by appropriate choice of fit/clearance and extremely precise manufacturing (narrow tolerance classes), possibly by reworking (careful grinding) the fitting surface of the clamping bolts during assembly.

The different **coefficients of thermal expansion** of the various steels are physical properties that are difficult to circumvent. (In addition, the thermal expansion of the *Vacodur sheet metal stack* was unknown.) The only solution lies in choosing a suitable ratio between the resilience of the bolt and the clamped parts and the pre-stressing forces, as shown in the *diagram* in Fig. 7.33.

Settling effects can be reduced by preconditioning of the rotor before the actual installation in the flywheel housing. The leveling of manufacturing-induced roughness peaks is a process that only occurs during operation and not immediately after installation. The number of load cycles up to which the setting process is completed cannot be estimated in advance, so it is necessary to observe the loss of preload during the conditioning process. During operation, cycles of thermal expansion as well as vibrations excited by unbalancing and magnetic forces are to be expected, which can cause these setting phenomena. These triggering mechanisms were simulated in a conditioning test bed, which is essentially a mechanical press. Figure 7.34 shows the finished assembly schematically and the

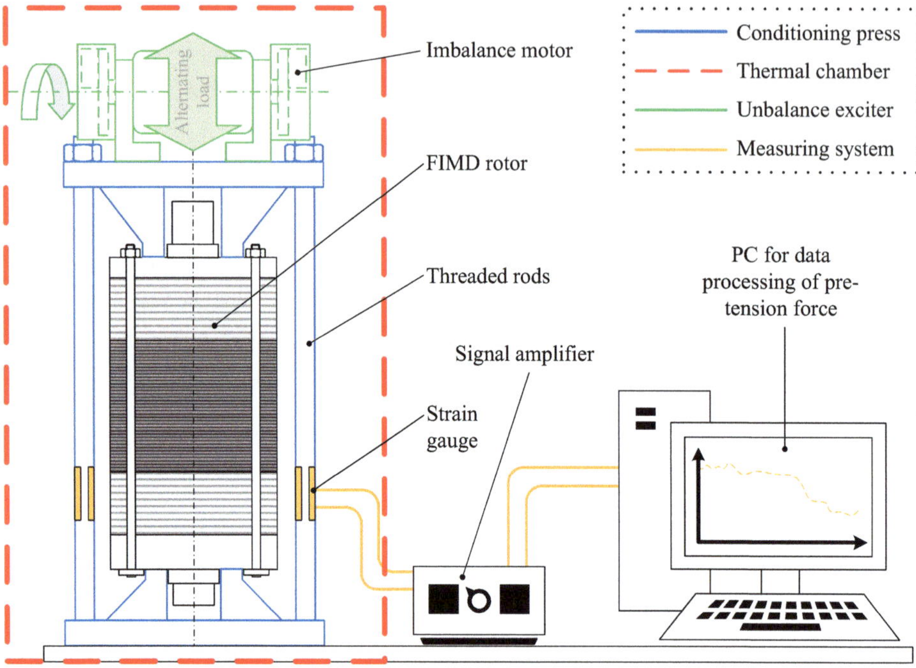

Fig. 7.34 Schematic layout and photo of the FIMD rotor conditioning unit

corresponding photo. It is to be noted that sinusoidal pressure loading in the longitudinal direction of the rotor was carried out before temperature cycling in the thermal chamber.

A. **Assembly and Conditioning Procedure**

Figures 7.35 and 7.36 show the rotor during assembly. The procedure can be described as follows:

1. Manufacturing of individual parts
   (a) Manufacturing turned/milled parts from raw material
   (b) Heat treatment
2. Manufacturing the active part of the electric machine
   (a) Production and heat treatment of Vacodur electrical sheets
   (b) Coating with full-face bonding layer and pressing
   (c) Wire EDM of the rotor contour
3. Stacking of the individual layers on only two clamping bolts
4. Inserting and adjusting the remaining clamping bolts
5. Hydraulic pretensioning of the bolts
6. Conditioning in the setting unit (as shown in Fig. 7.34.)
7. Hydraulic tightening of the clamping bolts and caulking of the nuts

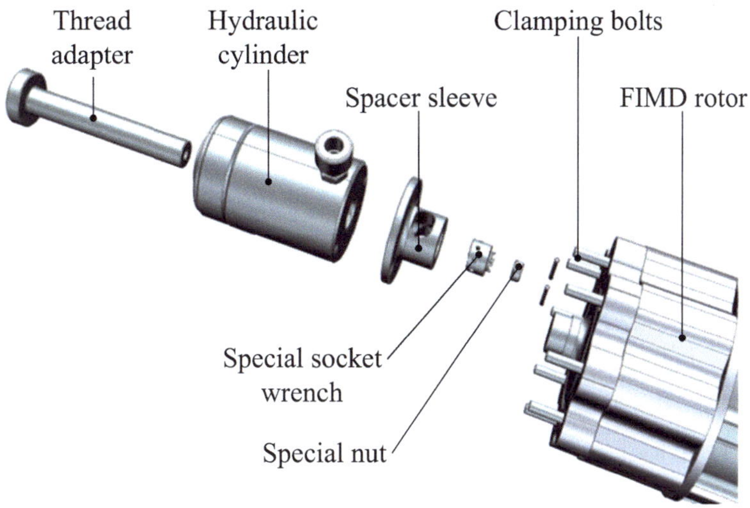

Fig. 7.35 Exploded view of the hydraulic bolt tensioning tool for the clamping bolts

Fig. 7.36 Rotor during assembly and force measurement of the hydraulic bolt tensioning unit

7.5.2 Burst Testing the *FIMD Rotor*

The layered structure of the rotor made it possible to subject a shortened version, containing only a few metal sheets of each material, to a burst test. However, due to the rotationally symmetrical distribution of the centrifugal forces, the results can be transferred with high accuracy to the actual rotor of the *FIMD prototype.*

Burst testing of a rotor this size is offered by only a few companies in Central Europe. In this case, the company *Schenck Rotec* in Darmstadt, Germany, was chosen to carry out the

Fig. 7.37 FIMD burst rotor on *Schenck Pasio 50* balancing machine

test in a so-called spin pit, model *Centrio100*. The speed level at which rotor failure is to be expected was determined using FE methods at about 45,000 rpm. In order to achieve such high angular speeds, the test rotor must offer an extremely high balance quality, as otherwise, shaft vibrations may occur which could damage the bearings of the spin pit's drive unit. A maximum vibration amplitude of 250 μm measured at the quill shaft was defined as the abort criterion. Therefore, exact balancing of the rotor before the burst test was essential.

In order to prevent the rotor from wobbling, a further machine-dynamic condition must be fulfilled. The ratio of the moments of inertia of the rotor J_1/J_2 must be less than 0.7 or greater than 1.25, where J_1 is related to the axis of rotation of the rotor and J_2 is related to the transverse axis through the center of gravity. A *Schenck Pasio 50* balancing machine was used for pre-balancing the burst rotor, which is shown in Figs. 7.37 and 7.38 during the balancing process.

Before the first balancing process, the rotor showed a "raw unbalance" of 300 g∗mm, which can be considered relatively good in comparison to the total rotor mass of 25 kg. After several dynamic balancing runs at 700 rpm, a residual unbalance of 4.2 g∗mm was achieved by attaching an M6 grub screw and placing a 1.9 g sliding block in the dovetail groove and the rotor endplate. Due to this high balancing quality, a very low shaft vibration was expected and the actual burst test could be started.

Figure 7.39 shows the burst test rig (spin pit) *Centrio100* by *Schenck Rotec* with its protective containment consisting of four solid, concentric thick-walled steel cylinders. In the center, see a simple, much smaller burst housing made of mild steel can be seen, which was used for a qualitative burst test. In this image, the burst rotor is already hanging from the flexible quill shaft in the center of the upper part of the machine. To reduce the power requirements for achieving such high rotational burst speeds and also to reduce the viscous damping on impact, the test is carried out in a vacuum.

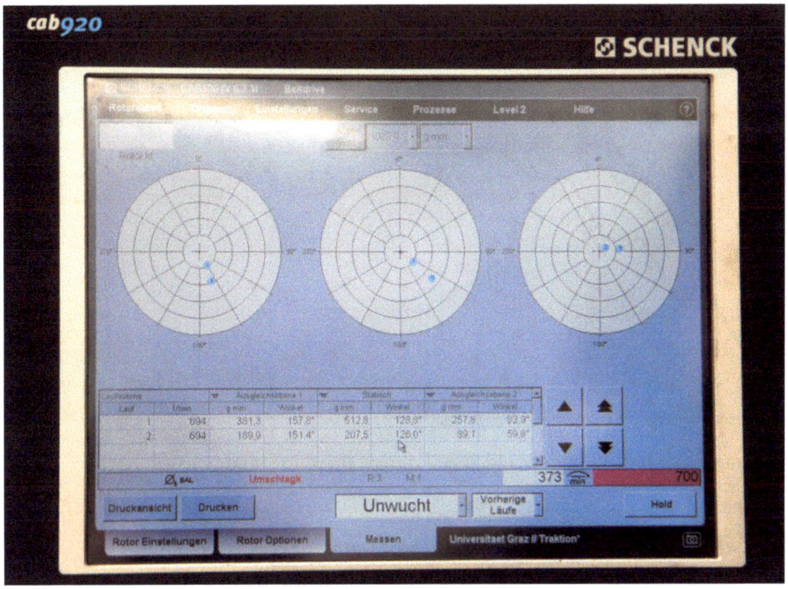

Fig. 7.38 User interface of the *Schenck Pasio 50* balancing machine while balancing the FIMD burst rotor

Fig. 7.39 Burst rotor mounted on the flexible shaft of the spin pit before the first run

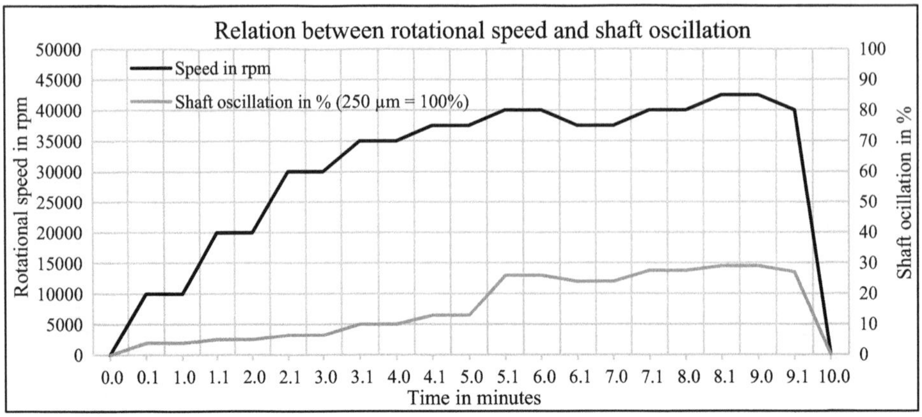

Fig. 7.40 Course of shaft vibration and speed during the first spin cycle

First Spin Cycle

Since the test flywheel is inevitably operated to its yield point and even further during a centrifuging testing, the balance quality changes over the speed due to settling phenomena and plastic deformation of the rotor material. The speed is therefore increased in discrete steps. Figure 7.40 shows recorded data from the first run of the overspeed test. It can be clearly seen that the increase in speed from 37,500 to 40,000 rpm resulted in a significant increase in shaft vibration (i.e., a significant deterioration in balance quality). However, since the balancing quality did not deteriorate any further if the speed was kept constant at 40,000 rpm, and a reduction in the speed also resulted in a reduction in shaft vibration, the experiment was not stopped immediately, but a further increase to 42,500 rpm was carried out.

In order to be able to visually inspect the rotor and possibly discover the cause of the significant deterioration in the balance quality, the test was interrupted for the time being. In addition, renewed balancing was necessary in order to achieve the planned burst speed of 45,000 rpm. The visual inspection and rebalancing showed the following results:

- The decrease in balance quality in the two planes (on each endplate) from 7.5 to 121 and 5.6 to 163 g∗mm
- Probable cause: Plasticizing of a rotor material
- Radial expansion of the flywheel mass plate (42CrMo4) by approximately 5/10 mm
- Rebalancing to 6.2 or 10.1 g∗mm per balancing plane

The radial expansion of the flywheel mass plate made of 42CrMo4 by approximately 5/10 mm is clearly visible in Fig. 7.41, as the outer contour of the assembled rotor was overturned during production and therefore should not show any sudden variation in diameter.

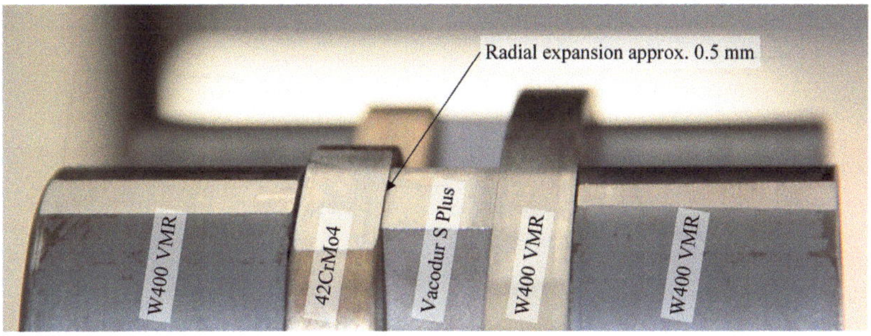

Fig. 7.41 Visual inspection of the FIMD burst rotor after a run up to 42,500 rpm

Table 7.9 Comparison of the mechanical properties of the different materials used in the *FIMD rotor*

Material	*Böhler W400 VMR*	*42CrMo4*	*Vacodur S Plus*
Number	1.2343	1.7225	–
Yield strength (approx.)	900 N/mm^2	700 N/mm^2	800 N/mm^2
Tensile strength (approx.)	1300 N/mm^2	1200 N/mm^2	1200 N/mm^2

The cause of the change in the balance quality was thus reaching the *yield point* of the flywheel mass plate made of 42CrMo4, as these have the most ductile properties of all rotor parts used. However, the *tensile strength* of the flywheel mass plates is higher than that of the electrical sheet stack made of *Vacodur S Plus*, as shown in Table 7.9.

Second Spin Cycle

The procedure for the second spin cycle was identical to the first one. The rotational speed was again increased in discrete steps, this time with the shaft vibration being significantly lower up to 42,000 rpm, and even decreasing as the rotational was speed increased further. Only when increasing from 42,000 rpm to 44,000 rpm, an increase in shaft vibration from approximately 3% to about 12% could be noticed. From this point on, however, the shaft vibration continued to increase even when the speed was kept constant and did not become stable again, a sign that the centrifugal forces were so high that the rotor components were permanently plastically deformed. At 45,000 rpm, the rotor broke, which was noticeable by a loud bang and a disproportionate deflection of the shaft vibration. The course of the measured values of the second spin cycle is shown in Fig. 7.42.

7.5.2.1 Qualitative Postmortem Analysis

After the rotor parts and fragments that were still rotating in the bursting chamber had spun down, the scenario was photographically documented and the data prepared for a qualitative, postmortem analysis.

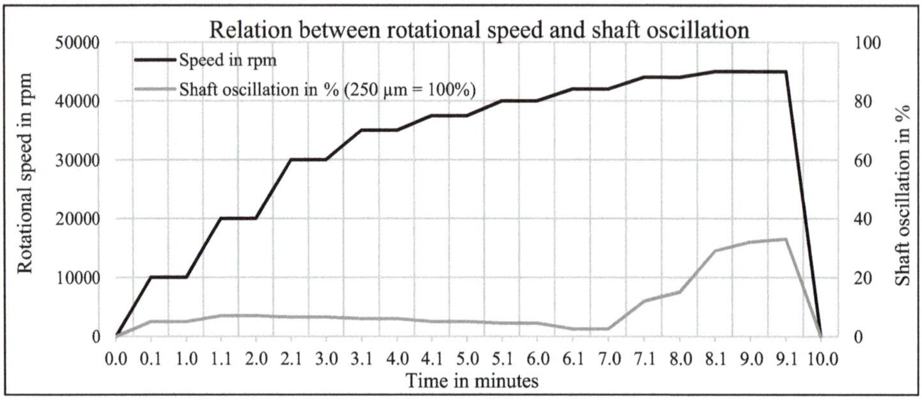

Fig. 7.42 Progression of shaft vibration and speed during the second spin test until rotor burst

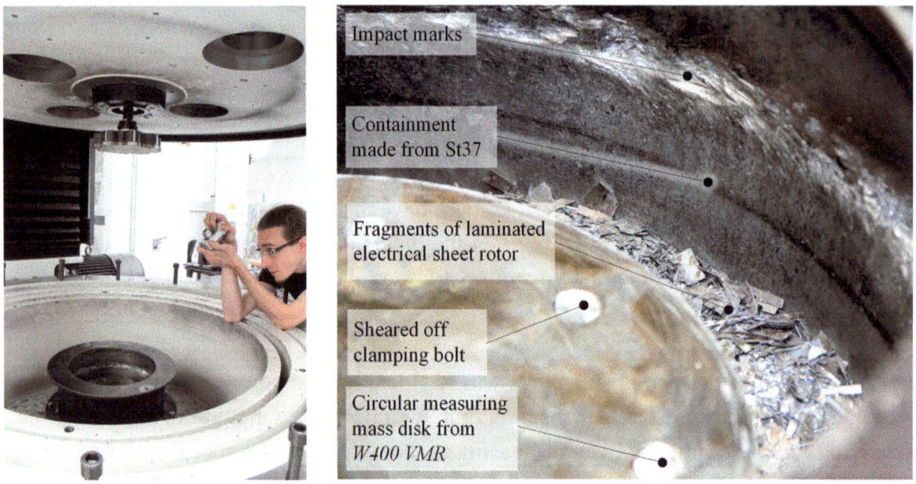

Fig. 7.43 Burst rotor after failure at 45,000 rpm

The rotor had been split in two in the horizontal plane, with the lower part consisting of the solid endplate and the circular measuring mass disk (both made of *W400 VMR*) falling into the touchdown bearing of the burst rig and being responsible for the long spin downtime (about 10 min) due to gyroscopic stabilization. The upper part of the rotor continued to hang on the drive shaft, with the ductile flywheel mass plate still connected to it. The stack of electrical sheets of the electric machine made of *Vacodur S Plus* had been completely pulverized. Figure 7.43 shows the test specimens immediately after opening the burst chamber.

All fragments and components found in the chamber were collected and examined in detail in order to reconstruct the failure mechanism as well as possible. Figure 7.44 shows

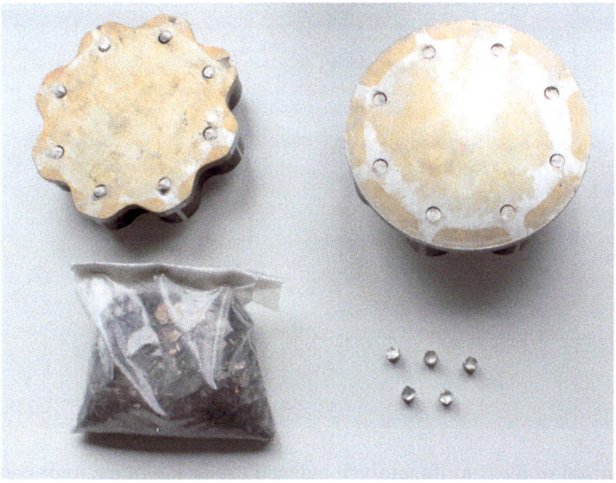

Fig. 7.44 Rotor parts remaining after the burst test

Fig. 7.45 Flywheel mass plate made of 42CrMo4 and measuring mass disk made of W400 VMR with remains of the clamping bolts

the parts remaining after the burst test. Fragments of the clamping bolts are visible in the lower right-hand corner of the picture, whereby only five of the eight cylindrical fragments were found in the burst chamber.

Although an almost play-free transition fit (H7/j6) was selected between the clamping bolts' shaft and the bores through the rotor layers, an extreme widening of the bore in the flywheel mass plate made of 42CrMo4 can be seen in Fig. 7.45 on the left. The same phenomenon could also be observed in the measuring mass disk made of *W400 VMR*, but to a much lesser extent, as shown in Fig. 7.45 on the right. This is another indicator of the higher ductility but lower strength of 42CrMo4 compared to Böhler's vacuum-cast hot working steel. The comparison of the two materials after this excessive centrifugal load in

Fig. 7.46 Comparison of the centrifugal force-induced expansion of the rotor parts from 42CrMo4 and W400 VMR after 45,000 rpm

Fig. 7.47 Remains of the clamping bolts that were sheared off at the level of the *Vacodur electrical sheet metal stack*

Fig. 7.46 confirms this: The radius of the 42CrMo4 steel sheets is 0.6 mm greater compared to the solid endplate made of *W400 VMR* (Fig. 7.47).

The end faces of the bolt remnants each showed strong indications of high shear forces and did not indicate failure due to tensile stresses (Fig. 7.48 left.) Furthermore, the originally cylindrical contour showed a slight bending of the symmetry axis. On the convex side of each screw fragment fragments of the brittle-hard, 0.1-mm-thick *Vacodur S Plus* electrical steel sheets are "baked-in." It can be assumed that these fragments were welded to the relatively soft bolt material *X6CrNiMoTi17-12-2* (No. 1.4571) during impact on the

Fig. 7.48 Macrographs of a sheared clamping bolt. Left: top view with clear shearing marks. Right: side view. On the right side of the fragment, "baked-on" fragments of *Vacodur S Plus* can be seen

burst containment wall. The circumferential speed at the time of the rotor failure was almost 520 m/s (1870 km/h).

Reconstruction of the Rotor Burst

All the evidence points to a sequence of failures that can be reconstructed as follows:

1. **First plasticizing of components at 42,000 rpm**

A significant decrease in the balance quality was observed by measuring the quill shaft vibration (compare Fig. 7.41).

2. **Bursting of the *Vacodur S Plus* electrical sheet stack at 45,000 rpm**
   - Despite the degressive growth of the shaft vibration, sudden failure of the rotor occurred, which was not detected in advance by vibration measurement—a classic brittle fracture, which even before postmortem analysis indicated that a fracture of the Vacodur sheet stack had occurred.
   - This spontaneous, brittle burst behavior was to be expected based on previous tensile tests (see Fig. 7.49).

3. **Shearing off the clamping bolts**
   - The clamping bolts, which due to the electromagnetic requirements had to be made of non-ferromagnetic and therefore relatively low-strength austenitic steel, could not withstand the centrifugal forces in the radial direction exerted by the fragments of the electric machine on the cylindrical inner bore surface. It was not only the lower

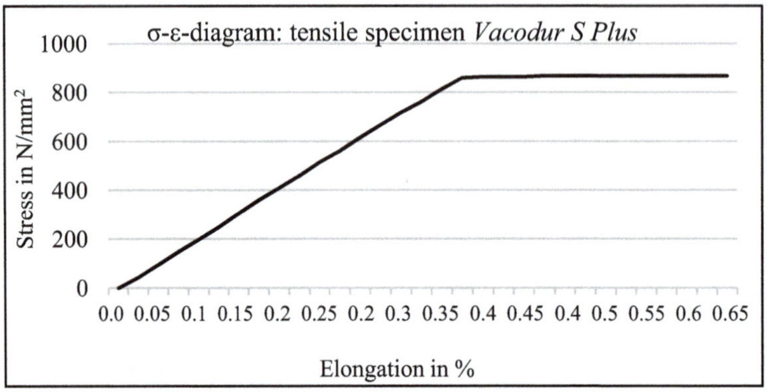

Fig. 7.49 Tensile test of a sample of *Vacodur S Plus*, annealed for maximum mechanical strength

tensile strength of the screws made of *1.4571* compared to all other rotor materials but also the inferior hardness compared to *Vacodur S Plus* that made it possible to form large shears as shown in Fig. 7.48 on the left. The fragments of the clamping screws all had exactly the same height (12 mm) as the sheet metal stack of the electric machine.

4. **Impact of the rotor fragments**

At a speed of around 500 m/s, the parts of the electric machine's sheet stack, which still contained the sheared clamping bold pieces, struck the inner test burst containment, which consisted of two concentric steel rings, each 8 mm thick.

Figure 7.50 shows the rotor bursting process in a cartoon image.

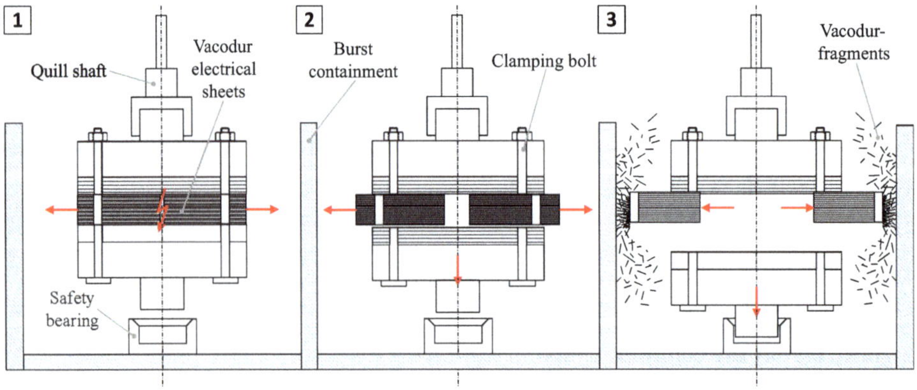

Fig. 7.50 Reconstruction of the rotor failure sequence: (1) brittle fracture of the electric machine's sheet stack, (2) shearing of the clamping bolts, (3) impact of the fragments of the electric machine on the burst containment and touchdown of the lower endplate in the safety bearing

7.5.3 Summary of Results: Fully Integrated Multi-Disk Rotor (*FIMD*)

The manufacturing and theoretical design of the fully integrated multi-disk (FIMD) rotor was subject to many uncertainties, including:

- Static overdetermination of the clamping bolts (fitting of shafts)
- Possible setting phenomena between the sheets and loss of pretension
- Most different thermal material properties
- Unknown material properties of the full-face bonding layer (backlack insulator) of electrical steel sheets
- Unknown total rotor stiffness
- Unknown vibration damping properties

Nevertheless, the rotor behaved as expected (both during the burst test and the actual commissioning). The following advantages of the design compared to solid steel rotors, and even compared to some composite rotors, could be demonstrated with the *FIMD* concept:

1. **Burst behavior:** A skillful choice of material, which assigns the highest quality (strongest) material to the largest and most massive components and the thinnest layers (= smaller rotor components) the most brittle or weakest material, guarantees a "good-natured" burst behavior. This means that

$$\sigma_{\mathrm{W400}\ \ VMR} > \sigma_{\mathrm{42CrMo4}} > \sigma_{\mathrm{Vacodur\ \ S\ \ Plus}} \tag{7.11}$$

Although the sheared segments of the clamping bolts with a diameter of 10 mm and a height of 12 mm weighed only approximately 7.4 g each, their impact in the burst containment was clearly visible (Fig. 7.51). Almost equidistantly, in each case at 45° angular spacing around the circumference, a distinct indentation can be seen, which was caused by the high kinetic energy of the bolt stubs (about 1000 joules or 0.3 Wh each). The stack of about 120 0.1-mm-thick *Vacodur S Plus* sheets, on the other hand, was "pulverized" during impact, whereby on the one hand part of the energy was converted into friction by creating new surfaces, and on the other hand the housing thus experienced a homogeneous "pressure load" and was not punctured at any points. The avoidance of large, high-energy fragments in the event of failure thus paid off. Further information on the design of burst containments and the interaction with the FESS rotor can be found in Chap. 8.

2. **Monitoring of operational safety**: The use of ductile steel flywheel mass plate allows early detection of a change in the balance quality or a possible overload due to plasticizing before brittle fracture of the hardened steels sets in. This is also proven by

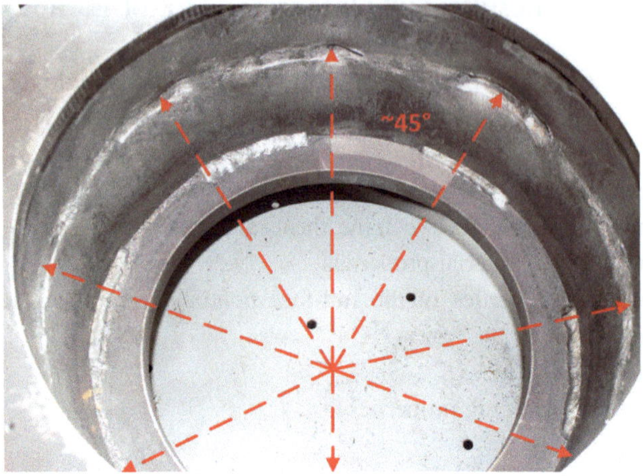

Fig. 7.51 Impact of the clamping bolt fragments in the test burst containment of the FIMD burst rotor

Figs. 7.40 and 7.42. A sequential failure behavior is obtained, whereby the thinnest rotor layers are assigned the lowest strength (electric machine sheet stack with 0.1 mm plates) and the most massive parts (solid endplates made of W400 VMR, 30 mm thick) the highest specific strength.

3. **Achieving and maintaining a high balance quality**: Both rotors, the full-size prototype rotor with a height of 472 mm and a mass of 84 kg and the shortened burst rotor with a height of 157 mm and a mass of 22 kg, did not show any change in balance quality due to settlement phenomena or similar during operation. Only plastic deformation at 42,000 rpm caused the balancing class to drop. There are two main advantages compared to wound composite rotors:
   - Addition of a third balancing plane close to the rotor center through radial holes in the flywheel steel plates is possible (see Sect. 9.6.1.).
   - No change in balance quality due to creep of highly pre-stressed fibers in the matrix.
4. **Good thermal properties:**

While conventional electric machines are usually cooled by axial airflow, flywheel systems have to manage with heat radiation and conduction alone due to the evacuated atmosphere. Therefore, the consideration of the thermal situation of the FESS in the design process is of special importance [56].

Two properties of steel materials are of particular benefit in this case:

- Higher temperature resistance than CFRP or other materials with epoxy resin matrix. The advantages of steel over composite materials mentioned in Sect. 7.2.1.2 have been fully exploited in the laminated multi-disk steel rotor discussed here. The strength of metals is retained up to higher temperature ranges, as shown in Fig. 7.52.

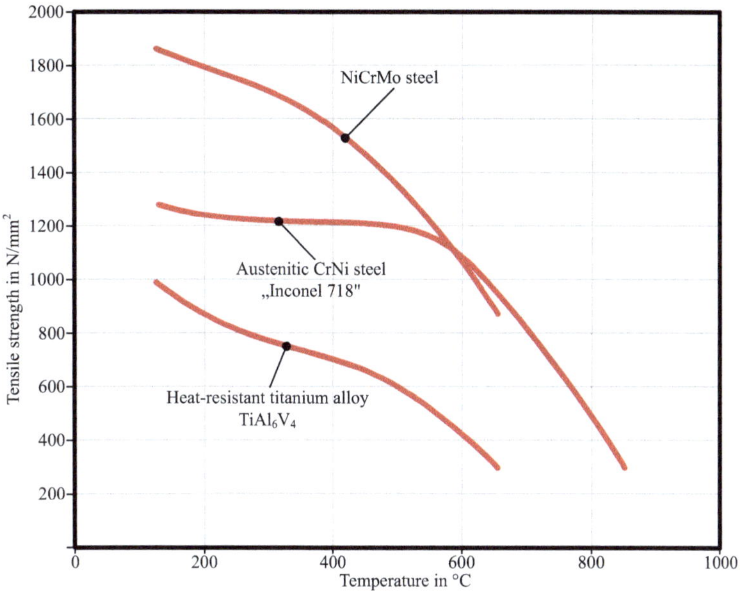

Fig. 7.52 Tensile strength of metals at temperature increase (data from [54] and [55])

- Good thermal conductivity of steel. As can also be seen in Sect. 6.2, plastic parts around the rotor of the electric machine act as thermal insulators. In the case of the fully integrated multi-disk rotor, the axially adjacent flywheel mass sheets conduct much better, and they even serve as temporary "heat sink" or thermally inert mass. This way, temperature peaks can be compensated for when high loads occur for short periods.

5. **Cost reduction**
   - The rotor can be produced by conventional manufacturing processes such as turning, milling, and drilling. The only exception in the case of the evaluated prototype was the active part of the electric machine, which was wire-eroded. However, for higher quantities or in series production, more cost-effective methods such as punching are possible.
   - The use of low-cost materials allows a theoretically achievable rotor material price of about 150 Euro. The specific costs of the Böhler W400 VMR are 11.00 €/kg, while those of the 42CrMo4 are about 1.79 €/kg.

References

1. J. Feldhusen und K.-H. Grote (2007) Dubbel – Taschenbuch für den Maschinenbau, 22nd Edition. Springer, Berlin, Heidelberg, Germany
2. P Selke und B. Assmann (2006) Technische Mechanik – Band 2: Festigkeitslehre, 16th Edition, 2006. Oldenburg Wissenschaftsverlag GmbH, Munich, Germany.
3. F. Strößenreuther (1996) Machbarkeitsstudie und Konzept einer stationären Schwungradanlage zur dezentralen, verbraucherorientierten Energiespeicherung (Diplomarbeit), Lehrstuhl für Dampf- und Gasturbinen, Aachen, Germany.
4. G. Genta (1985) Kinetic Energy Storage: Theory and Practice of Advanced Flywheel Systems. Butterworths, London, UK.
5. P. von Burg (1996) Schnelldrehendes Schwungrad aus Faserkunststoff, ETH Zürich, Schweiz.
6. S. Renner-Smith (1980) Energy Storage: Search for the Perfect Flywheel. Popular Science, Issue January 1980.
7. O.J. Fiske and M.R. Ricc (2005) Third Generation Flywheels for High Power Electricity Storage. LaunchPoint Technologies, Goleta, California, USA.
8. A. Kubo, H. Kameno and R. Takahata (2003) Development of a Compact Flywheel Energy Storage System. Koyo Engineering Journal, English Edition No. 163E.
9. J. Carter (2014) The use of the Gyrodrive hybrid system in bus, truck and off highway vehicles. GKN Hybrid Power, Grove UK.
10. J. Arseneaux (2011) 20 MW Flywheel Energy Storage Plant. Beacon Power LLC, Wilmington, Massachusetts, USA.
11. T. Dever (2013) Development of a High Specific Energy Flywheel Module and Studies to Quantify Its Mission Applications and Benefits. NASA, USA.
12. A. J. Deakin (2014) High performance and low CO_2 from a Flybrid® mechanical kinetic energy recovery system. Torotrak Group PLC. Preston, Lancashire, UK.
13. R.J. Hayes, J.P. Kajs, R.C. Thompson and J.H. Beno (1999) Design and Testing of a Flywheel Battery for a Transit Bus. SAE International Congress and Exposition, Detroit, Michigan, USA.
14. Robert Hebner, Joseph Beno and Alan Walls (2002) Flywheel Batteries Come Around Again. IEEE Spectrum, pp. 46-51, Issue April 2002. https://spectrum.ieee.org/energy/the-smarter-grid/flywheel-batteries-come-around-again
15. M. A. Pichot, J. M. Kramer, R. C. Thompson, R. J. Hayes and J. H. Beno (1997) The Flywheel Battery Containment Problem. 1997 SAE International Congress and Exposition, Detroit, Michigan, USA.
16. NEXUS Projects SL (2012) Durability of Composites – Fatigue. Martorell, Barcelona, Spain. http://nexusprojectes.com/durabilidad.aspx?lang=en. [Accessed August 17th 2016].
17. Anthony J. Colozza (2000) High Energy Flywheel Containment Evaluation. NASA, Brook Park, Ohio, USA.
18. S.K. Ha, K.K. Jin and Y Huang (2008) Micro Mechanics of Failure (MMF) for Continuous Fiber Reinforced Composites. Journal of Composite Materials, Bd. 42 (18) pp. 1873–1895 Issue July 2008.
19. J. Koyanagi (2011) Durability of filament-wound composite flywheel rotors. Mechanics of Time-Dependent Materials, Bd. 16, Nr. 1, pp. 71–83.
20. H. P. Luckett (1979) PS/What's News. Popular Mechanics, p. 75, Issue October 1979.
21. B. Nearing (2011) Flywheels fail at energy project. TimesUnion, Issue October 19th. 2011.
22. Universal Science (2012) Thermal Conductivity of Materials. http://www.universal-science.com/wp-content/uploads/2012/08/Thermal-conductivity-table.pdf. [Accessed January 8th 2016].
23. Böhler (2012) Werkzeugstähle Schnellarbeitsstähle. Lieferprorgamm BÖHLER – Stahl für die Besten der Welt, Nr. Issue May 2012 pp. 10–74.

24. T. Tian (2011) Anisotropic Thermal Property Measurement of Carbon-fiber/Epoxy Composite Materials. University of Nebraska, Lincoln, Nebraska, USA.
25. A. Dasgupta and R. K. Agarwal (1992) Orthotropic thermal conductivity of plain-weave fabric composites using a homogenization technique. Journal of Composite Materials, Edition 26, pp. 2736–2758.
26. R. D. Sweeting (2004) Measurement of thermal conductivity for fibre-reinforced composites. Composites Part A: Applied Science and Manufacturing, pp. 933–938.
27. R. C. Wetherhold and J. Wang (1994) Difficulties in the theories for predicting transverse thermal conductivity of continuous fiber composites. Journal of Composite Materials, pp. 1491–1498.
28. A. Storer (2015) What is the maximum temperature stability of carbon fiber composite and glass fiber composite? https://www.quora.com/What-is-the-maximum-temperature-stability-of-carbon-fiber-composite-and-glass-fiber-composite. [Accessed January 8th 2016].
29. P.E. Mason, K. Atallah and D. Howe (1999) Hard and Soft Magnetic Composites in High Speed Flywheels, ICCM-12 Paris, France.
30. GKN Hybrid Power (2014) Gyrodrive by GKN Hybrid Power – Driving Efficient Transport, Unit 1 Pentagon South, Abingdon Science Park, Barton Lane, Abingdon, Oxford OX14 3PZ, UK.
31. L.A. Bisby (2003) Fire behaviour of fibre-reinforced polymer (FRP) reinforced or confined concrete, (Dissertation), Queen's University, Kingston, Ontario, Canada.
32. Clean Motion Offensive (2011) Projektinhalt. Clusterland OberAustria GmbH, Hafenstraße 47-51, 4020 Linz, Austria. http://www.cleanmotion.at/index.php?id=19. [Accessed February 20th 2016].
33. Klima- und Energiefonds (2011) CMO – Clean Motion Offensive. Klima- und Energiefonds, Gumpendorferstr. 5/22, 1060 Wien, Austria. https://www.klimafonds.gv.at/unsere-themen/e-mobilitaet/leuchttuerme/cmo-clean-motion-offensive/. [Accessed February 20th 2016]
34. VAC – Vacuumschmelze (2013) Weichmagnetische Kobalt-Eisen-Legierungen (Datenblatt VACOFLUX and VACODUR). Vacuumschmelze, Hanau, Germany.
35. E. Lindsley (1973) Hybrid Car: Part-Time Engine + Part-Time Flywheel = Full-Time Transportation. Popular Science, Issue August 1973.
36. A. P. Armagnac (1974) Flywheel Brakes Store New Train's Energy for Electricity-Saving Starts. Popular Science, pp. 70–72, Issue February 1974.
37. D. Scott (1980) Hydrobus, Gyrobus use brake-generated energy. Popular Science, pp. 76–77, 1980.
38. D. Scott (1961) Fifth Wheel Runs Bus… Stops it Too! Popular Science, pp. 98–102, Issue May 1961.
39. R. C. Clerk, J. Adams and J. A. Howell (1970) Flywheel aided power surge. Commercial Motor Archive, 30 October 1970
40. W. Novy (2008) Start-Stopp – aber mit Schwung! Kietische Energiespeicher als Alternative zu Akkumulatoren und Kondensatoren. AUTOMOTIVE, pp. 64–66, Issue 11 2008.
41. P. Dietrich (1999) Gesamtenergetische Bewertung verschiedener Betriebsarten eines Parallel-Hybridantriebes mit Schwungradkomponente und stufenlosem Weitbereichsgetriebe für einen Personenwagen (Dissertation) p. 86. ETH Zürich, Switzerland.
42. Parry People Movers Ltd. (2009) PPM Technology. Parry People Movers Ltd, Overend Road, Cradley Heath, West Midlands, B64 7DD, UK. http://www.parrypeoplemovers.com/technology.htm. [Accessed August 20th 2016].
43. H. Schreck (1977) Konzeptuntersuchung Realisierung und Vergleich eines Hybrid-Antriebes mit Schwungrad mit einem konventionellen Antrieb. Fakultät für Maschinenwesen der Rheinisch-Westfälischen Technischen Hochschule, Aachen, Germany.
44. N. N. Gulia, (1986) Der Energiekonserve auf der Spur. Verlag Harri Deutsch, Thun, Germany.

45. Compact Dynamics (2008) KERS – Energy Recovery System (Version 08). Compact Dynamics, Moosstrasse 9, D-82319 Starnberg, Germany.
46. B. Schweighofer, M. Recheis, P. Fulmek and H. Wegleiter (2013) Rotor Losses in a Switched Reluctance Motor – Analysis and Reduction Methods. EPJ Web of Conferences, Volume 40, 2013. JEMS 2012 – Joint European Magnetic Symposia. https://doi.org/10.1051/epjconf/20134017008
47. E. Chiao (2012) Amber Kinetics DOE Peer Review. U.S. Department of Energy, Washington D.C., USA.
48. Grosschädl Stahl (2016) Lager-Preisliste, Stabstahl 42CrMo4 + QT. Graz, Austria.
49. Edelstahl Service Schulz (2016) Übersicht über die verarbeiteten Werkstoffe – Nichtrostende Stähle (austenitisch) – Sonderstähle. Edelstahl Service Schulz, Augustenstr. 10 a, 70178 Stuttgart, Germany.
50. D. Breslavsky (2011) European steel and alloy grades and numbers. National Technical University, KhPI, 21 Frunze Str., Kharkov 61002, Ukraine. http://www.steelnumber.com/en/steel_composition_eu.php?name_id=335. [Accessed July 22nd 2016].
51. BI-WAT GmbH. – Bad Ischler Wassertechnik und Edelstahldesign (2013) Edelstahl-Information | Chemische Beständigkeit. Marie-Luisenstraße 1A, 4820 Bad Ischl, Austria.
52. Thyssen Krupp Materials International (2008) Werkstoffblatt TK 34CrMo(S)4 bis 42CrMo(S)4, p.3. Thyssenkrupp AG, Essen, Germany.
53. A. Buchroithner, I. Andrasec and M. Bader (2012) Optimal system design and ideal application of flywheel energy storage systems for vehicles. 2012 IEEE International Energy Conference and Exhibition (ENERGYCON). Florence, Italy. DOI: https://doi.org/10.1109/EnergyCon.2012.6348295
54. P. M. Rudeloff (1909) Der Einfluß erhöhter Temperaturen auf die mechanischen Eigenschaften der Metalle. Polytechnisches Journal, Berlin, Germany. http://dingler.culture.hu-berlin.de/article/pj324/ar324182
55. C. Brummer (2013) Licht hilft beim Formen anspruchsvoller Materialien – Laserunterstütztes Metalldrücken verbessert Formänderungsverhalten hochfester Werkstoffe. Industrieanzeiger Future Trends, Issue 22. April 2013. Konradin-Verlag Robert Kohlhammer GmbH, Leinfelden-Echterdingen, Germany.
56. Grosschädl Stahl (2016) Lager-Preisliste, Stabstahl 42CrMo4 + QT. Grosschädl Stahl Graz, Südbahnstraße 10, A-8020 Graz, Austria.

Flywheel Energy Storage Housing

8

Nothing harms the economic success of a technology more than its reputation of being dangerous. Even though there are hardly any known accidents involving energy storage flywheels that actually resulted in personal injury, incidents such as the much-cited rotor burst in *Beacon Power*'s grid stability plant in Stephentown are sufficient to fuel mistrust of FESS technology [1]. But so far, only a few accidents in which the burst housing was penetrated and rotor fragments escaped have become public. Two prominent examples are, however:

- **2011, *Beacon Power*** (compare Fig. 8.1): Powder explosion of carbon fiber dust [2].
- **2015, *Quantum Technologies*:** Causes of errors were not published by the operators [3].

All other FESS rotor and housing damages found in literature are intentionally performed burst tests within the scope of scientific investigations. Since the burst behavior of isotropic (usually steel) rotors is completely different from that of anisotropic (fiber composite) rotors, a clear distinction must be made here.

8.1 Requirements Derived from *Supersystem Analysis*

The housing of the flywheel is a component that is essentially responsible for three main tasks:

- Interface of the connection between moving parts of the flywheel and the vehicle/surrounding
- Providing the required vacuum tightness
- Protective function against escape of fragments in case of rotor failure/crash

A. Buchroithner, *Flywheel Energy Storage*,
https://doi.org/10.1007/978-3-658-35342-1_8

Fig. 8.1 Carbon dust explosion at the *Beacon Power* flywheel plant in Stephentown, USA, in 2011 [2]. (Image rights: The Eastwick Press)

In the following, only the third subtask will be dealt with, as the first two points are self-explanatory and can be considered as largely solved.

Figure 8.2 illustrates the eight most important aspects of FESS housing design resulting based on considerations of the *supersystem*. In addition to the safety aspect (in the event of rotor burst and crash), which must always have the highest priority in the process of housing design, the following requirements must be met:

1. **Low weight:** The rather high specific energy of the rotor alone is usually only a fraction of the entire system, since the housing has accounts for the largest weight share.
2. **Good integration into the vehicle:** A corresponding interface/attachment to the vehicle must be designed, which is generally easier to implement in commercial vehicles due to the more generous installation space available.
3. **High stiffness and damping**: Favorable influence of the machine dynamics and acoustics by the housing structure and its properties is required.
4. **Suitability for integration of a cooling system:** It must be possible to integrate an appropriate cooling system for the electric motor generator and bearings in the housing.
5. **Low costs:** As ductile (structural) steels are usually used, which are very cost-effective, special attention must be paid to selecting a design suitable for cheap manufacturing in order to avoid high process-related costs.
6. **Attractive design and appearance:** Since the customer does not see anything of the "inside," i.e., the actual technology of the FESS storage unit, the external appearance plays a central role in terms of marketing.

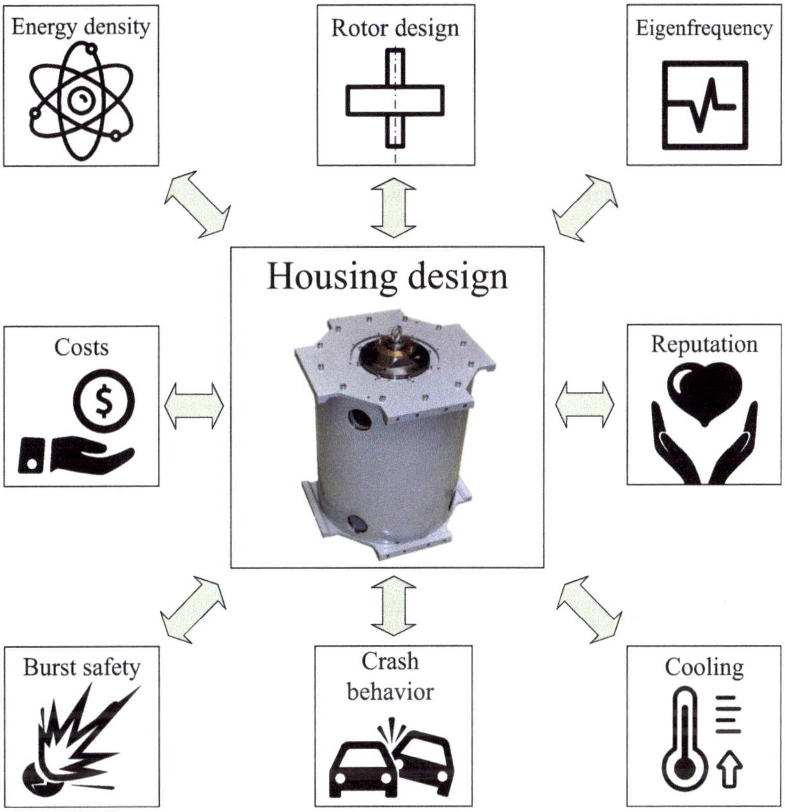

Fig. 8.2 The eight essential, mutually influencing parameters of FESS housing design

The design and calculation methods for flywheel safety containments available in the literature (*NASA* [5], *Lockheed* [6], *Genta* [7]) (see Sect. 8.6) show strong divergences and are partly based on specific, empirical input variables.

However, progress in the design of FESS burst containments is particularly important for the following two reasons [4]**:**

- **In order for FESS to successfully enter the market, public confidence in the technology must be created:**
- As the short calculation example in Chapter 2, Eq. 8.2, and the accidents mentioned at the beginning of the chapter show, the consequences of rotor burst can reach catastrophic dimensions.
- Experts from the *Oak Ridge National Laboratories* (US Department of Energy) also agree that:

> "An accident resulting from a containment failure by any one of the flywheel developers would be detrimental to the entire industry" [8].

This also applies to the attitude of large investors who, by withdrawing their financial resources, could extremely slow down the technical development of flywheel energy storage systems [1].

- **Increasing the specific energy by reducing the housing weight:**
  The achievement of the *threshold energy density* mentioned in Sect. 5.4 must be regarded as an essential goal in the development of mobile flywheel energy storage systems.

8.2 Safety Requirements for Mobile Energy Storage Devices

All energy storage systems must comply with certain safety regulations. Especially in the automotive industry, they are subject to particularly strict regulations, which are not always easy to comply with due to the limited installation space and the trend toward lightweight design. While supercaps and batteries have no moving parts and potential danger lies primarily in possible electric shock or fire due to a short circuit, a flywheel energy storage system requires a different, comprehensive safety concept. The main problem with FESS is that the entire kinetic energy can be released within a very short time. A simple numerical example shows what damage can occur if a flywheel with an energy content of 1.5 kWh fails. The energy content is simply compared with the kinetic energy of a 1.5 ton personal car:

$$1{,}5\ [kWh] = 1{,}5 * 1000 * 3600 = 5400000\ [J] = \frac{1}{2}\ m\ v^2 = 750\ v^2 \tag{8.1}$$

$$v = \sqrt{\frac{5400000}{750}} \approx 85\ \left[\frac{m}{s}\right] \mathrel{\hat{=}} 306\ km/h \tag{8.2}$$

Example
The energy content of a 1.5 kWh flywheel is therefore equivalent to the kinetic energy of a car traveling at over 300 km/h. The greatest danger is the breakage of the rotor and the high energy of the fragments due to the extreme rim speeds. It is therefore needless to say that a sophisticated condition monitoring system and emergency strategy is essential.

Essentially, *four possible scenarios* during vehicle operation can be distinguished and safety requirements can be defined accordingly:

1. **Normal driving operation**
   The FESS must exhibit safe behavior under the given operating conditions. It must be designed in such a way that the accelerations, frequencies, and temperature ranges

occurring in normal driving operation do not damage the flywheel in any way. Due to various influences, however, sudden technical failure can occur even during regular operation. This can be caused by:

- High cycle fatigue due to continuous load
- Thermal aging of plastics
- Bearing wear
- Stochastic errors of the control system
- Stochastic material defects (blowholes, etc.)

2. **Early detectable failure**

A monitoring system must be provided, which can distinguish "normal" from "disturbed" operation by measuring operating parameters such as acceleration and/or amplitude of the flywheel shaft, temperature, etc. If a critical limit is exceeded for an operating parameter (e.g., acceleration of a bearing), the system must be switched off in a controlled manner and without damage.

The following methods, among others, are available for condition monitoring:

- Vibration-based method: [9, 10]
- Measurement of the deformation: [11]
- Crack detection in the rotating flywheel by spectral analysis: [12, 13]

The abovementioned methods belong to the field of measurement technology and signal processing and will not be described in detail in this book. More information can be found in the listed literature references.

3. **The So-Called Repair Crash**

These are accidents at low speeds, for example, parking damage. Insurance companies carry out standardized crash tests (*AZT/RCAR tests*), according to which the repair costs of the car are determined (compare Figs. 8.3 and 8.4.) The aim of every automotive company is to keep the repair costs as low as possible in order to obtain favorable insurance premiums for the vehicle. If the flywheel were to be damaged by the *RCAR* repair test, this would be a knockout criterion for series use in the vehicle from an economic point of view.

The following applies to both front and rear tests: The impact speed is defined as $v_F = 15$ km/h, the *barrier angle* $\alpha = 10°$, and the rounding radius $R = 15$ cm.

4. **The Serious Accident**

In this case, the flywheel energy storage unit may be destroyed, but of course no fragments should escape from the safety housing and thus represent a further risk of injury. It would

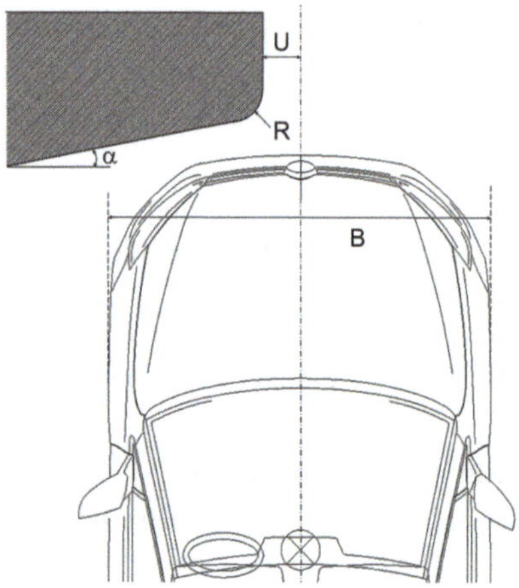

Fig. 8.3 New AZT/RCAR repair test: 10° front. (Image rights: ATZ Automotive GmbH)

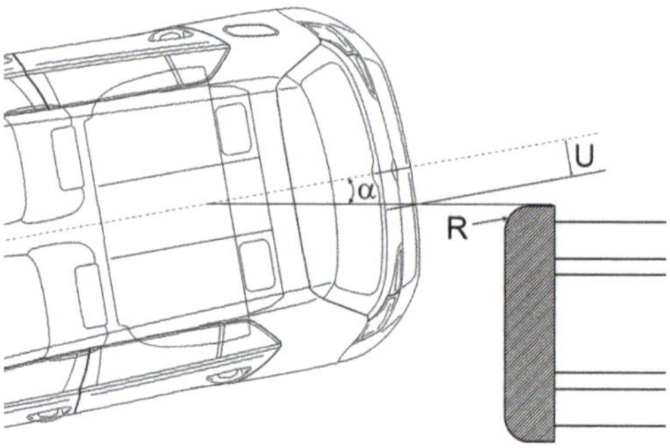

Fig. 8.4 New AZT/RCAR repair test: 10° rear. (Image rights: AZT Automotive GmbH)

Fig. 8.5 Sled crash test of a *Flybrid Systems* flywheel energy storage module. (Image rights: PUNCH Flybrid)

be an obvious approach to couple a "switch-off" or "self-destruction" strategy with the activation of the vehicle's restraint system.

The test procedures for the usual accidents occurring in traffic are manifold. In Europe, the regulations of the *EEVC* (*European Enhanced Vehicle-safety Committee*) apply in principle. However, the *EuroNCAP tests* are still very important for sales, as customers attach great importance to the star ratings achieved by the vehicle types when it comes to vehicle safety. Vehicles equipped with flywheel energy storage should therefore comply not only with *EEVC* but also with *EuroNCAP regulations*. (The exact test procedures and guidelines can be found in technical literature, for example [14].) In most cases, the crumple zone of a vehicle is designed in such a way that accelerations of approximately 10 g occur when the bumper is destroyed, 40 g when the bumper is then penetrating into the engine compartment, and 60 g shortly before reaching the splash wall in order to avoid intrusions into the passenger compartment. In addition, every vehicle manufacturer carries out in-house misuse tests, which must also be taken into account when designing the mechanical flywheel components.

So far, only one crash test (sled test) is known to have been performed with a FESS [21]. A flywheel from *Flybrid Systems* was successfully crash tested at an operating speed of 64,000 rpm. The flywheel module intended as *KERS* in Formula 1 was subjected to accelerations of more than 20 g in a protective housing without any adverse effect on its operational reliability [15]. The test setup is shown in Fig. 8.5.

The electrical protection regulations apply to batteries, *supercaps*, and also flywheel energy storage if the energy is transmitted electrically. There are already a number of technically recognized recommendations regarding this matter. Very comprehensive standards are the *ECE-R-100* and *DIN EN 1987*, which are only referred to but not discussed any further in this book.

8.3 Analysis of Existing Systems/State of the Art

Taking a closer look at examples of burst housings for flywheel energy storage systems available in the literature, one finds that:

- The design was usually very conservative, i.e., the housing is characterized by significant overdimensioning
- The influence of the housing weight is regarded as irrelevant for stationary FESS systems and the flywheels are usually additionally lowered in concrete pipes buried in the ground (i.e., the building foundation)

These two measures are of course unsuitable for mobile applications, as they considerably reduce the energy density of the energy storage system.

8.3.1 Example: Safety Housing for Composite Rotors of Stationary FESS

This specific and innovative housing concept was developed by *Boeing Phantom Works* and is called *S-Bracket Containment Structure*. The basic idea behind the selection of this concept was that a great deal of deformation energy can be absorbed by s-shaped steel fins. This type of housing was specially adapted to a wound, anisotropic fiber composite rotor. In this case the rotor burst, process is associated with an enormous increase in volume.

On the one hand, the s-brackets, which are welded to the inner steel cylinder of the housing, allow the rotor to expand, and on the other hand, the rotor is braked more and more with increasing diameter by the contact forces that occur. An isotropic steel rotor has not yet been tested in this type of housing, but it does not appear to make much sense due to the specific fracture kinematics (two to three large, sharp-edged fragments), since the brackets are intended to "gently" reduce the angular velocity of the composite rotor and do not represent an optimized structure for massive projectiles with ballistic trajectory.

Figure 8.6 shows *Boeing's* "s-bracket housing" concept. The wound composite rotor (carbon fiber) is clearly visible. Comparing the geometric dimensions of the rotor and the

Fig. 8.6 The s-bracket FESS containment [16] developed by *Boeing* especially for composite rotors. (Image rights: US Department of Energy)

housing and taking into account that CFRP has a much lower density[1] than steel, it becomes clear that the housing accounts for the majority of the FESS's total weight.

Important

- The example described above already shows how much the housing influences the volumetric and gravimetric energy density of the entire flywheel energy storage system.
- For mobile FESS, safe lightweight construction concepts must be developed and empirically verified.

8.4 Relevant Findings from Past Research Projects

In the years between 1960 and 1990, a number of low-speed steel flywheels were tested in vehicles. Some of the rotors had only a very sparse housing or even no housing at all. In contrast, the burst behavior of fiber composite rotors is far more complex due to their anisotropy and greater number of degrees of freedom in design choices. The complexity in the prediction of the fracture behavior is contrasted by the lower kinetic energy of the smaller fragments (particles) and the generally more "good-natured" fracture behavior. (However, it must be noted that the overall kinetic energy content is usually much higher in the case of composite rotors.) Essentially, there are three cases of failure in composite rotors, which are summarized in Table 8.1.

In rare cases, however, the unbalance can become so large during delamination that it exceeds the strength of the bearings and shaft, and the rotor is no longer guided and spins freely in the housing. This case and the spontaneous rotor breakage ("burst") thus form the worst-case scenario and basis for the housing design. It should also be noted that even a composite rotor usually has a metallic hub or shaft, which may break into three segments with significant high kinetic energy. An impact model of small composite fragments (particles) for the housing design was developed at the *University of Austin*, *Texas* [17], and is briefly described below.

Referring to Fig. 8.7, the letters a and b mark inner and outer *radius* of the rotor. The inner radius of the housing with *wall-thickness* h is marked c. At the time $t = 0$, the rotor rotates at the *angular velocity* ω, and it is assumed that it suddenly breaks into a large number of fragments with the *characteristic size* Δm. Each fragment moves unhindered along the *path* s until it hits the wall of the housing. The load on the housing was approximated by modeling a continuum and does not consider discrete rotor fragments. Introducing the identity $\Delta m = l\, r\, \rho\, \Delta\Theta\, \Delta r$, the *length* l of the rotor, and the *material density* ρ, an equation for the compressive load on the containment can be established [17]:

[1] In [4], a density of 1608 kg/m^3 is given for the composite rotor.

Table 8.1 Overview of delamination and transverse matrix cracks [17]. Courtesy of Joseph Beno, *Center for Electromechanics*, University of Texas

Failure scenario	Cause	Impact of failure
In the radial plane (R)		
Delamination: Radial separation of layers of composite Windings	Overspeed Epoxy matrix degradation Creep of imbedded fibers Exceeding the operating temperature	Maybe benign for some designs: Causes strong imbalance forces due to mass shift → system shutdown required
In the Thea plain (θ)		
Fracture (burst): Hoop fiber crack	Overspeed Fatigue fracture	Can reach catastrophic proportions → releases total kinetic energy in a short time
In the axial plane (Z)		
Crack in the matrix: Transverse to tangentially wound hoop fibers	Induced by high axial tensile stresses due to Poisson effect	Benign in limited occurrences: May cause imbalance forces → system can maybe be switched off in time

Fig. 8.7 Geometric definitions for the rotor burst calculation model according to [17]. (Image rights: Center for Electromechanics, University of Texas)

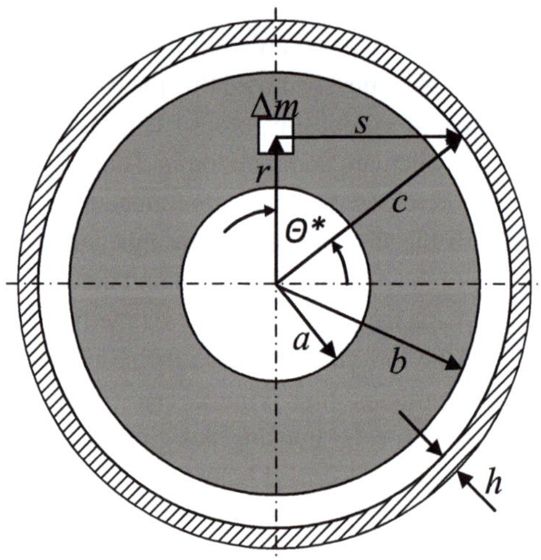

$$p = \frac{\rho * c^2 * \omega^4 * t^2}{(1 + \omega^2 * t^2)^3} * \left[1 - \frac{d\omega}{dt} \frac{1 + \omega^2 * t^2}{c * \omega^2 * t}\right]^2 \tag{8.3}$$

The *time t* describes the *impact duration*, based on the angular velocity of the particles and the *path length s*. Due to the high circumferential speeds, this time duration is usually in the μs range, as shown in the example in Fig. 8.8.

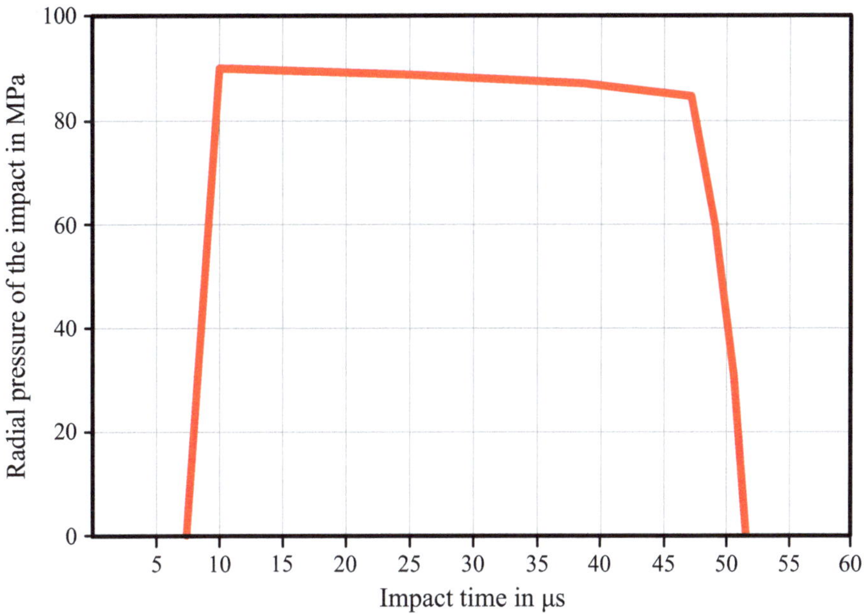

Fig. 8.8 Predicted uniform pressure loading on the burst containment as a function of time at impact of a fragment [15]. (Image rights: Center for Electromechanics, University of Texas)

8.4.1 Particle Kinematics

Practical experiments have shown that the housing must also offer high axial strength and rigidity, as particles are deflected by the radial housing walls. Only few results about particle kinematics in an evacuated flywheel housing are available, but according to [17], the following possibilities seem plausible:

- An axial pressure gradient in a fluidlike bed of debris is developed due to the accumulation of particles.
- "New" fragments meet fragments that already hit the containment (scattering of fragments at the wall from the impact of incoming fragments.)
- A kind of "particle wedge" is formed on the inner housing circumference, which drives the particles to the upper or lower housing cover plate (fracture of fragments along oblique planes as they hit the wall).

Regardless of the cause, it can be assumed that two equal pressure-induced forces act on the housing top and bottom cover plates. However, in addition to the impact forces of the particles on the housing walls caused by the centrifugal forces, a free torque due to the mass moment of inertia must also be expected. The assumption that the kinetic energy is internally dissipated in the case of a complete, fine pulverization of the rotor without tangential forces acting on the housing is only valid if the rotor runs within an extremely large housing. In mobile applications, however, the *gap thickness* ($c - b$ in Fig. 8.7) should only be selected on the basis of aerodynamic considerations (windage losses) and should be as small as possible for space-saving reasons. It can therefore be assumed that the still intact part of the rotor can transmit a torque to the inner wall of the housing via the freely moving rotor parts (fragments and particles). The resulting free torsional moment can also be calculated with the help of this model, assuming that the particles remain on the inner wall of the housing and a kind of "deposit layer" is formed, on which the particles that follow later impinge. In the worst case, the entire rotor will disintegrate at once, pulverize, and deposit on the inner circumference of the housing. In any case, two components contribute to the torque on the casing:

- Stopping of these particles, which after *path s are* decelerated at housing wall
- Impulse of following particles, which still carry part of the kinetic energy of rotor

The *pressure at the housing wall* N_p is calculated according to [17] as

$$N_p(t) = \rho\ (1 - \gamma)\ \omega_d^2\ \frac{c^3 - r_2^3}{3c} \tag{8.4}$$

with the *porosity of the debris bed* γ, the *inner radius* r_2 of the debris bed, and the *inner radius c*, whereby the latter can be calculated through mass conservation. Introducing the *coefficient of friction* μ (which must be determined experimentally and is usually in the range of 0.3–0.4 for fiber composite rotors) and the *length l*, the *torque* $T_p(t)$ *on the housing wall* can be calculated:

$$T_p(t) = 2\pi c^2 l \mu N_p(t) \tag{8.5}$$

In the event of failure, it would be desirable for the rotor to disintegrate into as smallest possible particles, all of approximately the same size, as this would convert a maximum of the kinetic energy of the rotor into fracture energy. Many small fragments escaping from the rotor (each with relatively low mass and hence low kinetic energy) would also stress the protective housing much more evenly and thus more controllably than individual large fragments with high kinetic energy [25]. Ballistic investigations of various protective containment materials have shown that, in addition to kinetic energy, the impact surface, hardness, and sharpness of the projectile are the main factors influencing the impact consequences. Furthermore, the geometric size of the fragments is also very decisive. In the following, the optimum fragment shape is derived using simple model considerations as an example:

The kinetic energy of a hoop-shaped rotor is

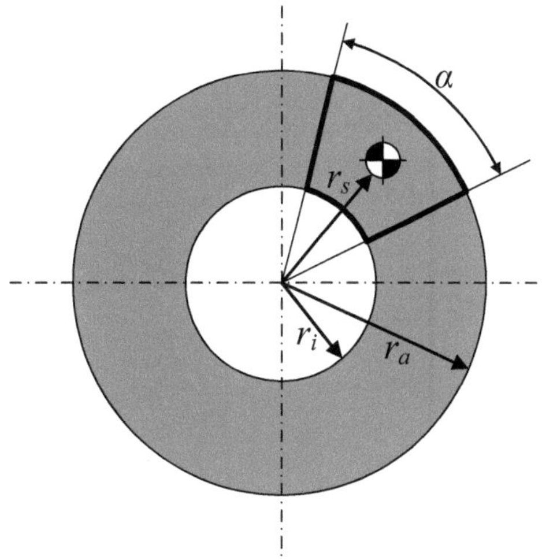

Fig. 8.9 Geometric division of the rotor into individual fragments according to [26]. (Image rights: Peter von Burg, ETH Zürich)

$$E_{kin} = \frac{1}{2} I\omega^2 = \frac{1}{4} m \left(r_a^2 + r_i^2\right) \omega^2 \tag{8.6}$$

If the actual fracture work is neglected, the law of conservation of energy can be used to derive the shares of the rotor energy split into pure translational and rotational energy of the fragments with *size a* (Fig. 8.9).

The translational energy is calculated from the instantaneous velocity of the center of gravity of the newly created fragment. The *center of gravity radius* r_c of such a fragment is

$$r_c = \frac{4}{3} \frac{\sin(\alpha/2)}{\alpha} \frac{r_a^3 - r_i^3}{r_a^2 - r_i^2} \tag{8.7}$$

The *translational energy* E_{kin_transB} of a fragment is thus proportional to the stored *energy* E_{kin}:

$$\frac{E_{kin_{transB}}}{E_{kin}} = 2 \frac{m_B \, r_{S_B}{}^2}{m \left(r_a^2 + r_i^2\right)} = \frac{8}{9} \frac{\left(\sin(\alpha/2)\right)^2}{\pi \, \alpha} \frac{\left(r_a^2 + r_a r_i + r_i^2\right)}{\left(r_a + r_i\right)^2 \left(r_a^2 + r_i^2\right)} \tag{8.8}$$

The ratio of the translational to rotational energy of the section of a thin hoop or disk as a function of the number of fragments is shown in Fig. 8.10. Thus, the x-axis "size of fragments in circumferential direction" shows not only their size but indirectly also their number. For a fragment size of 36°, for example, ten fragments are considered.

Ideally, the fragments should be as thin-walled as possible and cover the entire rotor circumference (detachment of hooplike rings), since in such a case, the translational energy share of both, the sum of all fragments as a whole, and the individual fragments can be minimized.

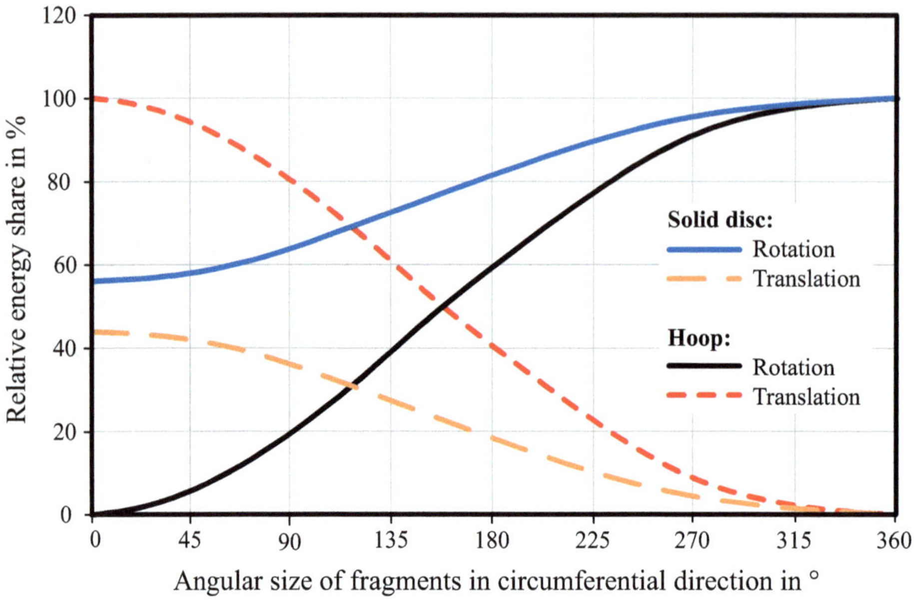

Fig. 8.10 Distribution of rotor fragment energy, based on [26]. (Image rights: Peter von Burg, ETH Zürich)

8.5 Practical Design of FESS Housings

Several theoretical housing concepts were presented in [17], but without mentioning quantifiable design guidelines. The simplest design, which is mostly used for stationary flywheel applications, is not suitable for mobile and especially racing applications—a housing that is so thick-walled and heavy that it can withstand a rotor breakage in any conceivable case of damage and also allows excessive plastic containment deformation. Due to space and weight restrictions, alternative approaches must be sought. Because of the relatively high torque acting on the housing in the event of failure, it is advisable to take design measures to absorb the free torque in a targeted manner and thus mitigate loads on the housing mounts. One possible solution is the use of "rotatable liners" made of Kevlar (shown in Fig. 8.11), which are set in rotation in the housing when the rotor fragments hit it. A similar concept was tested by *Lockheed* already in 1972. The safety housing shown in Fig. 8.12 contains fiberglass rings, which were set in rotation relative to the outer casing wall made of steel by the fracture (*tri-burst*) of a steel rotor. The flywheel with a diameter of about 500 mm failed at a speed of 16,750 rpm, which corresponds to an energy content of about 343 Wh [6]. However, according to the *Lockheed Missiles Company*, a solid steel ring was the lighter and safer containment variant.

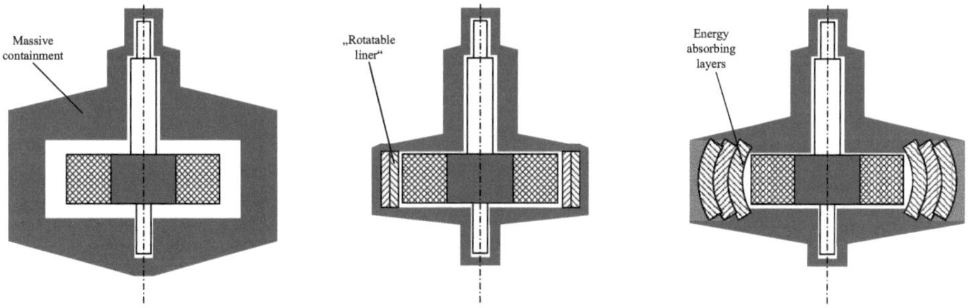

Fig. 8.11 Non-validated, theoretical burst containment concepts by the *Center for Electromechanics* of the University of Texas, Austin, USA [17]. (Image rights: Center for Electromechanics, University of Texas)

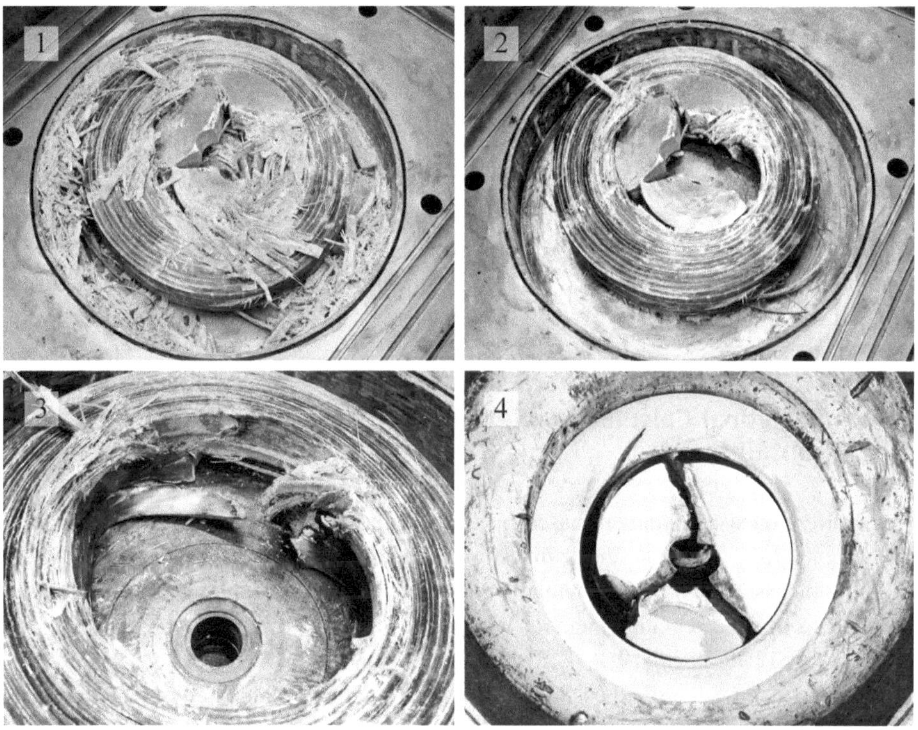

Fig. 8.12 FESS containment with "rotatable liners" made of fiberglass composite (Figs. 1–3) and solid steel ring (Fig. 4) [6]. (Image rights: Lockheed Martin)

Another possibility is the use of "energy absorbing layers" on the inner housing circumference, as shown in Fig. 8.13. The absorber layer is made of a soft material, which is attached to the inside of a rigid "backup structure." Although an *energy absorbing*

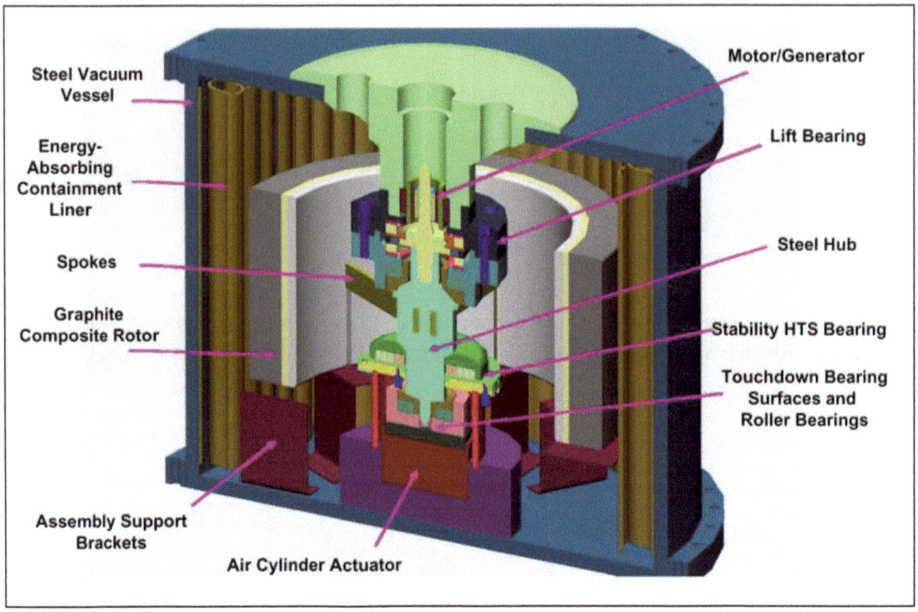

Fig. 8.13 Schematic diagram of a 10 kWh *high-temperature superconducting* (HTS) flywheel [18]. (Image rights: US Department of Energy)

containment liner is also mentioned in [18], no guidelines or publications on its design can be found.

8.6 Analytical Calculation Methods for Designing FESS Burst Containments

A cylindrical burst containment based on the calculation methods available in literature was designed for the spin pit for the investigation of FESS containments described in Sect. 8.9.2. In this case, test flywheels with defined burst speed and fracture geometry are used. The analysis of the calculation methods was partly supported by student theses at the Graz University of Technology [19].

The three major established calculation methods and their assumptions and main results are summarized below.

8.6.1 Calculation According to Lockheed Missiles Company [6]

The *Lockheed* company carried out a research project in 1972 with the aim of realizing a flywheel hybrid vehicle. In the course of this project, first considerations were made

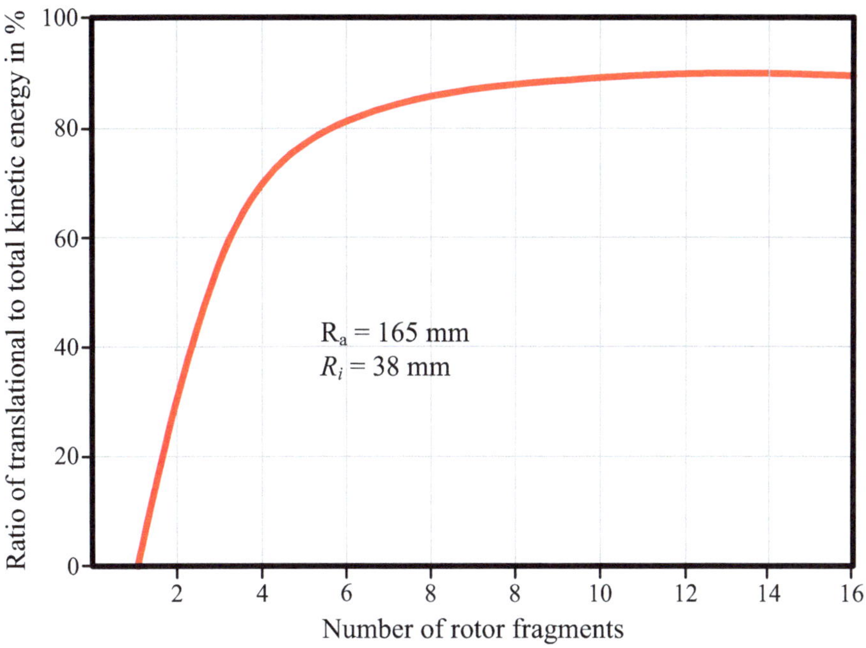

Fig. 8.14 Number of fragments and kinetic energy according to [6]. (Image rights: Lockheed Martin)

regarding the crash and burst safety of the FESS. Two methods were used for the design of the burst containment structure.

Method 1

This method is basically suitable for the calculation of a safety housing for steel flywheels, since it is assumed that the rotor breaks up into a small number n (usually 3) of fragments. The total energy in the system is made up of the rotational and translational share of the fragments. The rotational component decreases as the number of fragments increases, but the translational component increases, as shown in Fig. 8.14.

However, the stress distribution in the ring-shaped burst containment is (in very simplified way) subsequently assumed to be homogeneous. The method is based on the idea that the centrifugal force acting on the individual fragments at the time of bursting is subsequently introduced directly into the housing wall. The following applies to a scenario involving three rotor fragments:

$$F = \frac{\rho * h * w}{3} * \left(r_a{}^3 - r_i{}^3\right) \tag{8.9}$$

where:

F... centrifugal force of the flywheel over a radial cross section
ρ... density of the material used in kg/m^3
h... height of the flywheel rotor in m
ω... angular velocity of the flywheel during bursting in 1/s
r_o... outer radius of the flywheel in m
r_i... inner radius of the flywheel in m

It must be noted that neither energy absorption through plastic deformation nor the creation of new surfaces through crack formation in the rotor fragments or dissipation through friction between the fragments is considered in this method.

Method 2
In the second method, an internal pressure on the protective ring-shaped containment is calculated from the energy stored in the rotor. Although it is assumed that this pressure is evenly distributed across the entire inner containment surface, the usually short duration of the impact is ignored. This significant deviation from reality, the assumption that the internal pressure is constant over time, results in extremely massive wall thicknesses being calculated. The theoretical internal pressure is calculated by

$$p = \frac{\rho * \omega^2}{3} * \left(\frac{r_a{}^3 - r_i{}^3}{r_a}\right) \tag{8.10}$$

p... theoretical internal pressure in Pa
and causes the following stresses in the cylindrical containment:

$$\sigma_t = \frac{R_{ci} * p}{R_{ca}{}^2 - R_{ci}{}^2} * \left(1 + \frac{R_{ca}{}^2}{R_{ci}{}^2}\right) \tag{8.11}$$

$$\sigma_r = \frac{R_{ci} * p}{R_{ca}{}^2 - R_{ci}{}^2} * \left(1 - \frac{R_{ca}{}^2}{R_{ci}{}^2}\right) \tag{8.12}$$

σ_t... tangential stresses in the burst protection housing in N/mm2
σ_r... radial stresses in the burst protection housing in N/mm2
R_{ci}... inner radius of the containment in m
R_{co}... outer radius of the containment in m

8.6.2 Calculation According to *Giancarlo Genta* [7]

In 1985, the Italian *Giancarlo Genta* wrote a book entitled *Kinetic Energy Storage*, which is probably the most cited publication within the flywheel community. This comprehensive work, which examines the history of flywheel energy storage up to modern developments such as *high-temperature superconducting bearings* in a detailed and scientific way, also presents calculation guidelines for the design of FESS bust containments. The calculation procedure is a two-stage process:

In the first part, the method assumes that the fragments strike the protective housing and slide along the inside of the containment wall until they finally come to a standstill, slowed down by the resulting friction forces. Further assumptions are an *inelastic impact*, *rigid body fragments*, and *no significant change in the geometry* of the burst protection housing (no severe deformation). It is assumed that the rotor breaks into three equally sized fragments:

$$p_c = \frac{m * v^2 * \cos(\varphi)^2}{2 * \pi * R_{ci} * (R_{ci} - dg) * h} \tag{8.13}$$

- p_c... centrifugal pressure acting during impact in Pa
- m... mass of fragment in kg
- v... velocity of the fragments after rotor burst in m/s
- φ... angle between the resulting velocity and the tangential velocity of a fragment in rad
- R_{ci}... inner radius of the containment in m
- d_g... distance of the center of gravity of a fragment from its outer radius in m
- h... height of the flywheel in m

In the second part, the energy absorption capacity of the cylindrical burst containment is calculated and compared with the total energy content of a fragment:

$$E - 2 * \pi * R_{cm} * h * t_c * \left(\sigma * \left(\epsilon - \frac{\sigma}{2 * E}\right) - p_c * \epsilon * \left(\frac{R_{cm}}{t_c}\right)\right) \tag{8.14}$$

- E... energy absorption capacity of the first burst containment ring in J
- R_{cm}... mean radius of the first burst containment ring in m
- t_c... wall thickness of the first burst containment ring in m
- σ... yield strength in N/mm^2
- ε... elongation at break in %.
- E... modulus of elasticity of the containment material in N/mm^2

8.6.3 Calculation According to *NASA* [5]

In 2000, the *National Aeronautics and Space Administration* (*NASA*) published a report entitled "High Energy Flywheel Containment Evaluation." In this report, calculation methods for the design of flywheel safety housings are included, although they are subject to considerable simplification. A fictitious worst-case scenario is assumed:

The rotor breaks into three equally sized parts. The total energy of the flywheel is converted in equal shares into purely translational energy of the fragments; thus, energy absorption by *crack growth*, *deformation*, *friction*, etc. is neglected. Only the impact of a fragment is examined, because it is assumed that all three impacts are completely identical. The maximum stress occurs in the contact area between the fragment and the protective housing. An important factor in this calculation is the *impact time*. In the available literature, this impact time is given at 75 μs [17]. Based on the impact forces, the internal pressure on the housing can be calculated as follows:

$$F = \frac{m * v}{t_i} \tag{8.15}$$

F... impact force of a fragment on the housing in N
m... mass of a fragment in kg
v... velocity of a fragment before impact in m/s
t_i... impact time in s

$$P = \frac{F}{A_i} \tag{8.16}$$

A_i... area of impact in m^2

$$\sigma = \frac{h^2 * \left(\frac{U}{3}\right)^3 * P}{2 * \left(h^2 + \left(\frac{U}{3}\right)^3\right) * d^2} \tag{8.17}$$

σ... maximum occurring stress in protective housing in N/mm^2
h... height of the flywheel in mm
U... circumference of the flywheel in mm
P... pressure exerted by a fragment on the protective housing in MPa
d... wall thickness of the protective housing in mm

8.7 Application of the Calculation Methods and Comparison of the Results

In order to demonstrate the divergence of the calculation methods, each of them was applied to dimension the housing of a spin bit at the *Energy Aware Systems* Group, Graz University of Technology (see Sect. 8.9.2). Basis for the calculations was a test flywheel with a specifically designed burst speed and fracture behavior ("tri-burst"). These parameters were adjusted by means of water jet cut notches (see Fig. 8.15, Table 8.2).

The energy of the three fragments is to be absorbed by a burst protection cylinder made of mild steel (material no. 1.0038 or *S235*) with an **inner diameter of 170 mm and a wall thickness of 5 mm**. Based on these input data, the stresses in the cylinder were determined using the calculation methods described in Sect. 8.6 and are shown in Table 8.3.

Fig. 8.15 First version of the test flywheel designed for the rotor burst test rig at Graz University of Technology

Table 8.2 Characteristic values of the originally designed test flywheel for burst tests

Test flywheel data	
Material	C45
Diameter	130 mm
Height	30 mm
Central bore	30 mm
Mass	3 kg
Burst speed	35,000 rpm
Fragment energy	3 × 14.8 kJ
Rim speed	238 m/s

Table 8.3 Preliminary design of the burst containment cylinder for TU Graz spin pit according to analytical methods [19]

	Result		
Calculation method	Stress	Energy consumption	Qualitative statement
Lockheed method 1	413 N/mm^2	–	Housing may withstand impact[a]
Lockheed method 2	2067 N/mm^2	–	Housing does not withstand impact
Genta centrifugal pressure	10 N/mm^2	–	Housing withstands impact
Genta energy intake	–	1265 J	Housing almost withstands impact
NASA	325 N/mm^2	–	Housing withstands impact

[a]The tensile strength of material S235 varies from 360 to 510 MPa, depending on actual steel quality

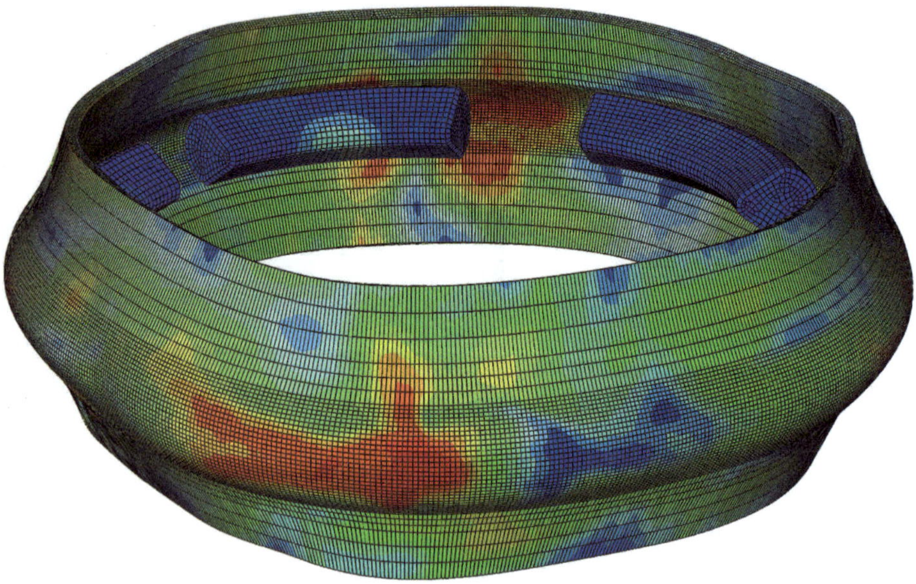

Fig. 8.16 Visualization of the deformations and von Mises stresses in a burst protection cylinder. The maximum stress is around 233 N/mm^2 [19]

It is quite easy to see that the results vary greatly. A comparison with a quick finite element calculation is deceptive, as linear-elastic behavior is usually assumed. However, this numerical method may still be of use, if considered as an intermediate step. Figure 8.16 shows the maximum deformation determined by the FEM calculation in the linear-elastic range, i.e., at local stress maxima below 300 N/mm^2. Based on this geometric deformation, the deformation energy was calculated, which is only a fraction of the total flywheel energy.

The reason for this significant discrepancy lies in the neglecting of local plasticizing of the protective ring in the rotor fragments.

8.7.1 Summary and Plea for Empirical Burst Containment Studies

The explanations in Sects. 8.3, 8.4, 8.5 and 8.6 have shown that there have been some empirical and theoretical investigations of burst containments for FESS in recent decades. The focus, however, has been on researching the failure mechanisms and fracture behavior of fiber composite rotors. In the process, some findings regarding particle kinematics have been obtained, but the resulting calculation methods for the dimensioning of the housing offer only inaccurate and extremely conservative approximations. Furthermore, the available information on protective housing design usually refers to the interaction of *a single specific rotor and housing design* and does not allow the derivation of generally applicable design guidelines! In summary, the following can be stated:

- Burst tests were carried out to investigate primarily the burst behavior of the rotors rather than the energy absorption capacity of the containment structures.
- The know-how about the fracture behavior of *steel* flywheels can essentially be summarized with the historical knowledge that isotropic rotors break into a few large fragments.
- When bursting fiber composite rotors, a homogeneous pressure load on the inner wall of the housing is usually assumed.
- The available analytical pre-design formulas usually consider fictitious "worst-case scenarios" and thus do not allow for housing optimization in the sense of lightweight design.
- General ideas for housing architectures are available in the technical literature, but quantifiable and comprehensible design specifications are lacking.
- Optimized lightweight FESS containments, in which innovative materials such as metal foams, honeycomb structures, or composite materials are used, are hardly known or not sufficiently described.
- Finite element simulations mainly provide satisfactory results if either homogeneous pressure loading is assumed or the geometry, position, and impact duration of the fragment are exactly known.
- Information on the behavior of burst containments in technical reports, such as [6], often refers to a very small number of tests and therefore lacks any statistical significance.
- So far, an unmistakable connection between fragment energy and energy absorption capacity of the housing, as well as fragment geometry and safety against containment penetration in the case of FESS steel rotors, has not been determined.

The arguments just presented and the fact that safety-relevant components such as the protective housing have to be tested by means of practical trials anyway speak in favor of an in-depth empirical investigation of the matter.

8.8 Qualitative Analysis and Overview of Previous Burst Tests

Quite often in literature, the burst behavior of fiber composite rotors is described as "good-natured," since breakage is usually caused by delamination of the tangentially wound fiber layers. This does not only cause detectable imbalance forces but also means the energy content of the individual fragments (particles) is lower than in the steel rotor. (This circumstance should not be confused with the fact that the kinetic energy of fiber composite rotors is generally much higher than that of steel rotors.) This means that the housing—ideally—experiences homogeneous pressure loading and is not stressed locally above the strength limit by sharp-edged, high-energy, and hard fragments [16].

Nevertheless, investigations have shown that under certain operating conditions or due to factors related to manufacturing processes, fiber composite rotors may also experience spontaneous failure [20]. Furthermore, the creation of fine composite material particles, such as carbon dust in case of rotor failure, which is dealt with in Sect. 7.2.1, is a problem that has not been completely solved to date.

A detailed survey of findings published in technical literature on the subject of housing design for flywheel energy storage and a subsequent qualitative analysis have led to the following conclusions:

- The number of published results of flywheel burst tests is not sufficient to perform a *statistically significant* qualitative analysis and to derive clear design guidelines (or at least recommendations).
- The so-called spin tests (or overspeed tests) were mostly carried out in over-dimensioned, bunker-like spin pits with extremely high safety factors. Investigations of explicit lightweight housings for mobile applications were realized only in [15], whereby the rotor of the FESS did not actually break during this crash test (compare Fig. 8.5).
- It was not until 2015 that a research project called *FlySafe* was approved by the *UK Research Council* and funded to the amount of £764,854 [22]. The project aimed to investigate potential failure mechanisms of modern high-speed composite flywheels [23]. The results, which were prepared in collaboration with *Ricardo Energy*, *PUNCH Flybrid*, *GKN Hybrid Power*, *Imperial College London*, and the *University of Brighton*, will probably not be published in great detail due to confidentiality agreements. Design guidelines based on these experiments are not yet available.

Table 8.4 Damage patterns of *steel flywheels* after overspeed tests

Organization	Year	Description	Speed	Spec. energy	Ref.
Lockheed Martin Missiles Company	1972	Cylindrical steel containment with 515 mm diameter and 7 mm wall thickness	22,820 RPM	640 Wh	[6]
Lockheed Martin Missiles Company	1972	Cylindrical steel containment with 12.7 mm wall thickness and *rotating liner* made of GFK (178 mm thick)	16,750 RPM	340 Wh	[6]
ETH Zurich	1996	Steel tube in concrete	n.a.	n. a.	[26]
IME, Graz University of Technology	2008	Rectangular steel housing with 10 mm wall thickness and plywood lining; flywheel made of ceramic composite	8000 RPM	2.5 Wh	–
Schenck Rotec/ Graz University of Technology	2014	Laminated steel rotor (see Sect. 7.5) and steel safety housing consisting of 2 concentric cylinders with 8 mm wall thickness each	45,000 RPM	280 Wh	–

Table 8.5 Damage patterns of *composite* flywheels after overspeed tests

Organization	Year	Description	Speed	Spec. energy	Ref.
Lockheed Martin Missiles Company	1972	Cylindrical steel containment with 515 mm diameter and 75 mm wall thickness	25,000 RPM	n.a.	[27]
Ricardo UK Ltd.	2016	Woven burst flywheel for repeatable burst tests	90,000 RPM	n.a.	[28]
ETH Zurich	1996	Wound rotor made of glass fiber rovings and epoxy matrix	n.a.	n.a.	[26]
Oak Ridge National Laboratory	1980	Fiber composite flywheel in steel housing	n.a.	n.a.	[24]
Center for Electromechanics, University of Texas		Carbon fiber flywheel in aramid fiber burst containment	35,200 RPM	280 Wh	[16]

Tables 8.4 and 8.5 provide an overview of the burst tests described in literature including relevant technical data. A distinction was made between isotropic (steel) and fiber composite flywheels. It must be noted that the number of published results on this topic is small and therefore does not allow a statistically significant statement.

Damage patterns of two burst tests, which were carried out by Graz University of Technology, are shown in Figs. 8.17 and 8.18.

Fig. 8.17 Burst test of the FIMD rotor. The remnants of the electrical sheets of the motor generator are clearly visible

Fig. 8.18 Burst test of a ceramic disk (C-Sic), performed by Assoc. Prof. Michael Bader of Graz University of Technology in 2008. On the right, the reassembled rotor fragments are shown. (Image rights: Michael Bader)

8.9 Empirical Investigation of FESS Burst Containments

The results of the survey of the state of the art in the field of burst housings (see Sect. 8.3) and the conflicts between the analytical design formulas described above illustrate the necessity of an empirical approach supported by methodical test bench trials. This necessity was already recognized in 2011 at Graz University of Technology, and the design of a spin pit specifically for systematic flywheel rotor burst testing was started in the course of academic theses. The

Fig. 8.19 Remains of a fiberglass flywheel that was subjected to a burst test at ETH Zürich in the 1990s. (Image rights: Peter von Burg, ETH Zürich)

aim was to investigate burst containments and to determine a correlation between kinetic energy of the fragments and energy absorption capacity of the housing. It can be seen from [17] that the burst containment cylinder, which encloses the circumference of a steel flywheel, absorbs almost all the energy in the event of a fracture. The cover plates of the casing therefore play a subordinate role in this specific case. This is particularly true in the case of isotropic (steel) rotors, since the tightness of the casing is also irrelevant because there is no risk of dust formation. (An example of the failure of a fiber composite rotor where dust formation occurred can be seen in Fig. 8.19.) The main task of the test rig, i.e., the long-term goal of the empirical containment investigations, can therefore be summarized as follows:

- An analytical relationship between fragment energy and energy absorption capacity of the housing ring is to be determined by means of systematic test series, which include a statistically significant number of burst tests. Influencing factors such as fragment kinematics and geometry as well as maximum (plastic) deformation are also to be recorded and analyzed.

8.9.1 Commercially Available Spin Pits and Testing Services

Some companies offer commercial spin pits (rotor burst test rigs) for purchase or as a service. Table 8.6 gives an overview of the most important players in this segment. Apart

Table 8.6 Commercially available *spin rigs* and *spin testing services*

Company/model	Portfolio	Specifications	Ref.
Shoemaker Engineering GmbH	Customized spin pits for purchase	(Specifications depend on customer requirements. Spin pits for automotive alternators have been realized)	[32]
Schenck ROTEC GmbH/"Centrio 100"	Spinning service and spin pits for purchase	Max. rotor diameter/length: 900/900 mm Max. rotor weight: 400 kg[a] Max. speed: 250,000 rpm[a]	[33]
Schenck ROTEC GmbH/"BI 1–7"	Spin pits for purchase	Max. rotor diameter: 200–2700 mm Max. rotor weight 10–6300 kg[a] Max. speed: 3000–250,000 rpm[a]	[33]
Test Devices Inc.	Spin pits for purchase	(No data available)	[34]
BSI-Barbour Stockwell incorporated Inc.	Centrifuging tests as a service	Max. rotor diameter/length: 2,000/1,500 mm Max. rotor weight 2.7–1800 kg Max. speed: 18,000–200,000 rpm	[35]
Aerovent	Spinning service	Max. rotor diameter: 1397/2997 mm	[36]
Oceanfront Engineers	Spinning service (fan wheel overspeed tests)	(No data available)	[37]
Lingling Balancing Machinery Co td/"OTS 10–1500"	Spinning service (compressor wheel tests)	Max. rotor diameter: 300–1,500 mm Max. rotor weight 10–1,500 kg[a] Max. speed: 9000–65,000 rpm	[38]
Piller TSC Blower Corp.	Spin pits for purchase	Max. rotor diameter/length: 1.066/889 mm Max. rotor weight: 454 kg Max. speed: 60,000 rpm	[39]
Element Materials Technology GmbH	Spinning service (overspeed tests)	Max. rotor diameter/length: 1,600/1,000 mm Max. rotor weight: 2500 kg Max. speed: 65,000 rpm	[40]

[a]Max. rotor weight and speed depend on the selected (modular) gearbox. At a maximum speed of 250,000 rpm, the maximum permissible rotor mass is reduced to 10 kg

from the companies listed in Table 8.6, some turbomachinery manufacturers (such as *MAN Turbo*, *Rolls-Royce*, *MTU*) and a few research institutes (such as *Ohio State University [29]*, *Naval Postgraduate School [30]*, or *NASA [31]*) have their own burst test rigs to perform R&D activities. However, despite the seemingly large offer of test sites and equipment, the available facilities do not serve the purpose as a *strategic development* tool for the following reasons:

Availability: Time slots for spin tests must be booked in advance. Availability depends on the current order situation and can therefore not be guaranteed.

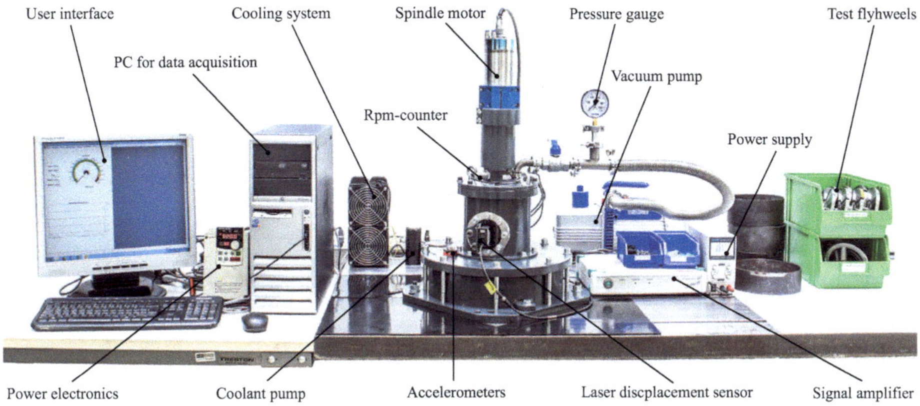

Fig. 8.20 Special test rig for the investigation of flywheel burst housings, *Graz University of Technology*

Costs: Depending on the rotor size/mass and the required measurement data, a single burst test costs between 1000 and 5000 Euro[2]. The purchase price of the Schenck-*ROTEC* model *Centrio 100 spin pit, for example*, is around 550,000 Euro.

Flexibility: In order to gain deep scientific insight in the complex rotor-housing system, a highly customizable test rig design regarding measurement equipment and vacuum feedthrough hardware is required.

Balance quality: Most commercially available spin pits require extremely accurate balancing of test rotors and do not allow burst testing of cheap rotors with low balancing quality.

Consequently, the facilities listed in Table 8.6 are usually only used for a single validation test and *not* as a strategic development tool during the entire FESS development process.

In order to investigate the suitability of burst housings especially for low-cost steel rotors and to derive design specifications or verify simulation models, the test rig shown in Fig. 8.20 was developed at *Graz University of Technology*. The process comprised several iteration stages and was supported by students' bachelor and master theses.

The aim of these empirical flywheel containment investigations is to determine an *analytical relationship between the kinetic energy of the rotor fragments and the energy absorption capacity* of a ductile housing structure. With the help of the resulting formulas and optimized simulation methods, it shall be possible to design flywheel housings safely and optimized regarding low weight and cost.

[2]Price depends on rotor size, material, burst speed, and required preconditioning (balancing, heating, etc.) and may well exceed 5000 € in some cases. Information is based on personal experience with European companies offering spin tests.

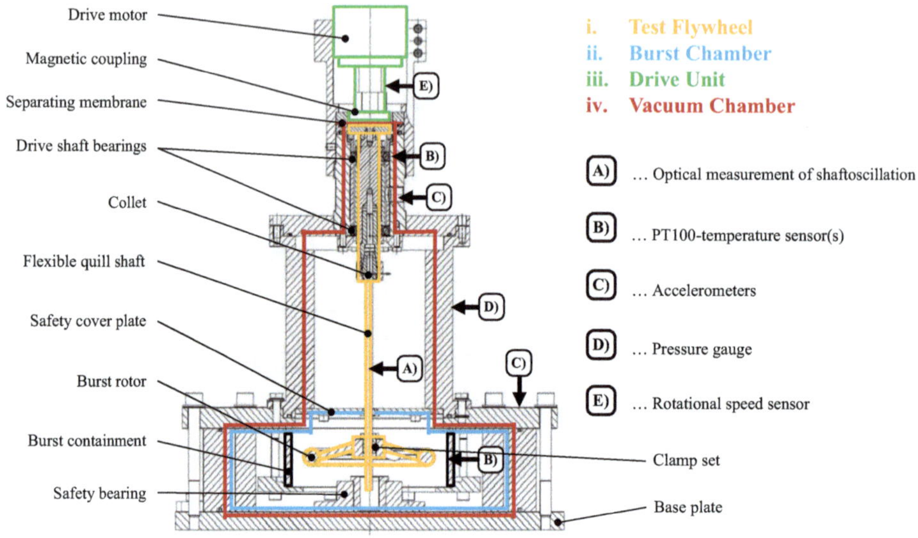

Fig. 8.21 Schematic structure of the burst test rig including measurement technology

8.9.2 Structure of the Burst Test Bench

The test stand was downscaled for cost and space reasons and can be divided into four main units as shown in Fig. 8.21:

- **Test flywheel (orange):** The test flywheel, which is to be destroyed in the course of the test, is mounted on a cantilevered, flexible shaft (quill shaft) by means of a clamping set. On the one hand, resonant frequencies and unbalance influences can be kept low, and on the other hand, the transmission of high force peaks to the spindle bearing arrangement during the bursting process can be avoided. The flexible, 6-mm-thick shaft is guided into the bursting chamber through a protective cover to prevent fragments from entering the bearings or hitting measuring equipment.
- **Burst chamber (blue):** The burst containment to be tested is located in this chamber. The outer walls of the burst chamber are made of 25-mm-thick solid steel. An emergency touchdown bearing is attached to the bottom of the burst chamber, which prevents the flexible shaft from excessive deflection and plastic deformation when passing through resonance.
- **Drive unit (green):** The spindle motor (asynchronous machine with frequency converter from *Mechatron*) delivers a maximum of 2.2 kW and 42,000 rpm. The torque is transmitted to the flexible shaft by means of a magnetic coupling, which makes it possible to hermetically separate the burst chamber from the environment.
- **Vacuum chamber (red):** Evacuating the burst chamber is important for two reasons: firstly, the aerodynamic losses would require high drive torques to reach rotor burst speeds, and secondly, the viscous damping and aerodynamic friction influence the

results of the burst tests. The pressure level can be reduced to around 0.5 mbar by means of a two-stage rotary vane pump.

It must be remembered at this point that the experimental test flywheels are only used as "projectiles" for the examination of the burst housings and represent sacrificial parts. In this case, the strength of the housing, not the rotor, will be examined.

Originally, the test flywheels were turned and milled parts, allowing the burst speed and the fragment geometry to be adjusted pretty precisely by means of adequate shaping. However, as the production of these relatively complex parts cannot be done in one step, high costs (~150 €/piece) had to be accepted. A favorable alternative is offered by suitable large series parts such as pulleys or handwheels made of gray cast iron. The brittle fracture behavior of gray cast iron must be seen as a further advantage, as ductile rotors can cause large imbalance forces through plasticizing, even before they burst. In some cases, this may lead to a detachment of the interference fit, which connects the rotor to the quill shaft before centrifugal force-induced breakage occurs.

8.9.3 Method and Experimental Procedure

The actual burst test, i.e., the rotational acceleration of the test flywheel until it bursts due to excessive centrifugal forces, takes only about 15 seconds for the test flywheel shown in Fig. 8.22, depending on whether there is contact of the shaft and the safety bearing during resonance pass-through and the moment of inertia of the flywheel. During the acceleration phase, speed, bearing acceleration, and shaft vibration are monitored and recorded. The evaluation of the test is much more time consuming and consists of the following steps:

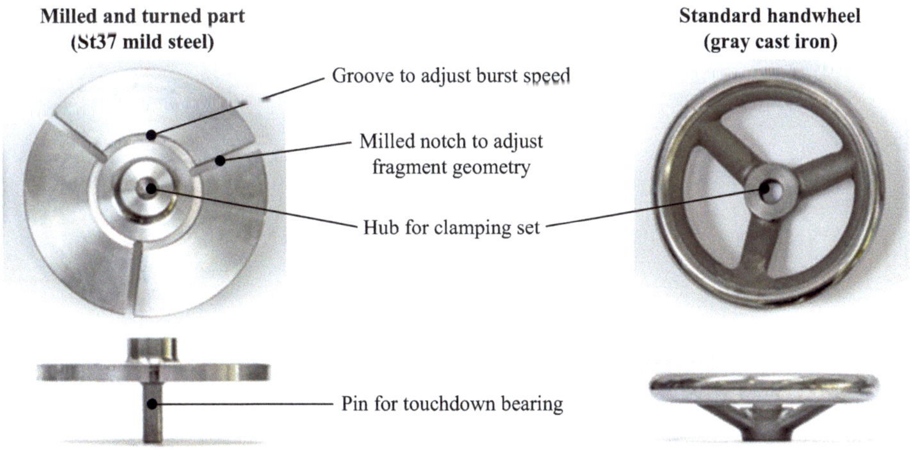

Fig. 8.22 Experimental flywheels with a diameter of 160 mm used for the burst housing tests

- **Documentation and reconstruction**

After the in situ photo documentation, the rotor fragments were reassembled, and the impact points on the burst housing were assigned to the individual fragments. The initial fracture can be determined by means of a precise analysis of the grinding marks on the rotor and housing and with the aid of color markings. In most cases, it can also be determined whether the rotor continued to fracture during impact or whether all cracks were caused by centrifugal forces (Fig. 8.23).

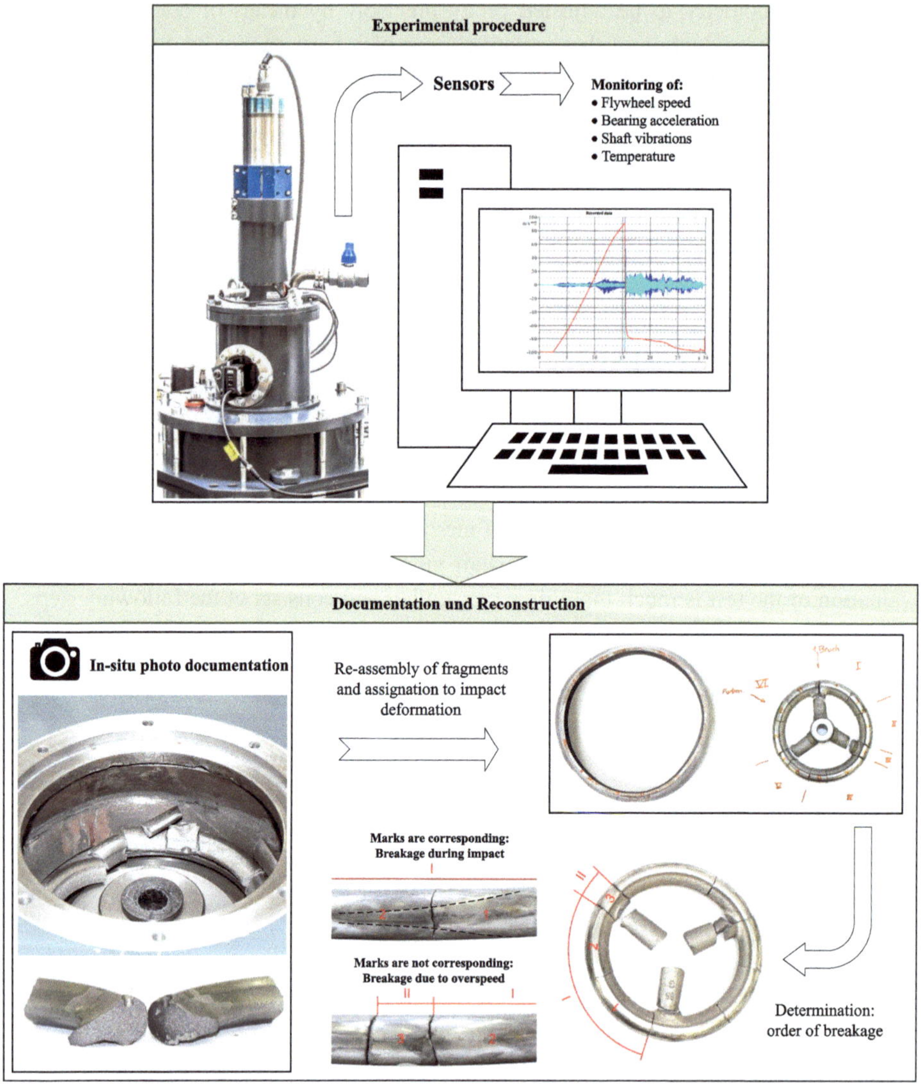

Fig. 8.23 Step 1 in the burst test: documentation and reconstruction

The plastic deformation energy introduced into the burst housing is a measure of the energy absorption capacity. If the containment cylinder has been penetrated, it is no longer possible to draw conclusions about the energy absorption, since part of the kinetic energy of the rotor fragments was destroyed by impact in the burst chamber. The plastic deformation energy is determined approximately by comparing the burst housing geometry *before* and *after* the impact. For this purpose, the circumference of the containment cylinder is measured at several heights and approximated by a fourth-order polynomial (Fig. 8.24).

In a further step, a more precise method was implemented by using a 3D scanner. The burst containment cylinders are scanned before and after the burst test, and a highly accurate 3D model is created. By comparing the two digital models, the surface strain can be determined numerically and thus the deformation work can be calculated (Fig. 8.25).

- **Calculation Evaluation**

The aim is to determine a *relationship between the kinetic energy of the rotor fragments and the energy absorption capacity of the burst housing*. In this context, it is essential to determine the rotational and translational components of the fragment kinematics (see Sect. 8.6, Fig. 8.26).

The enlargement of the housing surface due to plastic deformation is either approximated by summing up several truncated cone surfaces or determined numerically by means of 3D scanning. The strain (ε) can be determined by comparison with the original cylinder surface. If a stress-strain diagram of the containment material is available, the deformation energy can be determined as the area under the σ-ε-curve. However, it must be noted that stress-strain diagrams are normally determined by a quasi-static tensile test and dynamic effects of the ballistic impact are neglected (Fig. 8.27).

However, it must be noted that the dynamic effects of the *ballistic impact*,[3] which normally last only a few milliseconds, play an important role. An inelastic behavior of the containment material that depends on the strain rate is called *viscoplasticity* [41]. Neglecting these effects (e.g., *strain hardening* due to the high deformation rate [42]) in the mathematical evaluation can lead to high inaccuracies in the energy balance of the burst test. However, this error can be reduced by using a *Johnson-Cook material model* [43], but actual measured material data at these high deformation speeds is not available for any desired material.

8.9.4 Energy Balance

After the deformation energy of the burst housing has been determined, other terms of the energy balance must be examined. These are, for example, heating due to dissipation or crack formation in the flywheel before and during impact. The energy share of crack formation in the flywheel can be determined via a notch-impact test of samples made from the same material as the rotor.

[3]The test flywheel with a diameter of 140 mm has a rim speed of around 220 m/s at 30,000 rpm. Within only a few millimeters (or centimeters at the most), the rotor fragments are slowed down to zero at impact.

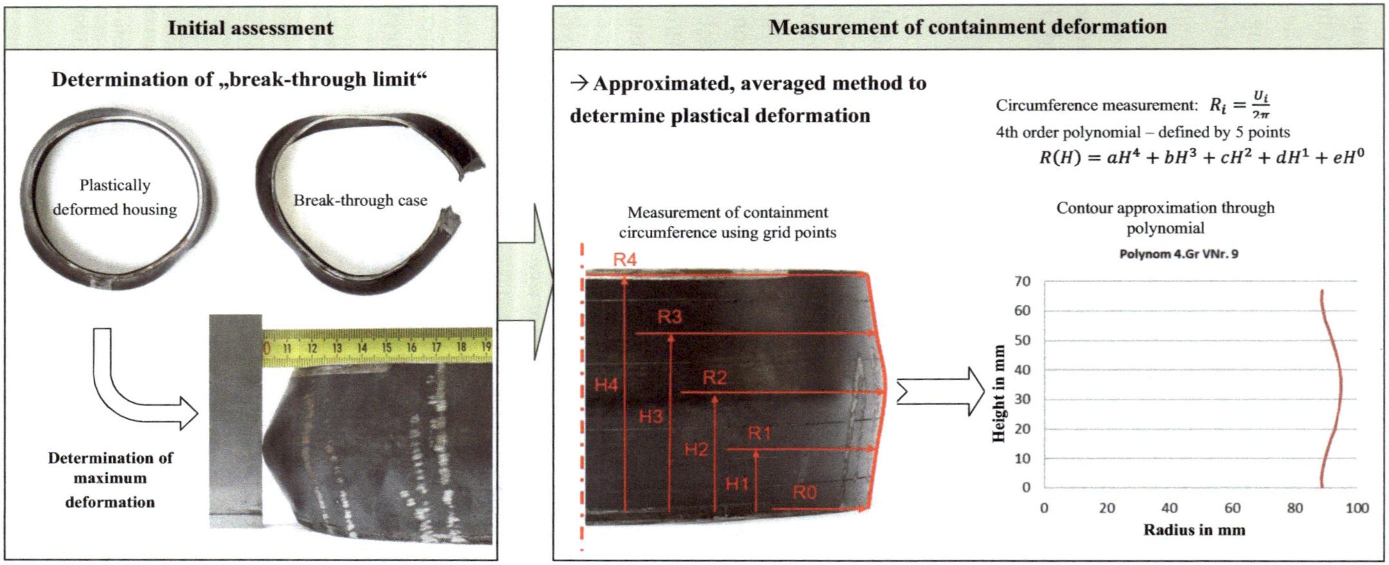

Fig. 8.24 Step 2 in the burst test: assessment and measurement of the housing deformation

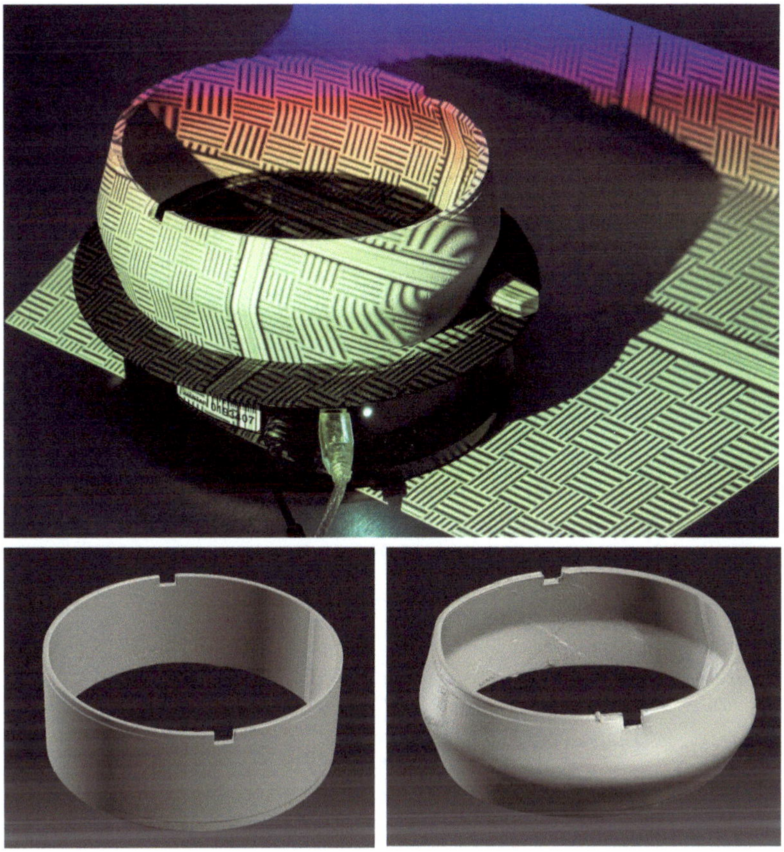

Fig. 8.25 3D scanning of the burst rings (top) and resulting CAD models before (bottom left) and after (bottom right) the burst test

Temperature entry in the housing on the other hand was measured directly by applying temperature sensors to the housing as shown in Fig. 8.28.

Figure 8.29 shows that a maximum temperature increase of around 10 °K is reached immediately after impact, and then the temperatures slowly drop again due to heat conduction. Since the actual spin pit is evacuated, there is no convective cooling of the burst containment, and radiation can be neglected at these low temperatures.

It can be concluded that the energy introduced into the containment in the form of heat (E_t) can be calculated as follows:

$$E_t = m_h * c_p * \Delta t \tag{8.18}$$

In this case, m_h is the mass of the cylindrical burst housing, c_p is the specific heat capacity of the housing material, and Δt is the maximum measured temperature difference. A more detailed description of the burst test rig and the development methodology was published in [43].

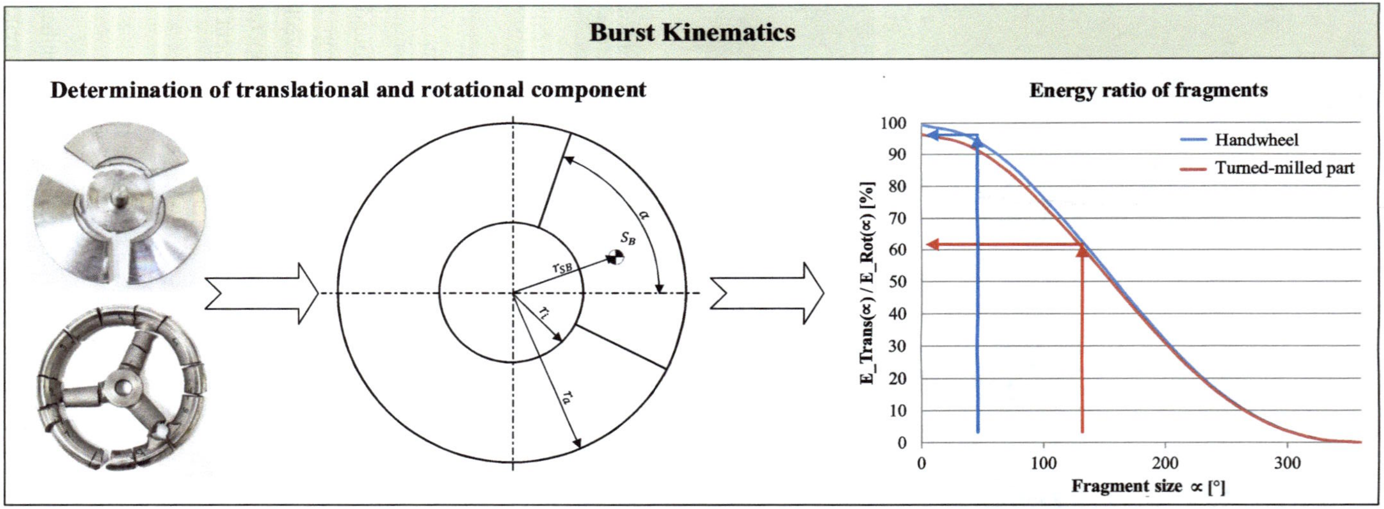

Fig. 8.26 Step 3a in the burst test: determining the fragment geometry and energy

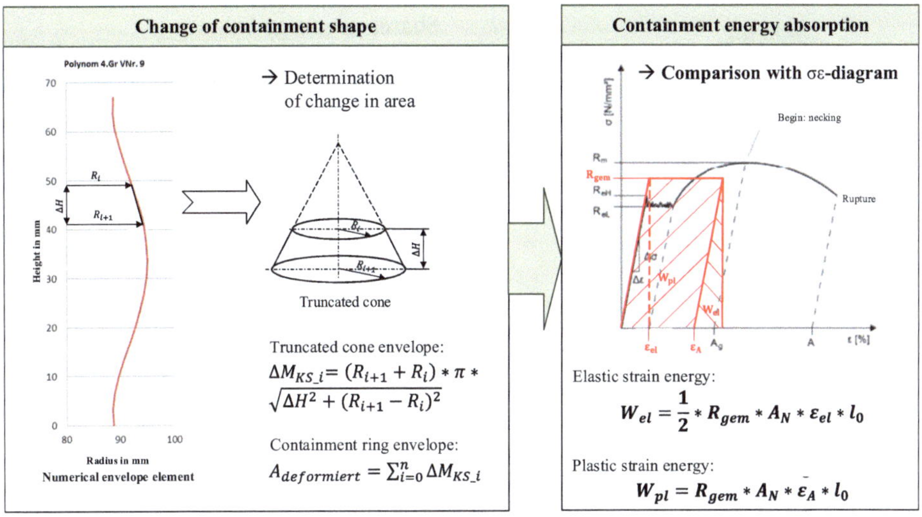

Fig. 8.27 Step 3b in the burst test: determining the deformation work dissipated in the housing using approximation methods

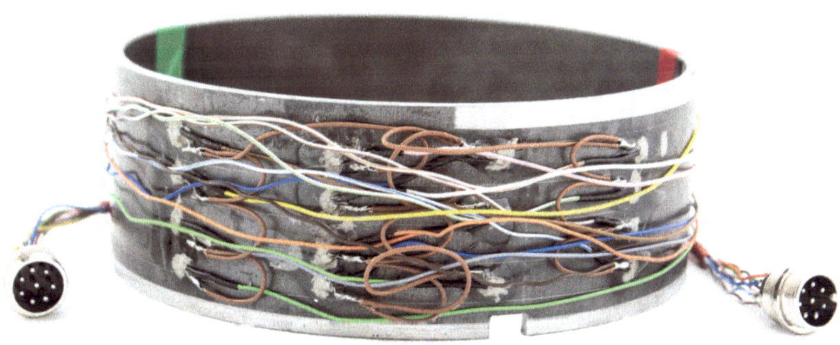

Fig. 8.28 Burst housing with 15 temperature sensors on the outer circumference

Fig. 8.29 Temperature rise of sensors 2.1–2.4 (middle row) in Fig. 8.28 immediately after the burst test

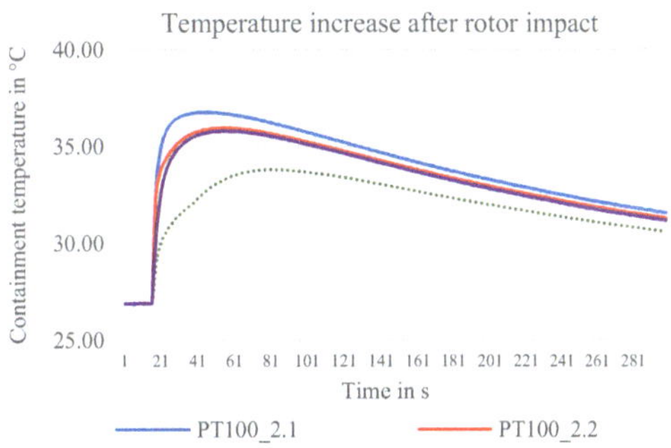

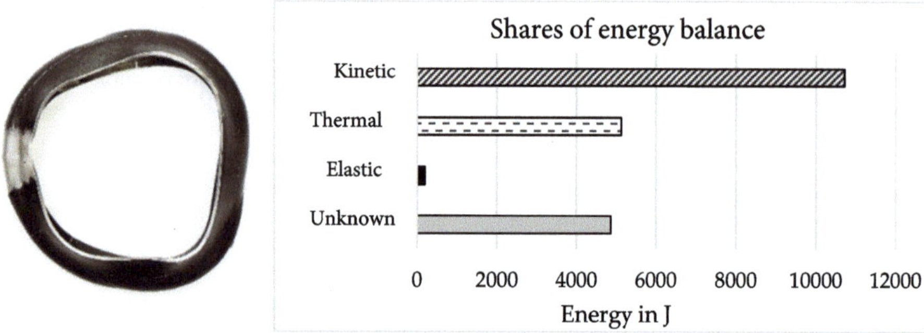

Fig. 8.30 Burst containment cylinder with applied temperature sensors and associated measurement diagram showing the temperature rise after the burst test

8.9.5 Summary of Previous Findings

Thanks to the low manufacturing and operating costs, the burst test rig described in Sect. 8.9.2 allows a statistically significant number of burst tests of flywheels and related burst containments to be carried out. It can therefore be used as a strategic development tool to study and validate the safety of flywheels in particular and of high-speed machines in general, with a high degree of reliability.

Figure 8.30 shows a burst containment after the test and a diagram of the associated energy balance, which quantifies the various energy shares. It should be noted that the *plastic deformation energy* consists largely of *thermal energy* and a presumably smaller residual amount of *elastic stress*. The relatively large unknown fraction can be explained by the following possible effects:

- Twisting/slippage of the cylindrical burst housing in the test bench
- Friction between rotor fragments and cover plates of the test rig or other components
- Elastic deformation of the containment, which causes the fragments to bounce off and ricochet
- Elastic deformation of the test rig
- Acoustic emission of the whole setup

Approaches to quantify these unknown energy shares are subject of current research.

Nevertheless, this simple experimental setup already allows a plausibility check of the energy balance of rotor fragments and energy absorption of the housing at good repeatability. This is particularly important for the validation of simulations of safety-critical components.

Table 8.7 summarizes some selected burst tests of containments with 3–6 mm wall thickness. The current energy balances show that about 20–30% of the translational rotor fragment energy is converted into deformation energy of the housing.

Table 8.7 Summary of some burst tests of 6–3 mm wall thickness

Wall thickness 6 mm		
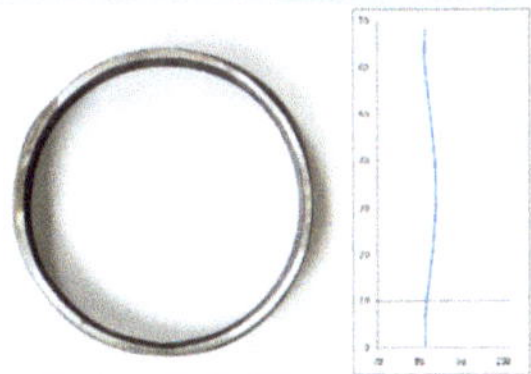	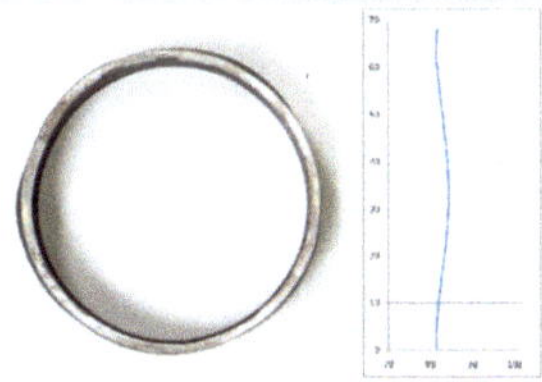	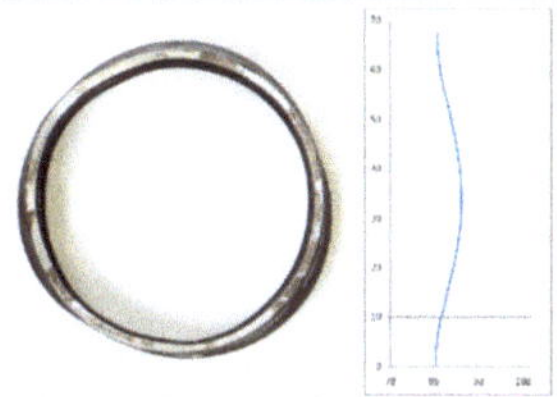
E_Flywheel_trans = 5839 J #4 bursting speed = 22,630 rpm W_deformation = 956 J 16.4%	E_Flywheel_trans = 6080 J #6 bursting speed = 24,460 rpm W_deformation = 1055 J 17.4%	E_Flywheel_trans = 7758 J #11 bursting speed = 28,638 rpm W_deformation = 1943 J 25.1%
Wall thickness 5 mm:		
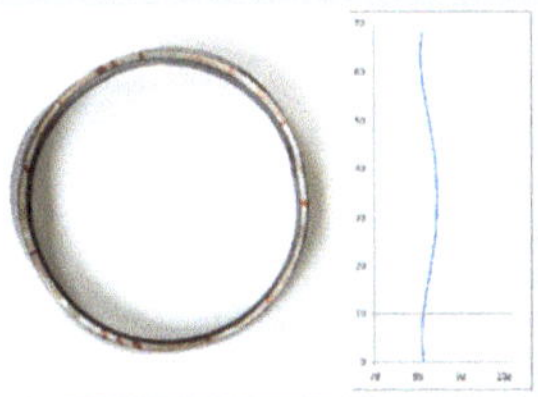	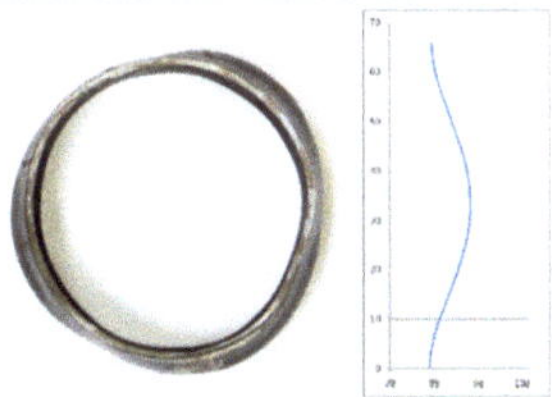	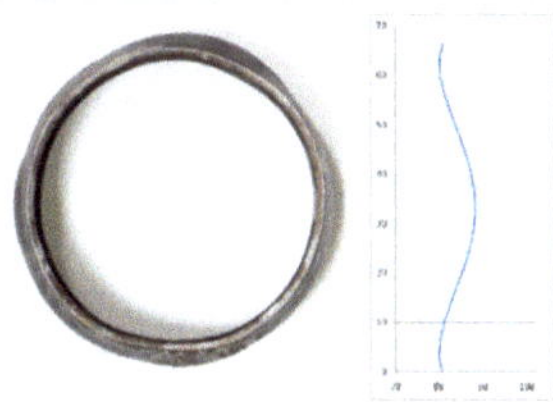
E_Flywheel_trans = 5446 J #3 bursting speed = 22,338 rpm W_deformation = 1012 J 18.6%	E_Flywheel_trans = 7824 J #12 bursting speed = 28,761 rpm W_deformation = 2308 J 29.5%	E_Flywheel_trans = 10,047 J #5 bursting speed = 29,431 rpm W_deformation = 2598 J 25.9%
Wall thickness 4 mm		
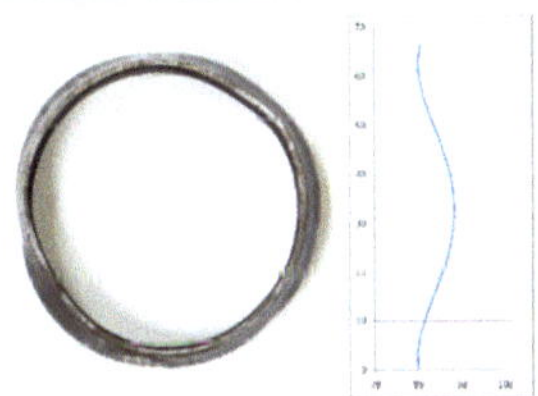	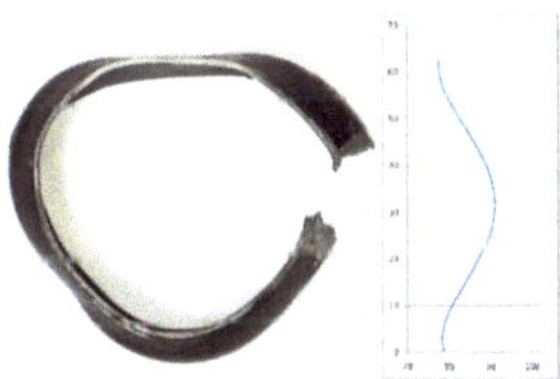	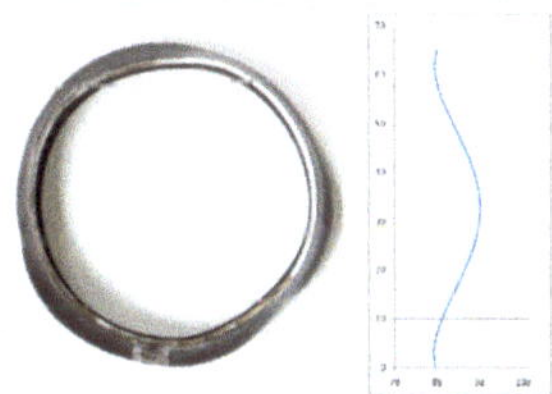
E_Flywheel_trans = 6946 J #7 bursting speed = 24,811 rpm W_deformation = 2069 J 29.8%	E_Flywheel_trans = 7763 J #13 bursting speed = 28,615 rpm W_deformation = 2432 J 31.3%	E_Flywheel_trans = 9799 J #8 bursting speed = 28,210 rpm W_deformation = 2699 J 27.5%
Wall thickness 3 mm:		

(continued)

Table 8.7 (continued)

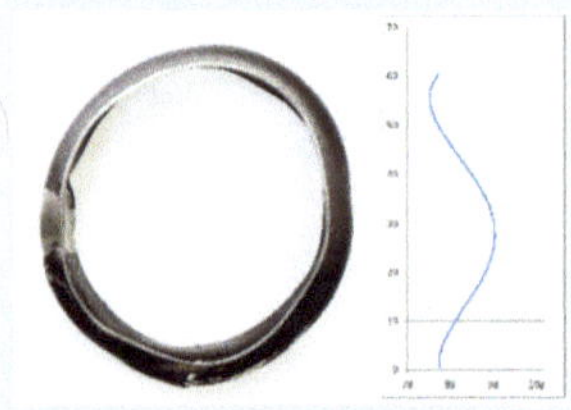	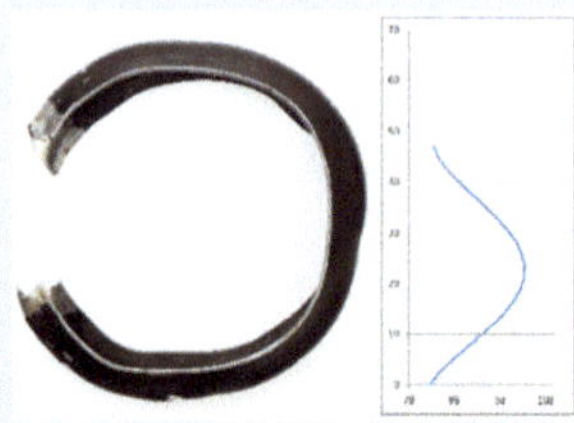	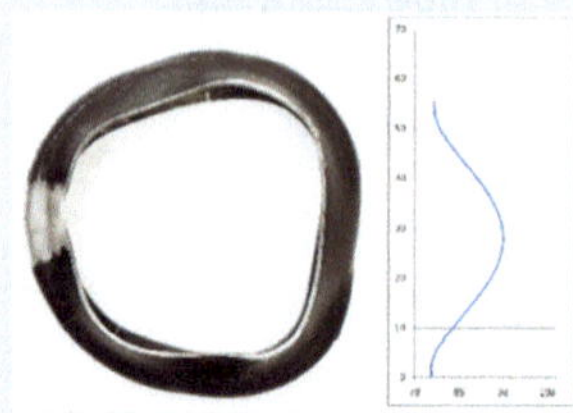
E_Flywheel_trans = 9104 J #14 bursting speed = 26,479 rpm	E_Flywheel_trans = 10,554 J #16 bursting speed = 28,644 rpm	E_Flywheel_trans = 11,944 J #15 bursting speed = 30,739 rpm
W_deformation = 2229 J 24.5%	W_deformation = 148 J 31.4%	W_deformation = 532 J 34.5%

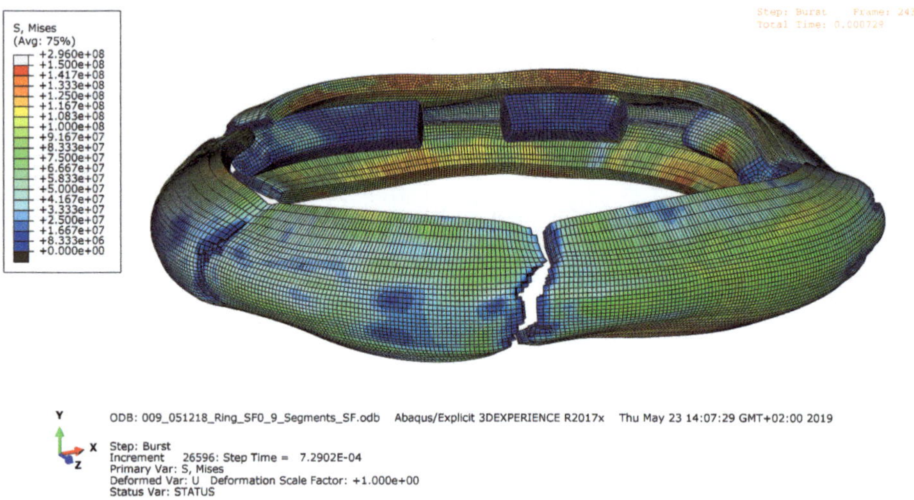

Fig. 8.31 ABAQUS simulation of a burst test with an aluminum housing (AlMgSi0.5) that could not withstand the impact forces

Simulation results of two burst tests conducted with the software tool ABAQUS are shown in Figs. 8.31 and 8.32.

Fig. 8.32 Simulation of a burst containment made of quenched and tempered steel 34CrNiMo6, also using ABAQUS and a Johnson-Cook material model. Despite strong plastic deformation, the housing withstands the impact loads

References

1. http://www.Wikinvest.com (2012) Safety failures by the company's flywheel products or those of its competitors could reduce market demand or acceptance for flywheel services or products in general. http://www.wikinvest.com/stock/Beacon_Power_%28BCON%29/Safety_Failures_Flywheel_Products_Those_Competitors_Reduce_Market. [Accessed May 14th 2016].
2. D. A. Flint (2011) Mishap at the Beacon Power Frequency Flywheel Plant. The Eastwick Press. Petersburg, NY, USA. https://eastwickpress.com/news/2011/07/a-mishap-at-the-beacon-power-frequency-flywheel-plant/. [Accessed May 19th 2017].
3. T. Mento and B. Ruth (2015) Injuries Reported in Explosion at Poway Business. KPBS Public Broadcasting, Issue Thursday 11th of June 2015.
4. A. Buchroithner and G. Jürgens (2017) Schwungradenergiespeicher – Eine Chance für die Automobilindustrie und darüber hinaus. VDI Fortschritt-Berichte 38. Internationales Wiener Motorensymposium, Reihe 12, Nr. 802, Bd. 2, Vienna, Austria, VDI Verlag, 2017, pp. 432–451.
5. A. J. Colozza (2000) High Energy Flywheel Containment Evaluation. NASA, Brook Park, Ohio, USA.
6. R. R. Gilbert, G. E. Heuer, E. H. Jacobsen, E. B. Kuhns, L. J. Lawson and W. T. Wada (1972) Flywheel Drive Systems Study – Final Report. Environmental Protection Agency, USA.
7. G. Genta (1985) Kinetic Energy Storage: Theory and Practice of Advanced Flywheel Systems, Butterworths, London, UK.
8. J. Hansen and D. O'Kain (2011) An Assessment of Flywheel High Power Energy Storage Technology for Hybrid Vehicles. Oak Ridge National Laboratory – Managed by UT-Battelle for the Department of Energy, Oak Ridge, TN 37831, USA.

9. Jerzy T. Sawicki, Xi Wu, George Y. Baaklini and Andrew L. Gyekenyesi (2003) Vibration-Based Crack Diagnosis in Rotating Shafts During Acceleration Through Resonance Proceedings of the 10th Annual International Symposium on Smart Structures and Materials.
10. G. Litak and J. T. Sawicki (2009) Intermittent behaviour of a Cracked Rotor in the resonance region. Chaos, Solitons & Fractals. Volume 42, Issue 3, November 15th 2009, pp. 1495–1501.
11. A. L. Gyekenyesi, J. T. Sawicki, R.E. Martin, W.C. Haase and G. Baaklini (2005) Vibration Based Crack Detection in a Rotating Disk – Part 2-Experimental Results. NASA/Glenn Research Center, USA.
12. D. Newland (1996) An Introduction to Random Vibrations, Spectral & Wavelet Analysis, 3. Issue 1996.
13. J. Kim et al (2003) Crack Detection in Operating Rotors using Direct Harmonic Wavelet Transformation. Mechanical Engineering Department, Imperial College of Science, London, UK.
14. H. Steffan (2010) Skriptum Vehicle Safety II, Graz: VSI – Vehicle Safety Institute, Graz University of Technology, Austria.
15. Green Car Congress (2007) Flybrid Flywheel Hybrid System Passes First Crash Test. 2007. http://www.greencarcongress.com/2007/10/flybrid-flywhee.html. [Accessed September 19th 2010].
16. M. Strasik (2003) Flywheel Electricity Systems with Superconducting Bearings for Utility Applications. Boeing Technology/Phantom Works, Seattle, USA.
17. M. A. Pichot, J. M. Kramer, R. C. Thompson, R. J. Hayes and J. H. Beno (1997) The Flywheel Battery Containment Problem. 1997 SAE International Congress and Exposition, Detroit, Michigan, USA.
18. M. Strasik and I. Gyuk (2002) Flywheel Electricity System – Project Fact Sheet: Superconductivity for Electric Systems Program and Industry Partners. United States Department of Energy, USA.
19. R. Rieger (2014) Erstellung eines Prüfstandes für Sicherheitstests an Berstschutzgehäusen schnell drehender, isotroper Schwungscheiben und Definition des Prüfumfangs. Institut für Maschinenelemente und Entwicklungsmethodik, Graz University of Technology, Austria.
20. G. Portnov, A.-N. Uthe, I. Cruz, R. P. Fiffe and F. Arias (2005) Design of Steel-Composite Multirim Cylindrical Flywheels Manufactured by Winding with High Tensioning and in situ Curing – 2. Numerical Analysis. Mechanics of Composite Materials, pp. 241–254, Issue May 2005.
21. Green Car Congress (2007) Flybrid Flywheel Hybrid System Passes First Crash. http://www.greencarcongress.com/2007/10/flybrid-flywhee.html. [Accessed September 19th 2010].
22. Research Councils UK/Innovate UK (2014) FlySafe – Flywheel-hybrid safety engineering. UK Research and Innovation, Polaris House, Swindon, UK. https://gtr.ukri.org/projects?ref=101306.
23. R. Nicolas (2015) A new flywheel test rig developed by Ricardo. http://www.car-engineer.com/a-new-flywheel-test-rig-developed-by-ricardo/. [Accessed July 22nd 2016].
24. Oak Ridge National Laboratory (2013) Oak Ridge National Laboratory Review. Oak Ridge National Laboratory's Communications and Community Outreach. http://web.ornl.gov/info/ornlreview/rev25-34/chapter8.shtml. [Accessed August 25th 2016].
25. P. von Burg (1996) Schnelldrehendes Schwungrad aus Faserkunststoff. ETH Zürich, Switzerland.
26. S. Renner-Smith (1980) Battery-saving flywheel gives electric car freeway zip. Popular Science, Issue October 1980.
27. J. Wheals, J. Taylor and W. Lanoe (2016) Rail Hybrid using Flywheel. Den Danske Banekonference, Tivoli Congress Centre, Kopenhagen, Denmark.
28. A. C. Hagg and G. O. Sankey (1974) The Containment of Disk Burst Fragments by Cylindrical Shells. Journal of Engineering for Power, no. 96, pp. 114–123.

29. Naval Postgraduate School (1999) Mechanical and Aerospace Engineering – Turbopropulsion Laboratory and Gas Dynamics Laboratory. https://my.nps.edu/web/mae/turbo. [Accessed Oktober 24th 2018].
30. NASA Glenn Research Center (2018) Special Projects Laboratory – Facility Description. National Aeronautics and Space Administration. https://www1.grc.nasa.gov/historic-facilities/special-projects-laboratory/facility-description/. [Accessed August 31st 2018].
31. Schuster-Engineering GmbH (2009) Schuster-Engineering GmbH, Am Söterberg 2, D-66620 Nonnweiler/Otzenhausen. http://schuster-sondermaschinen.de/6.html. [Accessed October 25th 2018].
32. Schenck RoTec (2005) Spinning Service. SCHENCK RoTec GmbH, Landwehrstraße 55, D-64293 Darmstadt. https://schenck-rotec.com/services/balancing-and-spinning-service/spinning-service.html. [Accessed October 25th 2018].
33. Test Devices Inc. (2015) Test Devices Inc., 571 Main Street, Hudson, MA, USA. https://www.testdevices.com/equipment/productionproof-burst-rigs/. Accessed October 25th 2018].
34. BSI – Barbour Stockwell Incorporated (2018) Barbour Stockwell, Inc. 45 Sixth Road, Woburn, Massachusetts 01801, USA. http://www.barbourstockwell.com/spin-test-services.html. [Accessed October 25th 2018].
35. Aerovent Industrial Ventilation Systems (2018) Aerovent, 5959 Trenton Lane North, Minneapolis, Minnesota 55442-3237, USA. https://www.aerovent.com/fan-testing-services/overspeed-testing/. [Accessed October 25th 2018].
36. Oceanfront Engineering (2018) Oceanfront Engineering, Fallbrook, CA 92028, USA. http://oceanfrontengineering.com/gallery/rotating-equipment/. [Accessed October 25th 2018].
37. Shanghai Lingling Balancing Machinery Co. (2015) Shanghai Lingling Balancing Machinery Co. Ltd., Build 2, Yong Lian Industrial Park, No168 HuaJi Road201108 Shanghai, China. http://www.linglingbalance.com/english-web/cp/cssyjj.html#OTS-300%E3%80%80%3E%3E. [Accessed October 25th 2018].
38. Piller TSC Blower Corporation (2015) Environmental XPRT. https://www.environmental-expert.com/services/overspeed-spin-test-214127. [Accessed October 25th 2018].
39. Element Materials Technology (2018). https://www.element.com/aerospace/aero-structures. [Accessed October 25th 2018].
40. P. Perzyna (1966) Fundamental Problems in Viscoplasticity. Advances in Applied Mechanics, Volume 9, p. 244–368.
41. J. F. Young, S. Mindess, A. Bentur and R. J. Gray (1998) The Science and Technology of Civil Engineering Materials, 1st Edition. Prentice Hall.
42. G.R. Johnson and W.H Cook (1983) A Constitutive Model and Data for Metals Subjected to Large Strains, High Strain Rates, and High Temperature. Proceedings of the 7th International Symposium on Ballistics, pp. 541-547, 19–21 April 1983.
43. A. Buchroithner, P. Haidl, C. Birgel, T. Zarl and H. Wegleiter (2018) Design and Experimental Evaluation of a Low-Cost Test Rig for Flywheel Energy Storage Burst Containment Investigation. Appl. Sci. 2018, 8, 2622. https://doi.org/https://doi.org/10.3390/app8122622.

9 Bearings for Flywheel Energy Storage

9.1 Analysis of Existing Systems and State of the Art

In the field of flywheel energy storage systems, only two bearing concepts have been established to date:

1. **Rolling bearings**, spindle bearings of the "High Precision Series" are usually used here.
2. **Active magnetic bearings**, usually so-called HTS (high-temperature superconducting) magnetic bearings.

A typical structure consisting of rolling bearings and an axial magnetic bearing (so-called lifting magnet for rotor weight compensation) is shown in Fig. 9.1. Alternative concepts such as friction bearings or aerostatic bearings are not used because of the requirements mentioned in Sect. 9.2. One of the few exceptions is the flywheel designed by *Kinetic Traction Systems*, which uses a *hydrodynamic pin* bearing as axial bearing.

Many of the stationary flywheel energy storage systems use active magnetic bearings, not only because of the low torque loss, but primarily because the system is wear- and maintenance-free, a characteristic that plays a central role, especially in continuous operation. Nevertheless, rotors with magnetic bearings require an additional pair of conventional bearings that act as *safety or touchdown bearings*. Three scenarios can make intervention of these touchdown bearings necessary:

(a) An intended shutdown of the system, powering down the magnetic bearings
(b) Failure of the power supply to the magnetic bearings or malfunction of the control system
(c) Exceeding the maximum permissible magnetic bearing load

A. Buchroithner, *Flywheel Energy Storage*,
https://doi.org/10.1007/978-3-658-35342-1_9

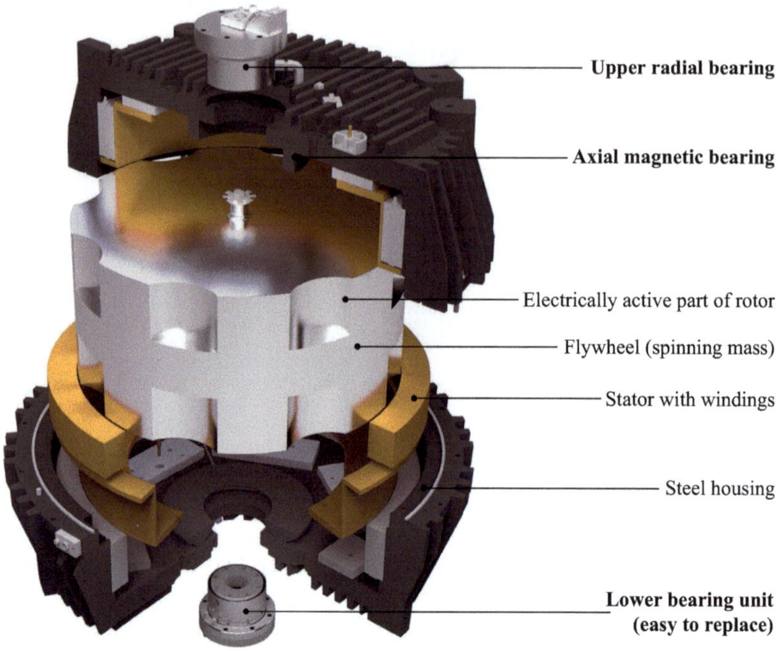

Fig. 9.1 General architecture and bearing system of a stationary flywheel energy storage unit (*Active Power HD625 UPS*). (Image rights: Piller Group GmbH)

In case (b) and (c), it may occur that the rotor hits the touchdown bearings at maximum rotational speed, which can lead to *forward* and *backward whirl* (a high-energy whirling motion of the rotor in the safety bearing) [1]. Due to the high energy stored in the FESS, the rotor and the touchdown bearing can suffer considerable damage in this case. In order to avoid this scenario, magnetic bearings must be used that can manage all expected bearing loads. However, this would lead to an increase in installation space and costs, which is especially critical in mobile applications.

On the basis of these arguments, only rolling bearings for FESS applications (partly in combination with permanent magnetic axial bearings, so-called lifting magnets) will be discussed in more detail below.

9.2 Requirements Derived from the *Supersystem Analysis*

Hardly any other machine element combines so many functions and influences the properties of the overall flywheel energy storage system as profoundly as the bearing. The following three properties must be particularly emphasized at this point:

1. **Service life:** The theoretical advantage that FESS can in principle achieve considerably higher numbers of charge/discharge cycles than chemical batteries highly depends on the choice and design of the bearing system.
2. **Friction:** "Achilles' heel" of FESS, high self-discharge, is primarily caused by friction losses in the bearings.
3. **Cost:** In order to significantly improve the two abovementioned properties (cycle life and self-discharge), active magnetic bearings are, at first glance, the obvious choice. However, even magnetic bearings are by no means loss-free and involve considerable design and financial effort, which must be regarded as a critical obstacle to the economic success of FESS in series production.

Naturally, there are several other aspects that must be considered when designing the bearing system for FESS application. The eight most important ones are shown as pictograms in Fig. 9.2.

9.2.1 Determination of Bearing Loads

Irrespective of whether magnetic or rolling bearings are used, the calculation of the bearing loads in flywheel energy storage systems represents a particular challenge. It is obvious that rolling bearing design always involves a thorough analysis or at least consideration of the *supersystem*, since the speed collective and external loads have a considerable influence on the design calculations and subsequently on the bearings' service life.

However, the rotational speed collective (proportional to the *state of charge* of the flywheel) depends on the energetic operating strategy or the use of the vehicle and the driver's behavior. The predictability of the load collective for passenger cars is a difficult task and can be determined much better for commercial vehicles.

The forces acting on the bearing can be divided into two groups:

1. **Forces caused by vehicle driving dynamics (*supersystem*)**
   (a) Linear acceleration through speeding up and decelerating the vehicle as well as cornering
   (b) Angular accelerations due to pitching, yawing, rolling
   (c) Gyroscopic reactions by pitching, yawing, and rolling
2. **Forces emanating from the flywheel (*subsystem*)**
   (a) Imbalance forces due to manufacturing tolerances (see also Table 9.9)

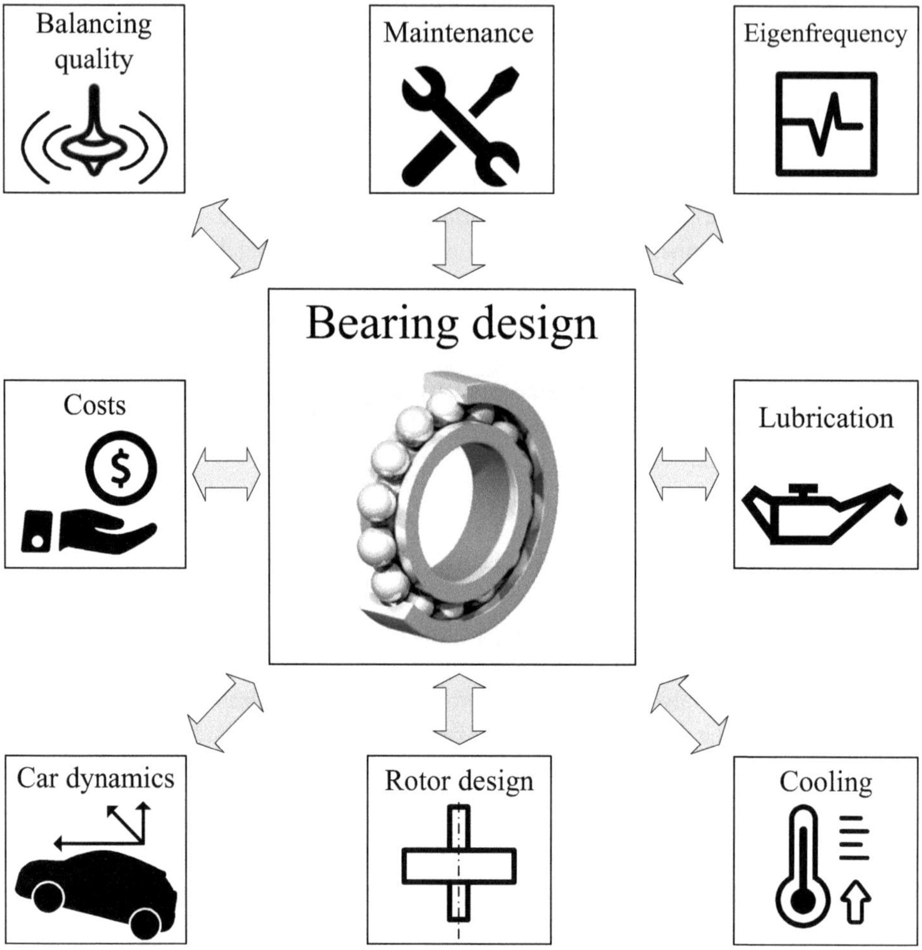

Fig. 9.2 The eight paradigms of bearing design for flywheel energy storage systems

The quantification or sorting according to the order of magnitude of these forces cannot be formulated in general terms. This is particularly due to the fact that the achievable balance quality depends on a large number of design and manufacturing parameters of the rotor (compare Chap. 7). However, for a representative case (*CMO rotor* with 11 kg mass, balance quality class G = 2.5, and max. 60,000 rpm), a sorting was made, which is shown in Fig. 9.3. The driving dynamics were determined by real-world measurements in the urban and rural area around the university city of Graz in Austria. More detailed information on this topic is given in Sect. 9.5.2.1.

The reason for the apparent dominance of the imbalance forces is the fact that these forces have the largest time component during operation of the FESS; they occur continuously as soon as the flywheel rotates, while all other load components only occur when the vehicle

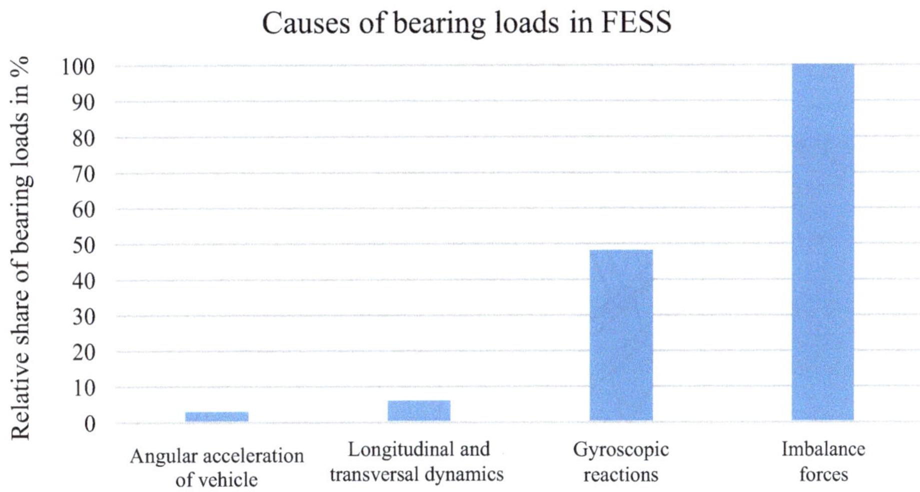

Fig. 9.3 Representative, relative shares of the bearing loads of a FESS in a vehicle

experiences significant deflection. However, the highest *absolute* bearing loads are caused by gyroscopic reactions or the dynamics of the vehicle. According to the current theories of fatigue strength calculation, these short-term loads can cause pre-damage that significantly reduces the bearings' service life. For this reason, a preliminary estimate of the gyroscopic bearing loads is essential and is explained in more detail in the following section.

It can be summarized that the following load types define the bearing life of flywheel energy storage units and must therefore be kept as low as possible:

1. Gyroscopic reaction forces.
2. Imbalance forces.

A detailed consideration of these two types of bearing loads follows in Sects. 9.3 and 9.5.

9.3 Gyroscopic Reaction Forces in Flywheel Energy Storage

9.3.1 The *Supersystem* of FESS Bearings: Analysis of Environmental Parameters

In order to design a bearing concept that meets all the requirements determined so far, a detailed requirement profile must be defined on the basis of the *supersystem analysis*. However, not only the energetic properties of the FESS, which define the rotational speed spectrum and its cumulative frequency, but also gyroscopic reactions that occur due to the

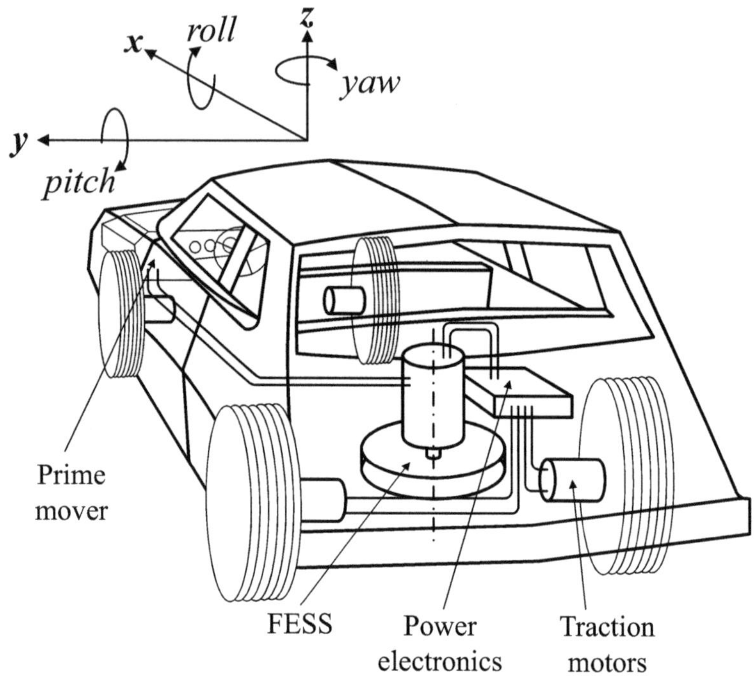

Fig. 9.4 Coordinate system and directions of movement of a vehicle with flywheel energy storage

vehicle's driving dynamics must be taken into account (see Sect. 9.5.2.1). Figure 9.4 shows a hybrid vehicle with flywheel energy storage and its degrees of freedom of movement.

9.3.2 Influence of FESS-Specific Operating Conditions on Bearing Design

The special operating conditions of a mobile flywheel energy storage unit have a strong effect on the design of the rolling element bearings. Some of the specific characteristics and their implications for the bearing technology are:

- **Vacuum:** The bearings must run in a hermetically sealed vacuum chamber, which affects the heat transfer and the lubrication system.
- **Vehicle dynamics:** The whole FESS is subject to extensive vehicular motion excerpting direct forces on the flywheel and resulting in gyroscopic reactions.
- **Extremely high angular velocities:** As shown in Chap. 7, the only way to achieve high specific energies is to increase the rotational speed. (Speeds of 20,000–80,000 rpm or rim speeds beyond 500 m/s are common in FESS.)

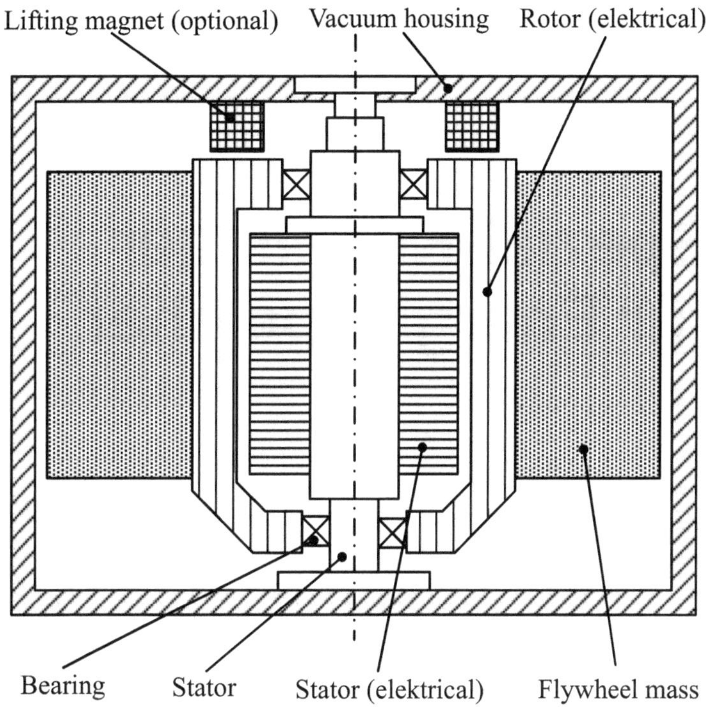

Fig. 9.5 Design of a FESS for a commercial vehicle: outrunner rotor, fully integrated design

- **Restrictions with regard to space and maximum weight:** As lightweight design is one of the most promising measures to reduce fuel consumption, the weight of the FESS must be kept as low as possible.

Figure 9.5 shows the schematic of a possible design of a vehicular FESS using an outrunner architecture, which was proposed at the beginning of a research project at Graz University of Technology. The purpose of the optional lifting magnet is to reduce gravitational (axial) bearing loads.

During a more detailed investigation of the proposed topology, which also included a simulation of the thermal behavior, two major shortcomings affecting bearing design were identified:

1. Maintaining a high balancing quality of a large carbon fiber (cf) rotor, subject to matrix aging [3],], is crucial in order to keep mass imbalance forces low (see also Sect. 7.2.1). The imbalance forces occurring must be taken into account when designing the bearing.
2. The carbon fiber bandages behave like a thermal insulator (Sect. 7.2.1) which inhibits the conductance and radiation of waste heat from the rotor of the electric machine. Therefore, thermal conductivity of heat across the ball bearings is essential and results in demanding bearing design requiring detailed analysis [4].

9.4 Complexity and Importance of FESS Bearing Design

As Fig. 9.6 shows, even small changes in the load collective can have a considerable effect on the rolling bearing's service life. The plot shows an estimate of the bearing life expansion factor over the partial load compensation caused by an assisting parallel active magnetic bearing (AMB) system. The theoretical advantages of this arrangement (shown in Fig. 9.7) have already been published in [4], but high costs have hindered introduction into the market so far.

The bearing life estimation is based on a measured load spectrum during the standard driving operation (see Sect. 9.5.2) and does not include misuse or destructive scenarios since their likelihood is hard to estimate. Figure 9.6 also clearly shows that a longer wheelbase helps to mitigate bearing loads. The longer wheelbase implies slower pitching and also a less sporty driving. Therefore, shorter and lighter vehicles would require stronger additional active magnetic suspension than longer and heavier vehicles. This means that accurate determination of the actual loads and detailed bearing life calculation is necessary to ensure reliability and long service intervals of FESS. Bearing life was calculated based on *FAG's modified bearing lifetime calculation*, which is a standard procedure in the industry.

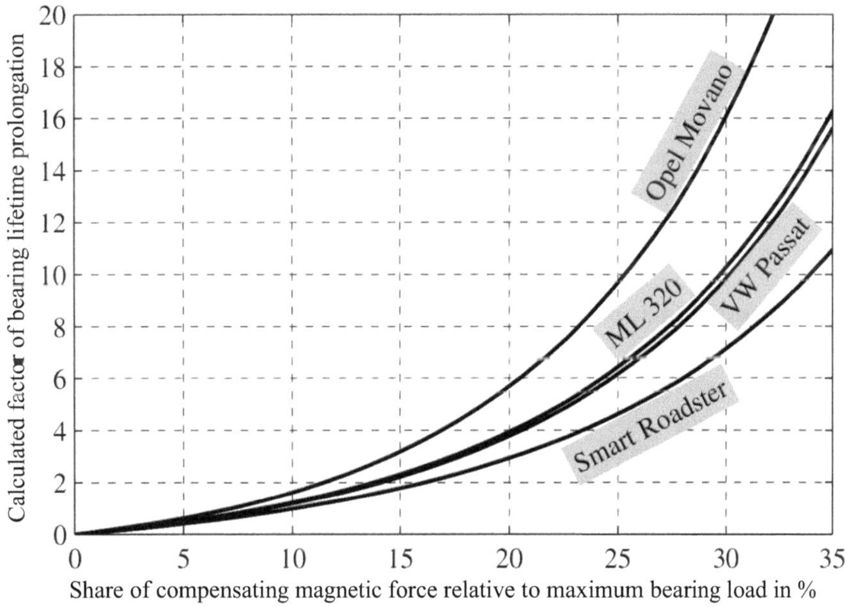

Fig. 9.6 Increasing the service life of a rolling bearing by reducing loads using a parallel magnetic bearing [4]. (Image right: Manes Recheis)

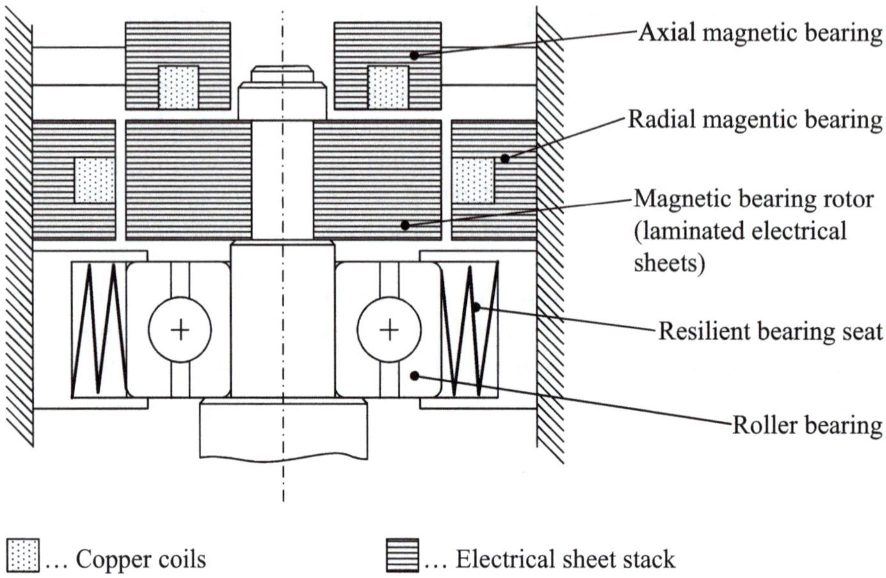

Fig. 9.7 Cartoon image of the combination of rolling element and active magnetic bearing

The longevity of the bearings or the achievement of extended service intervals plays a decisive role, especially in the important segment of commercial vehicles. High customer satisfaction through high reliability of the FESS is the key to market success and requires a thorough analysis of the highly nonintuitive gyroscopic reactions of the rotating flywheel in a moving system.

9.5 Determination of Gyroscopic Bearing Loads

The main bearing loads in an automotive flywheel energy storage system are the *gyroscopic reaction forces*, the *mass forces due to linear or angular acceleration*, and the *imbalance forces* of the rotor. Although in some cases the latter can greatly reduce bearing life, they will not be discussed in this section, as rotor design and balancing options are discussed separately in Sect. 9.6.

9.5.1 Step 1: Analytical Estimation

As mentioned before, the kinematics of a gyroscope and its reactions are complex and nonintuitive. Thorough understanding of the theory behind gyroscopic motion is necessary in order to design a proper FESS-to-vehicle mount and choose the correct bearing configuration. Solving the classic *Euler equations of gyroscopic motion*, which describe the

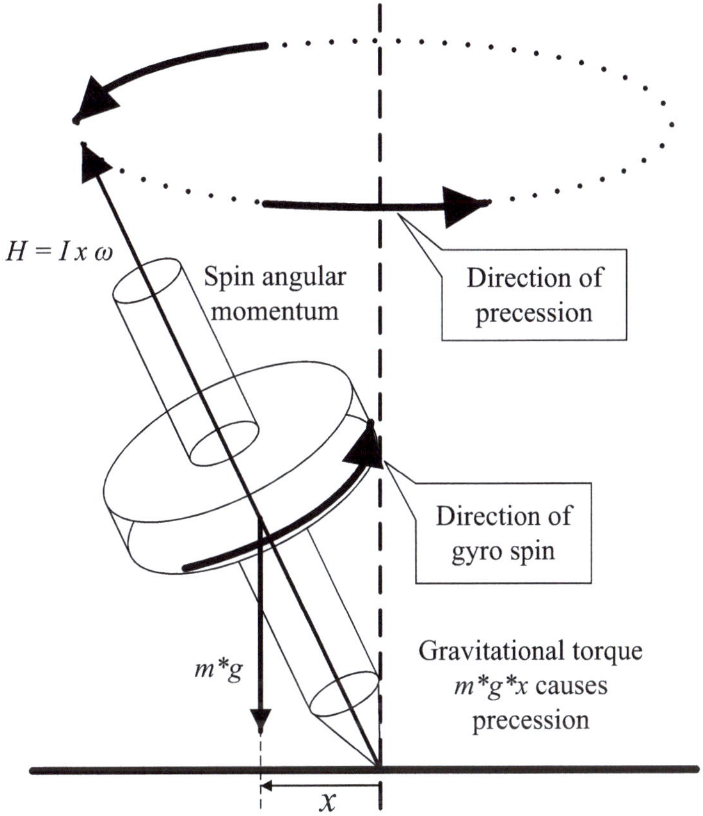

Fig. 9.8 Direction of angular momentum, precession, and nutation shown on a free gyroscope

Table 9.1 Vehicle and gyroscope motion variables according to Fig. 9.4

Denomination of angle	Description
α	Rotation around the X-axis (rolling)
β	Rotation around the Y-axis (pitching)
γ	Rotation around the Z-axis (yawing)
γ_k	Angle of FESS' own rotation axis

behavior of a free gyroscope as shown in Fig. 9.8 reveals the influencing parameters and offers a good first estimation of bearing loads due to gyroscopic reaction forces. The *Euler equations* and physics of a gyroscope in general are described in more detail in [5], whereas the following section summarizes the most important steps and findings.

Table 9.1 describes the relevant axes of rotation, based on the definition in Fig. 9.4.

Even though the rotating flywheel wants to keep its spin axes stable, the orientation of the spin axis changes in response to external torque. Consequently, gyroscopic reaction forces result in bearing loads and require proper mounting, and must be taken into account in the design.

What is relevant to bearing design is the resulting *gyroscopic moment* $\vec{M}_k$, which can be regarded as the time derivative of the *angular momentum* $\vec{H}$ (see Fig. 9.8):

$$\vec{M}_k = \frac{d}{dt}\vec{H} + \vec{\omega} \times \vec{H} \tag{9.1}$$

The angular momentum is the product of the *moment of inertia* I_i of the gyro or flywheel and its *angular velocity* $\vec{\omega}$:

$$\vec{H} = \begin{pmatrix} I_1 * \omega_1 \\ I_2 * \omega_2 \\ I_3 * \omega_3 \end{pmatrix} \tag{9.2}$$

However, the angular velocities (after coordinate transformation to a body-fixed coordinate system) can be described as

$$\vec{\omega} = \begin{pmatrix} \omega_1 \\ \omega_2 \\ \omega_2 \end{pmatrix} = \begin{pmatrix} \dot{\alpha} * \cos(\beta) * \cos(\gamma + \gamma_k) + \dot{\beta} * \sin(\gamma + \gamma_k) \\ -\dot{\alpha} * \cos(\beta) * \sin(\gamma + \gamma_k) + \dot{\beta} * \cos(\gamma + \gamma_k) \\ \dot{\gamma}_k + \dot{\gamma} + \dot{\alpha} * \sin(\beta) \end{pmatrix} \tag{9.3}$$

The above listed equations are the basis for the analytic calculation of gyroscopic reactions. It turns out that these differential equations can be *simplified significantly* if the following requirements are fulfilled:

- **Symmetric gyroscope:** The moment of inertia around two body-fixed axes must be identical.
- **Fast gyroscope:** The angular velocity of the flywheel ($\dot{\gamma}_k$) must be much greater than any component of $\vec{\omega}$.
- **Constant angular velocity:** The angular acceleration of the FESS, $\ddot{\gamma}_k$, around its own axis of rotation must be zero.
- **Generalization of the direction of action:** No vectors but only scalar values are considered (since these are the relevant quantities for bearing design).
- **Small angular displacement of disturbance:** *Pitching*, *yawing*, and *rolling* usually comprise angles <5° and therefore allow further simplification.

Table 9.2 Advantages and disadvantages of simplified, analytical determination of gyroscopic FESS bearing loads

Advantages	Disadvantages
• Quick and easy estimation • Vehicle dynamics can be estimated by values found in literature • Results of simplified equation are quite accurate (see Sect. 9.5.4)	• Neglects load collective • Neglects nonlinear vehicle motion • Neglects stiffness of mount • Neglects variation of rotational speed of FESS depending on state of charge

9.5.1.1 Results of the Analytical Assessment

If the simplifications just mentioned are applied to Eqs. 9.1, 9.2, and 9.3, the following scalar approximation formula is obtained:

$$M_{kf} = H * \omega_i \tag{9.4}$$

where the following applies:

$$H = I * \dot{\gamma}_k \tag{9.5}$$

Deflecting the gyroscope around an axis normal to the spin axes will change the orientation of the vector H. The FESS reacts with a *gyroscopic moment* M_k, which alters the momentum of spin according to the conservation of spin. As long as the angular deflection due to disturbances persists, the overall angular momentum (or spin) is increased.

Essentially, if values such as the moment of inertia and rotational speed of the FESS and maximum deflection velocity (pitch, roll, or yaw rate of the vehicle) are known, one can calculate the gyroscopic moment using formula Eq. 9.4. If the bearing spacing is known, the maximum bearing loads due to gyroscopic reaction can be estimated with sufficient accuracy (compare Sect. 9.5.4). Table 9.2 summarizes the possibilities and shortcomings of the approximation formula.

Table 9.2 shows that the simplified, analytical solution of *Euler equations* is an easy to use tool. However, it only allows a simple best-case/worst-case estimation, and further detailed analysis of the bearing loads is required.

9.5.2 Step 2: Numerical Simulation

The next and more detailed step is to compare the results of the analytical preliminary design with the results of a multi-body simulation (MBS). For this purpose, the simulation software *ADAMS* (*Automatic Dynamic Analysis of Mechanical Systems*) was used in the course of a master's thesis [6] at Graz University of Technology. The structure of the model is shown in Fig. 9.9 and is essentially composed of the following three elements:

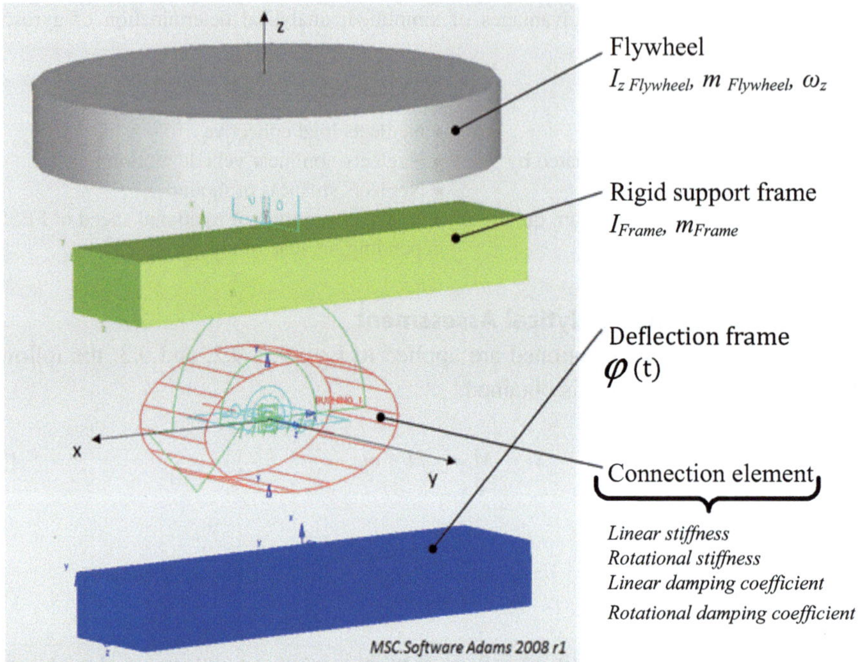

Fig. 9.9 Layout of the flywheel multi-body simulation model in ADAMS [6]. (Image rights: Andreas Brandstätter)

1. A *flywheel*, characterized by moment of inertia, rotational speed, and mass
2. A *rigid support frame*, which represents the FESS housing and is characterized by mass and moment of inertia
3. A *massless deflection frame*, which represents the vehicle chassis

The *rigid mounting frame* and the *deflection frame* are connected via a *viscoelastic element*, which was assigned stiffness and damping based on actual force-displacement measurements. The most important input data for the simulation are summarized in Table 9.3.

Investigating a basic load case (driving onto an angular ramp) acting on a gyroscopic mass, the simulation results were nearly identical to a time-dependent numeric simulation based on the derived *Euler equations* in *MATLAB*.

9.5.2.1 Load Collective and Peak Values

Feeding the *Euler equations* with actual vehicle measurement data enables identification of peak values in driving operation and even estimation of a probability density function of bearing loads, which will be called *load collective* from now on for better readability. In

Table 9.3 Table 1: Input data for numerical simulation, including measured stiffnesses of elastomer elements

Properties of the *viscoelastic connecting element*								Properties of rigid bodies			
Linear stiffness		*Rotational stiffness*		*Linear damping coefficient*		*Rotation damping coefficient*		*Property*	Symbol	Value	Unit
								Main moment of inertia of the flywheel	I_z	1,18	Kg*m^2
								Secondary moment of inertia	I_{xy}	4,6	Kg*m^2
N/mm		Nm/rad		Nm*s/mm		Nm*s/mm		Mass of the flywheel	$m_{Flywheel}$	58	Kg
c_x	140	c_{Rx}	20.2	d_x	4480	d_{Rx}	100	Angular velocity	ω_z	523	Wheel/s
cy	140	c_{Ry}	20.2	dy	4480	d_{Ry}	100	Moment of inertia of the rigid support frame	I_{Frame}	2,5	Kg*m^2
c_z	880	c_{Rz}	6727	d_z	4480	d_{Rz}	215	Mass of the rigid support frame	m_{Frame}	65	Kg φ

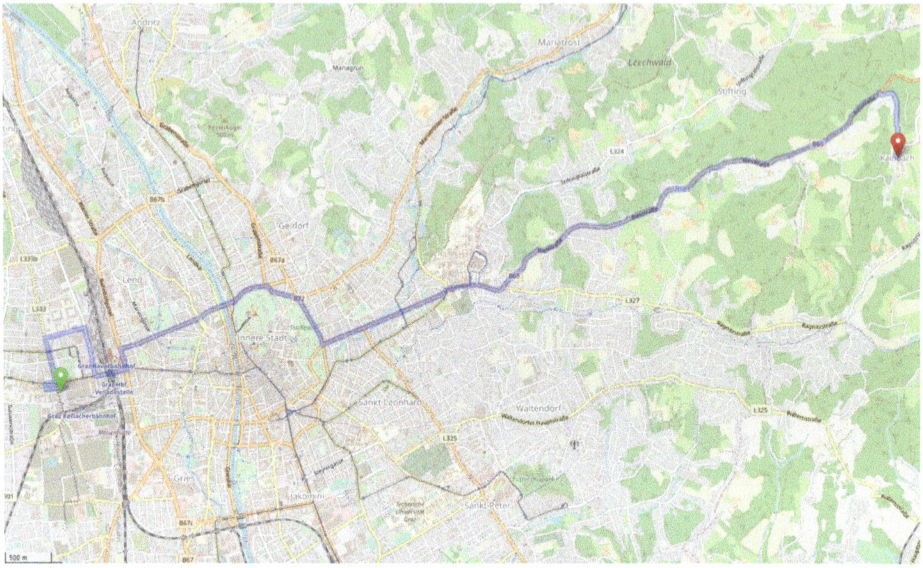

Fig. 9.10 Illustrated route (in red) of the urban, interurban, and highway vehicle measurements around Graz, Austria. (Image rights: OpenStreetMap)

this case, the collective was approximated, since there was no infinite number of samples and the results strongly depend on the measured track. The GPS track, which was chosen for acceleration measurements, includes urban inner-city traffic as well as cross-country roads and is shown in Fig. 9.10. The linear accelerations were measured using *Disynet* and *PCB Piezotronics* tri-axial acceleration sensors, and the angular speeds were recorded using a *Texys* tri-axial angular speed sensor. The car's ground speed was acquired using a *Racelogic VBox* GPS speed sensor. The test-drive route (as pictured in Fig. 9.11) has a total length of 58.7 km and contains 11.6 km (20%) single lane urban roads, 9.2 km (16%) of dual- and multi-lane roads, 16.4 km (28%) interurban roads, and 21.5 km (37%) of motorway. The measurements were carried out by the *Institute of Automotive Engineering* (*Institut für Fahrzeugtechnik*) at *FH Joanneum*, Graz, under the direction of Dr. Karl Reisinger (Table 9.4).

Even *Euler's equations*, which are greatly simplified with V0 to V6, are proven to be a valid instrument to estimate a bearing force probability density function based on actual vehicles driving measurement data. The simplifications are valid to estimate average loads and peak loads. They remain within a boundary of $\pm 3\%$ compared to the *full Euler equations* by reducing the calculation effort by a factor of 5 at the same time. However, they do change the probability density and therefore should not be used to estimate bearing life. The simplifications presented in subsection Sect. 9.5.3, however, enable a five times faster calculation time. To calculate the bearing loads from gyroscopic forces and acceleration from actual recorded driving data, the *University of Applied Sciences* in Graz has equipped several cars with acceleration sensors, angular speed sensors, a fast GPS tracking

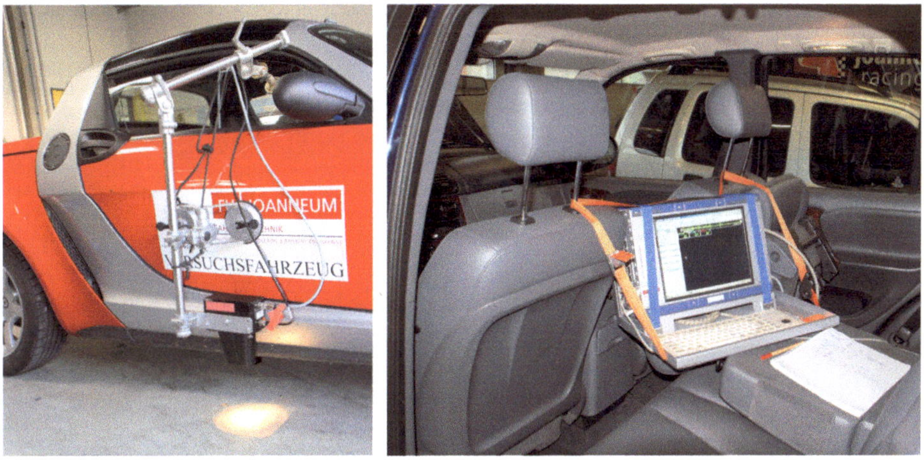

Fig. 9.11 *Smart Roadster* equipped with measuring equipment. Right: associated mobile data acquisition system. (Image rights: Graz University of Applied Sciences (Fachhochschule Joanneum), Graz)

Table 9.4 Simplifications of the *Euler equations* and their impact on *calculation time*, *peak load accuracy*, and *quality of average bearing loads* based on actual driving measurements [7]

No.	Description of simplification	Calculation time	Peak load quality	Accuracy of average bearing loads
V0	Complete set of the *Euler equations*	100%	100%	100%
V1	Rotationally symmetric rotor	69%	100%	100%
V2	Angular velocity $\dot{\gamma}_k$of the flywheel >> interfering angular velocities $\dot{\alpha}$, $\dot{\beta}$, and $\dot{\gamma}$ of the vehicle	49%	99%	98%
V3	Neglecting the charging/discharging of FESS	41%	99%	98%
V4	Neglecting the angular acceleration of the vehicle	36%	98%	97%
V5	Generalized vehicle speed in one direction	22%	97%	97,5%
V6	Cosine of roll angle β is set to 1	19%	97%	97,4%

device, a ground speed sensor, and a *Dewetron 3010* data acquisition system as shown in Fig. 9.11.

The cars were chosen strategically: a small sports car with stiff suspension (*Smart Roadster*), a long sedan class car (*VW Passat*), an SUV class car (*Mercedes Benz ML*) and a delivery truck with long wheelbase (*Opel Movano*). An overview of the maximum occurring vehicle accelerations is given in Table 9.5.

Table 9.5 Summary of the measured maximum values of vehicle accelerations and angular velocities in urban and motorway traffic

Vehicle model	Horizontal acceleration a_x, a_y m/s^2	Vertical acceleration a_x, a_y m/s^2	Roll rate—X ω_x °/s	Pitch rate—Y ω_y °/s
Mercedes ML 320	12.2	22.2	16.7	18.1
VW Passat	11.5	25.4	12.7	19.5
Smart Roadster	12.6	21.9	28.2	25.4
Opel Movano	10.6	19.2	16.9	12.2

From the raw sensor data, the angular acceleration and deflection of the vehicle were calculated and fed into the *Euler equations*. With the subsequent signal processed data, it was possible to show that the simplifications *V0–V4* presented in Table 9.4 have almost no influence on the estimated distribution of the bearing loads. Therefore, the simplified equations from V4 allow the calculation time to be cut in half, especially with a high sampling rate (>1 kHz) and hours of measurements and multiple channels to be processed. The estimation of the flywheel bearing's load probability density function was based on these actual car measurements.

9.5.2.2 Estimation of Heavy Misuse Bearing Loads

In order to also cover cases of vehicle misuse or small accidents like the so-called fender bender, acceleration sensor data from crash tests was additionally analyzed. Misuse tests were also recorded. One scenario is running over a speed bump with a vehicle speed of 45 km/h in a skewed angle. Another scenario represents three fast rounds in a small roundabout with a speed of 30 km/h to cover intense cornering. It needs to be mentioned that the effect of the gyroscopic torque on the vehicle movement during driving operation is neglected. To take this into account, it would have been necessary to put a rotating flywheel with the same momentum in the vehicle while recording data. A vehicle dynamics simulation prior to these measurements already showed that the gyroscopic reaction, from flywheels like this, is relatively small and can therefore be neglected. Note that the energy stored in this projected flywheel is 100 Wh at 60.000 rpm, which results in a small angular momentum compared to the vehicle size and mass. However, this assumption is not valid for all flywheels [8]. Heavy, slow-running flywheels such as the steel rotor with a diameter of 1 m and a mass of 500 kg used in the *Parry People Mover* [9] have a significantly higher angular momentum.

The result of the estimation of bearing loads in misuse tests is shown in Fig. 9.12. F_{total} describes the maximum force, consisting of F_{linear}, the bearing load due to linear vehicle acceleration, and the *gyroscopic reaction* due to the vehicle's pitch, yaw, and roll motion.

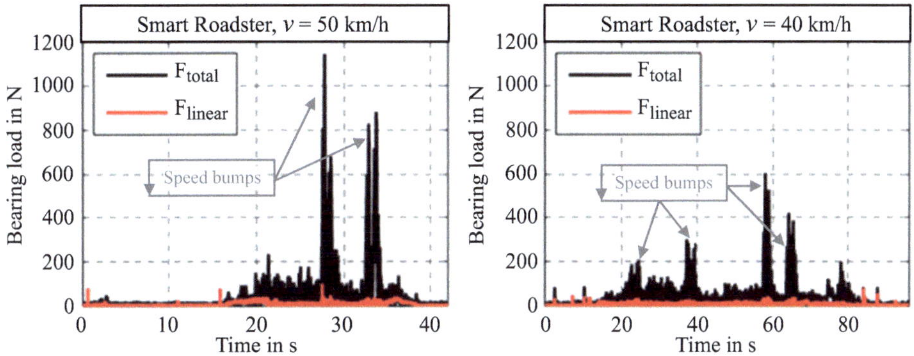

Fig. 9.12 Total calculated bearing loads of a 100 Wh flywheel in a *Smart Roadster* when driving over two "speed bumps." In the left diagram, the speed is 50 km/h, and in the right diagram, 40 km/h [7]. (Image rights: Manes Recheis)

9.5.2.3 Results of the Numeric Simulation

The maximum gyroscopic reaction forces almost precisely matched the results of the very simple analytical calculation, which is a good motivation to use the simple estimate at an early project stage. The absolute difference was found to be as low as 3% in most of the cases. It is important to point out that on the one hand, the peak values can be estimated rather quickly, and the whole bearing force probability function on the other hand is altered significantly by using the simplifications. When comparing the results of the numeric simulation to actual test bed measurements described in Sect. 9.5.3, it was found that the maximum gyroscopic moment was 13% below the experimental data presented in the next section. Reasons may be found in measurement inaccuracies and the fact that viscous damping of the elastomer elements depends on oscillation speed, which was neglected in the simulation.

9.5.3 Step 3: Empirical Verification

The shortcomings of the numeric simulation, which are listed in Tables 9.5 and 9.6 and in Sect. 9.5.2.3, clearly indicate the need for empirical investigation. The purpose of the test bench shown in Fig. 9.13 is not only to analyze the effect of a nonlinear, resilient flywheel mount but also to validate the analytic and empiric calculations.

The basic principle chosen for the test bench is the measurement of gyroscopic reactions due to forced deflection (tilting) of a flywheel. A steel disk with a diameter of 400 mm and a moment of inertia of 1.18 kgm^2 can be accelerated to 6000 rpm, resulting in an angular momentum of 741.4 kgm^2/s. This allows the simulation of the gyroscopic behavior of most flywheels relevant for automotive applications (including small, fast spinning ones) as long

Table 9.6 Advantages and disadvantages of determining FESS bearing loads using numeric simulation supported by vehicle acceleration data measurement during real driving operation

Advantages	Disadvantages
• Effect of real load collective can be estimated	• Difficult to judge plausibility due to nonintuitive behavior of the matter
• Real driving dynamics are taken into account	• Gyroscopic effect of FESS on vehicle dynamics is neglected
• "Proof-of-concept" simulations are easily possible	• Does not include nonlinear behavior of elastomer, viscous damping, etc. of FESS bearing and vehicle mount

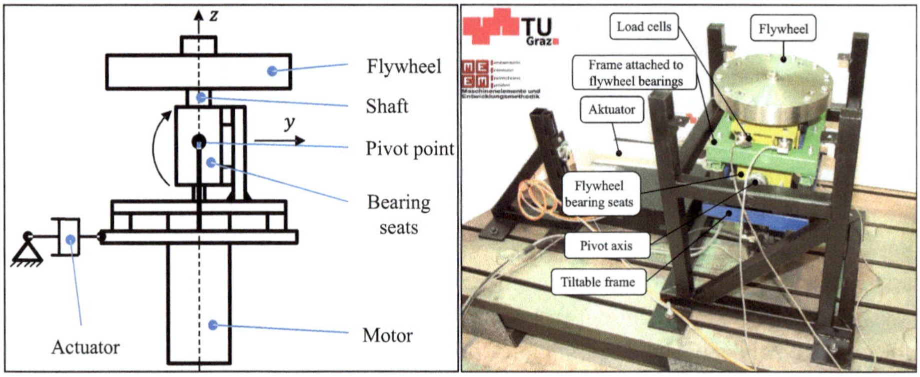

Fig. 9.13 Sketch (left) and photograph (right) of test bed for determination of flywheel bearing loads with adjustable FESS-to-vehicle mount stiffness [6]. (Image rights: Andreas Brandstätter)

as their angular momentum $\vec{H}$ does not exceed 741.4 kgm^2/s. (The physical and mathematical background is explained in Sect. 9.5.1.1.) The angular displacement around the tilt axes (x-axes) is provided by an electro-mechanic linear actuator by the company *Festo*. The reaction forces are measured by load cells, and the corresponding reaction torque around the x- and y-axes is calculated on the basis of the known geometric data. The deflection cycles were measured beforehand by applying sensors to relevant vehicles during real-life operation (as described in Sect. 9.5.2.1.) processed and fed into the actuator. To start the experiment, easy-to-interpret and simple cycles, like the crossing of a speed bump or driving onto a ramp (as shown in Fig. 9.14), were tested.

9.5.3.1 Results of the Empirical Verification

Looking at the gyroscopic torque response of the gyro (FESS) to the deflection in Fig. 9.15, it is interesting to see that the torque response around the y-axes followed the characteristics calculated by the approximation formula, but seemed to be superimposed by some kind of oscillation. It turns out that the low-frequency oscillation of around 12 Hz was caused by the gyroscopic reaction of the flywheel and the elastic deformation of the mount/test bench. Also the moment around the x-axes, which is not taken into account by the approximation formula, needs to be considered. The high frequent oscillations of

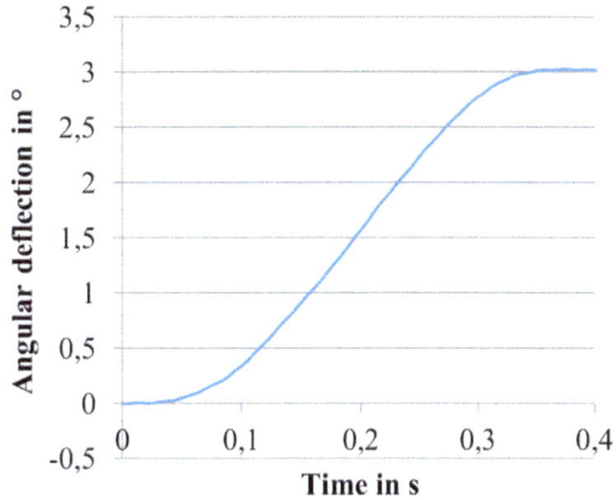

Fig. 9.14 Vehicle motion (angular displacement) when driving onto a ramp. (Image rights: Andreas Brandstätter)

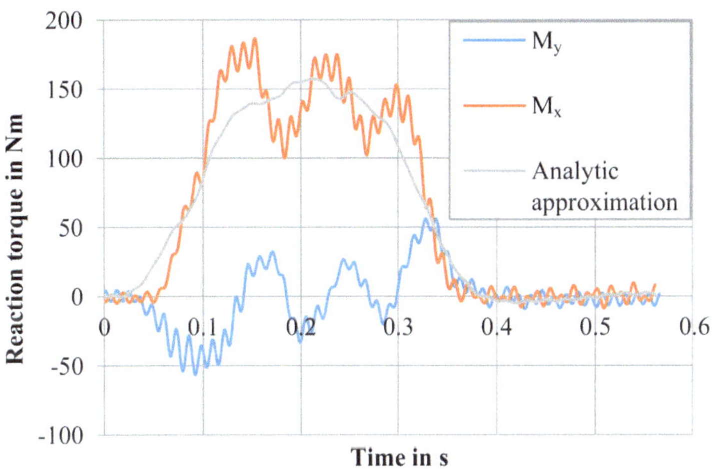

Fig. 9.15 Gyroscopic response of the FESS corresponding to the vehicle motion shown in Fig. 9.14. Measured torque around X (blue) and Y (green) and composite absolute value calculated via the approximation formula Eq. 9.4. (data from [6]). (Image rights: Andreas Brandstätter)

100 Hz and higher were caused by mass imbalance forces, representing a rotating force vector, which generated torque around x and y. Even though parts and components of the test bench cannot be seen as perfectly rigid bodies, the validity of Eq. 9.4 was confirmed once more [2] (Fig. 9.16).

It is difficult to mount the entire FESS to the vehicle using a resilient mount, which would filter vibrations. A soft mount with low coefficient of damping to reduce mechanical loads allows gyroscopic micro motion due to *precession* and *nutation*. The reaction forces at the bearings may actually end up being higher than as if the FESS were attached rigidly to the vehicle. This phenomenon was also proven by the test bed and is shown in Fig. 9.17.

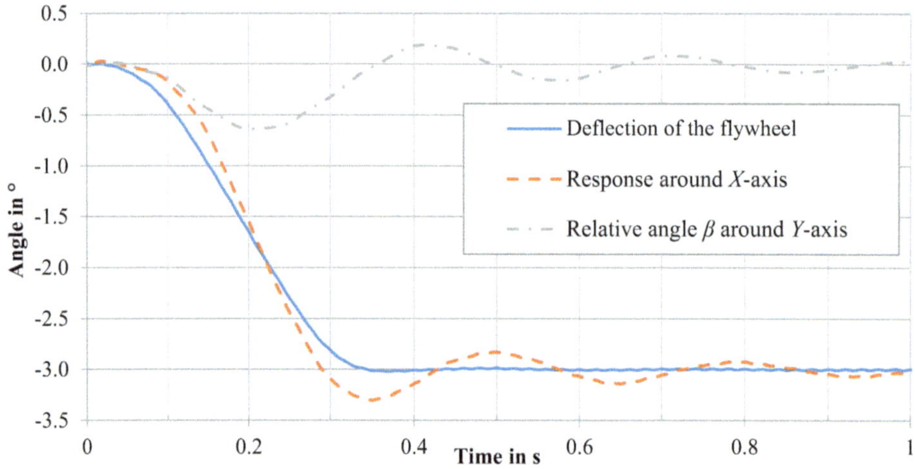

Fig. 9.16 Response of the FESS to forced deflection. One axis is actuated, but the elastic mount allowed three-dimensional motion due to gyroscopic reactions [6]. (Image right: Andreas Brandstätter)

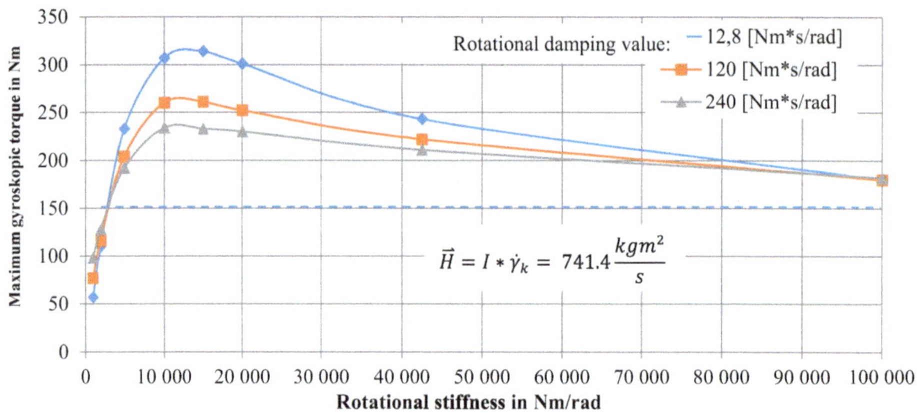

Fig. 9.17 Gyroscopic reaction torque of a FESS with an equivalent angular momentum of 741.4 kgm^2/s measured on the test rig as a function of the vehicle mount stiffness [7]. (Image rights: Andreas Brandstätter)

The reason for the amplification of bearing loads can be found in the superposition of the angular deflection velocity during the cycle (vehicle motion) and the gyroscopic reaction of the flywheel. Mounts with varying stiffness were applied to the test flywheel, and it was found that a connection stiffness of 12 kNm/rad resulted in the highest bearing loads for the configuration of this specific test bench. Therefore, the ideal FESS vehicle mount can be found at either end of the mount stiffness scale. Fig. 9.17 shows that either an extremely

Table 9.7 Advantages and disadvantages of empirical determination of FESS bearing loads

Advantages	Disadvantages
• Influence of vehicle suspension on bearing loads can be investigated	• Difficult to judge plausibility due to nonintuitive behavior of the matter
• Incorporates real vehicle dynamics	• Gyroscopic effect of FESS on vehicle dynamics is neglected
• Nonlinearity of elastomers, fluid damping, etc. are taken into account	• Time consuming and expensive hardware setup required

soft connection, like a gimbal, or a rigid mount is the best option for mitigation of gyroscopic bearing loads.[1]

9.5.3.2 Evaluation of the Empirical Verification

The surprising results presented in Fig. 9.17 justify the effort put into designing and building a test bench. The tests conducted with the help of the configuration shown in Fig. 9.13 showed clear tendencies regarding the stiffness of FESS mounts and influencing parameters.

Once the moment of inertia and angular velocity of the projected flywheel are known, results of the test bench can be scaled using the ratio of angular momentum as a scaling factor. This way, also the ideal mounting stiffness for the FESS in the vehicle can be found (see Table 9.7).

9.5.4 Conclusion Regarding Gyroscopic FESS Bearing Loads

As there are hardly any publications on this topic, it was necessary to conduct own investigations at the *Institute for Machine Elements and Development Methodology* (*Institut für Maschinenelemente und Entwicklungsmethodik*). Since the bearing—especially the rolling bearing—is one of the key elements in FESS, a detailed determination of the loads is of utmost importance. The presented process of load determination consists of an *analytical*, *numerical*, and *empirical step*, whereby a comparison of the results showed a good agreement. Although it turned out that the highly simplified form of *Euler's equations* provides sufficiently accurate results for maximum loads, the numerical simulation also confirmed that the complexity of gyroscopic mechanics and the nonlinearities of various assemblies made an empirical investigation on the test bench necessary. Even if, as mentioned above, the maximum bearing loads can be easily estimated with very good accuracy (~3%), the cumulative frequency, i.e., the influence of the load collective, is

[1]Design aspects may also influence the choice of FESS mount. A gimbal can only be used with electro-mechanic flywheels, while purely mechanic systems require a rigid mount because their power is transferred via shafts and gears.

Table 9.8 Comparison of the results of the different methods for determining the gyroscopic moment of FESS in automotive applications

Method	Analytic	Numeric	Empiric
Tool/ implementation	Simplified *Eulerian equations*	*ADAMS model* in combination with real vehicle dynamics measurement data	Test bed using a hinged steel flywheel
Load case	Ramp according to Fig. 9.14	Ramp according to Fig. 9.14	Ramp according to Fig. 9.14
Considered FESS mount	–	Elastomer 55 ShA	Elastomer 55 ShA
Maximum gyroscopic torque	219.4 nm	226.3 nm	242 nm
Relative error	−9.34%	−6.5%	–

ignored in the case of a simple, analytical preliminary design, and the result of the bearing life calculation is distorted. Table 9.8 summarizes the essential characteristics of the respective methods.

Vehicle-to-FESS mounts with various stiffness and coefficients of damping were investigated on a test bench, and the counterintuitive results show that either a totally rigid connection to the vehicle or a full gimbal mount results in the lowest bearing loads.

9.6 Imbalance Forces in Energy Storage Flywheels

While the gyroscopic reactions described in Sect. 9.3 occur *due to physical principles*, as they are inevitably caused by the angular momentum of the flywheel mass and the dynamics of the vehicle, imbalance forces are bearing loads, which *theoretically* can be avoided. The cause is a *manufacturing-related eccentricity* between the geometric axis of rotation and the center of gravity axis of the FESS rotor. The imbalance can be caused by geometric manufacturing errors and/or anisotropic material properties. Table 9.9 provides an overview of the possible types of imbalance forces in FESS rotors.

The in-depth background, physical basics, and details of rotor dynamics and balancing technology are not dealt with in this book, but reference is made to standard works such as "Balancing Technology" by *Hatto Schneider* [11]. The terms *unbalance* and *balancing quality* are mathematically described here as the main relationships. The *unbalance U* is defined by the product of the *unbalance mass u* and its distance from the *axis of rotation r*:

$$U = u * r \tag{9.6}$$

The balance quality G is obtained:

Table 9.9 Types of unbalance in FESS rotors and essential characteristics, created on the basis of [10]

	Static unbalance	Dynamic unbalance	Couple unbalance
Illustration	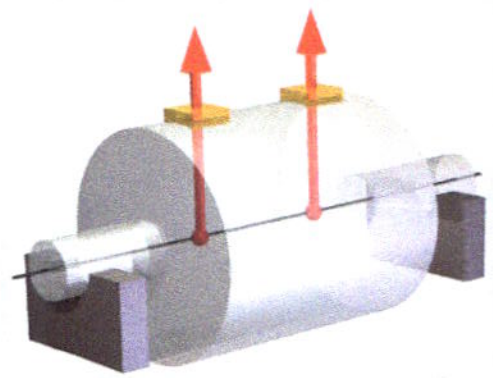	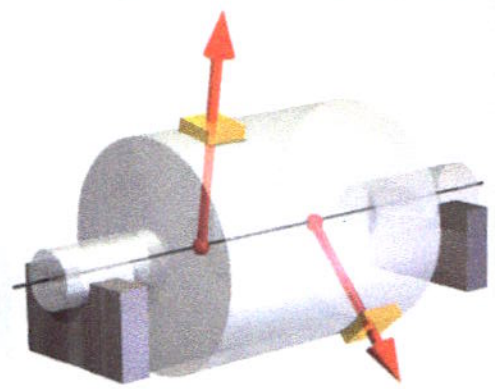	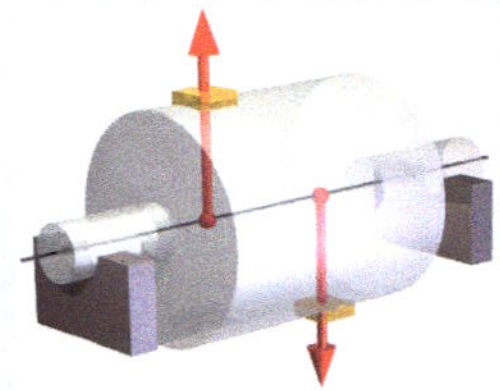
Cause	Two unbalances can have the same direction and angular position. The same condition results with a single unbalance, twice the order or magnitude, which acts at the center of gravity	The real rotor theoretically has an infinite number of unbalances, which are randomly distributed. They neither have the same magnitude nor a defined angular position	Two unbalances can have the same magnitude, but their angular position may also be offset by exactly 180°
Effect	It causes a shift of the center of mass from the geometrical center, whereby the rotor oscillates parallel to its axis of rotation during operation	Dynamic unbalance occurs in all rotors and includes both static unbalance and couple unbalance, whereby one or the other component may predominate	The running rotor performs a tumbling motion around its vertical axis (perpendicular to the axis of rotation), because the two imbalances result in a torque
Measure	Static unbalance can be corrected by measures taken at the plane of the center of gravity (attachment or removal of rotor mass)	Two compensation planes are required to completely correct the dynamic unbalance due to the couple unbalance component	A counter-torque is required to correct the couple unbalance, i.e., two correction unbalances with a certain (axial) distance between them

$$G = \omega * U * m_{rotor} \tag{9.7}$$

where ω is the angular velocity and m_{rotor} is the rotor mass.

The only way to reduce the imbalance forces and bearing loads of a FESS is to (dynamically) balance the rotor, i.e., to remove or add rotor mass so that the main axis of inertia (center of gravity) and the geometric axis of rotation coincide. However, since imbalance bearing loads are caused by a force vector rotating at the rotational frequency of the rotor, which acts on the roller bearing, it is possible to reduce these reaction forces by introducing *a resilient bearing seat*. Reducing bearing stiffness results in *supercritical operation*. Since—as the following practical examples will show—a certain residual unbalance cannot be completely avoided, a combination of the two methods can often be appropriate:

1. Dynamic fine balancing of the rotor (see Fig. 9.18)
2. Resilient bearing seats and supercritical operation

It can be seen that the *interdependencies between the FESS components*—in this case the rotor and the bearing—do not allow for an isolated view, but require a *holistic*, *transdisciplinary approach*. Rotor manufacturing, design, and assembly have a critical influence on the expected imbalance forces and must be considered in the early stages of rotor design, as the following case study shows.

Fig. 9.18 Dynamic balancing of a 5-kWh carbon fiber composite flywheel rotor (Image rights: FWT Composites & Rolls GmbH, Austria)

9.6.1 Balancing and Balancing Options of the *FIMD Rotor* Case Study

FIMD Due to the small air gap of the electric machine *FIMD flywheel*, only very small deflections of the rotor are permissible. This is why initially an attempt was made to design a rigid bearing support for *subcritical rotor operation*, meaning that the imbalance forces had to be kept as low as possible by balancing the rotor as extremely accurately. For this purpose, three balancing planes are provided, in which weights in the form of grub screws or freely movable sliding blocks in a dovetail groove can be attached (see Fig. 9.19).

The rotor was intended to be balanced in a *two-step procedure*:

1. **External balancing on balancing machine**

The bearing journals of the rotor are placed on the rollers of a balancing machine, and the rotor is brought to speed by means of a wraparound belt drive. The deflection of the compliant suspension of the rollers is determined, and the unbalance can be calculated based on this measurement data (compare Fig. 9.20).

2. In situ **balancing of the rotor**

The mounting process of the rolling bearings or the assembly of the entire FESS changes the balance quality as well as the multi-body dynamic vibration behavior (changed stiffness

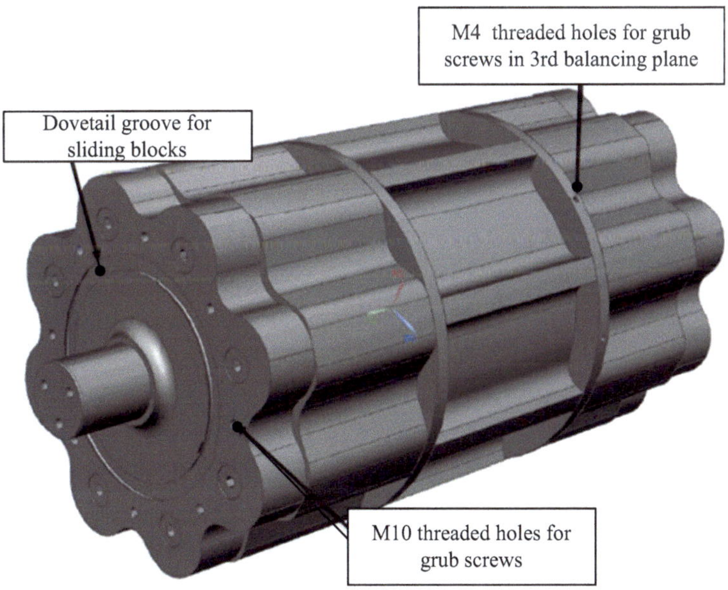

Fig. 9.19 Possibilities for attaching balancing weights to the FIMD rotor

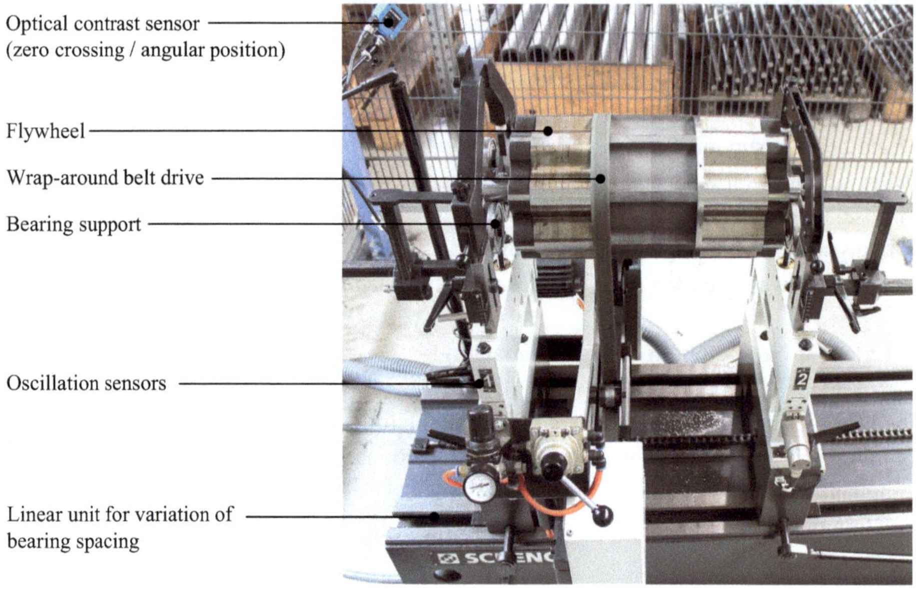

Fig. 9.20 FESS rotor on balancing machine (*Schenck Rotec HM 40*)

Fig. 9.21 Capacitive displacement sensors (type *CS 05*) in housing

and modal masses) of the system. This is caused by plastic micro-deformations and setting phenomena. As a result, balancing may become necessary when the rotor is installed in the housing. Two capacitive sensors (*Micro-Epsilon CS05*)—see Fig. 9.21—offset by 90° to each other and installed in the housing detect the contour of the circular "measuring mass

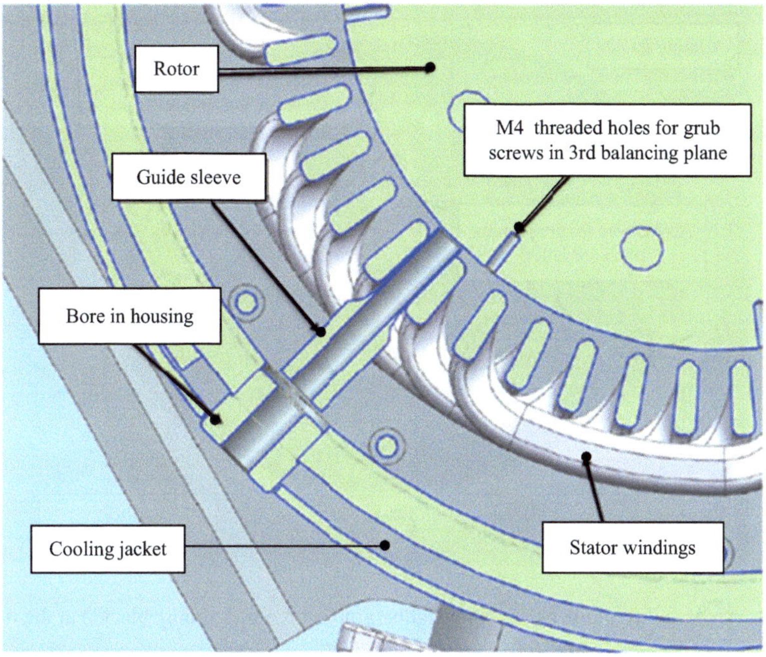

Fig. 9.22 Opening for grub screws as balancing weights in the middle compensation plane

disk" (see Sect. 7.5.1). First, the out-of-roundness of the measuring mass disk is determined at a sufficiently low speed. At higher speeds, this measurement is superimposed by the unbalance-related shift in the center of gravity. The geometric contour error must be compensated by calculation (subtraction of the out-of-roundness from the measurement signal recorded at higher speed).

The angular position and amplitude of the deflection are determined by software. In this case, the software provided for this purpose was purchased from the company *Data Physics.*

While Fig. 9.22 on the right shows the access to the central balancing plane of the rotor through an opening in the housing, the possibility of attachment/removal of balancing weights at the end faces (through the bearing shield) is shown in Fig. 9.23.

9.6.1.1 Problems of Subcritical Rotor Operation

Due to the expected highly dynamic load cycle and the application of the FESS in a vehicle, the focus was placed on the endurance of bearing high loads (gyroscopic reactions and shocks, as discussed in Sect. 9.5) and not on reducing the bearing loss torque for minimized self-discharge (see Sect. 10.2). The bearing arrangement—shown in Fig. 9.24—is comprised of a *7008 ACD Duplex Set* at the top of the shaft and a single angular contact ball bearing of the same bearing type *7008 CD* at the bottom. These are *SKF high-precision bearings* of the "Super-Precision Bearings" series, i.e., angular contact ball bearings with

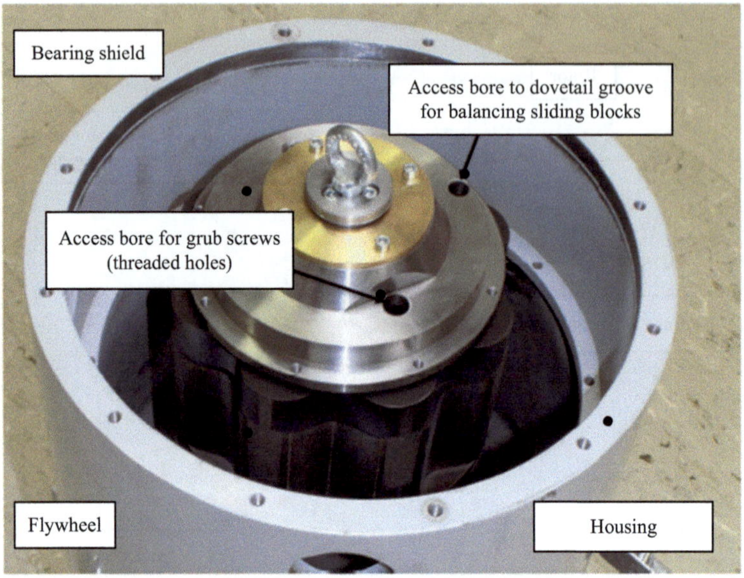

Fig. 9.23 Accessibility of the balancing weights (grub screws and sliding blocks) at the rotor faces

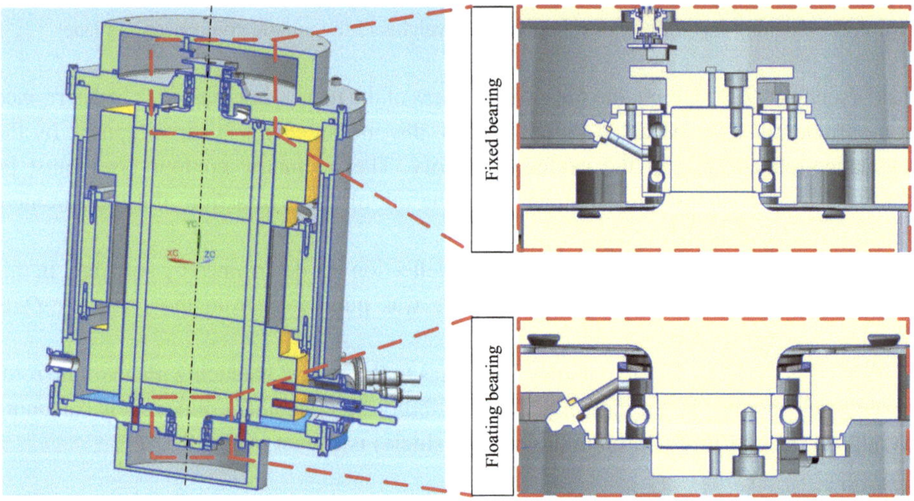

Fig. 9.24 FIMD rotor bearing concept

ceramic rolling elements (silicon nitride balls) for higher load and speed ratings and particularly good running tolerances.

While the "Duplex Set"—in this case a *back-to-back* arrangement, as shown in Fig. 9.25 on the right—has an internal, axial preload due to the minimal geometrical height

Main dimensions	Load factors	Max. speed lubricated with	Denomination
	dynamic static grease oil-air SKF		
d D B C C0			
mm kN U/min -			
40 68 15 15.9 10.4 20000 34000 7008 ACD/HCP4A			

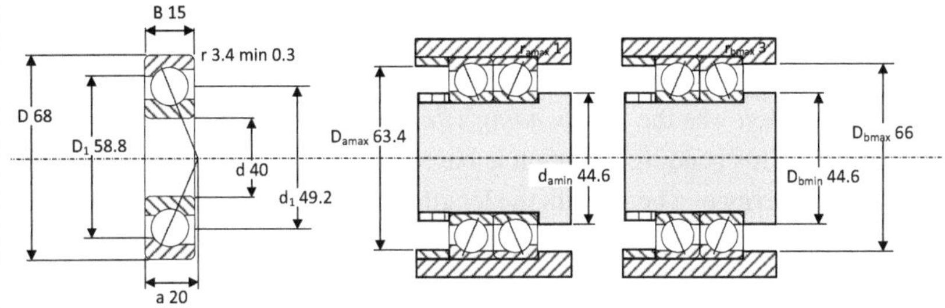

Fig. 9.25 Properties of the rolling bearings of the FIMD flywheel [12]. (Image rights: SKF)

differences of the bearing rings, the lower bearing must be axially preloaded by means of a spring assembly. For this purpose, a stack of cup springs from the company *Schnorr* was selected, which should guarantee an axial preload force of approximately 150–400 N depending on thermal expansion of the rotor.

As also shown in Fig. 9.25, the bearing bore diameter of 40 mm is relatively large for high-speed applications and therefore reduces the limiting speed to about 20,000 rpm when used with grease lubrication. Due to the light and high-strength silicon nitride rolling elements, the bearing can tolerate speeds up to the planned 37,500 rpm after consultation with the manufacturer, but the grease service life and overall service life are drastically reduced. Since this project was a prototype, this restriction was accepted.

Design details such as choice of fit, lubrication, etc. will not be discussed here. Instead, relevant results of the commissioning of the prototype and effects on the bearing system design will be presented.

9.6.1.2 Estimation of the Natural Frequency of the *FIMD Rotor Bearing System*

The complicated design of the multi-disk rotor (see also Sect. 7.5) made estimating the critical bending eigenfrequency a difficult task. Major uncertainties were:

- Total stiffness of the rotor due to the many joints/interstices
- Friction damping in the joints/interstices
- Mechanical properties of the full-face bonding layer (*Backlack*) between the electrical sheets
- Real stiffness of the bearing system (including bearing seats) due to manufacturing inaccuracies

Three different methods were therefore used for the calculation. However, it must be noted that only the FEM simulation also considered the bearings including their positions, while the analytical calculation and the natural frequency measurement only considered the *rotor system*.

1. **Analytical calculation**

The basis of the calculation was the *free bending vibration* of the rotor (considered as an oscillating continuum). This bending vibration is based on the consideration of the kinetic (E_k) and potential (E_p) energy of a beam with the length L, the mass allocation $\mu(x) = \rho A(x)$, and the bending stiffness $EI(x)$ according to [13]:

$$E_k = \frac{1}{2}\int_0^L \mu(x)\dot{w}^2(x, t)dx \tag{9.8}$$

$$E_p = \frac{1}{2}\int_0^L EI(x)w''^2(x, t)dx \tag{9.9}$$

where w denotes the deflection of the beam as shown in Fig. 9.26.

The complete derivation of Eqs. 9.8 and 9.9–9.13 can be found in the standard work *Vibrations* by *Magnus*, *Popp*, and *Sextro* [13]. A fourth-order differential equation can be stated, which describes the movement of the beam:

$$\mu(x)\ddot{w}(x, t) + [EI(x)w(\mathrm{x},\ \mathrm{t})] = 0 \tag{9.10}$$

The solution of the differential equation for beam vibrations requires a constant mass allocation $\mu(x) = \rho A =$ const. and constant bending stiffness $EI =$ const. Furthermore, the abbreviation

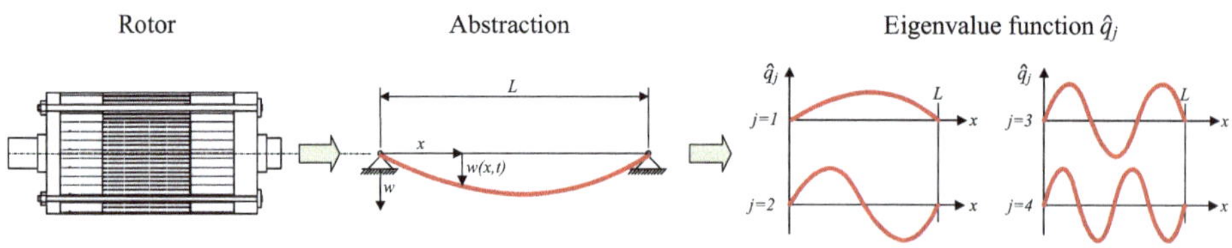

Fig. 9.26 Abstraction of the rotor with geometric dimensions of the lateral oscillation and eigenfunctions of a one-dimensional continuum supported on both ends

$$k^4 = \omega^2 \frac{\mu}{EI} \tag{9.11}$$

is introduced. The eigenfrequencies ω_j for the infinitely many eigenvalues λ_j of the transcendental equation result in formula (9.13):

$$\lambda_j = k_j L = j\pi \tag{9.12}$$

$$\omega_j = \lambda_j^2 \sqrt{EI/(\rho A L^4)} \quad \text{for} \quad j = 1,2,3,\ldots \tag{9.13}$$

The geometric dimensions A, I, and L can be easily determined, as can the density ρ for the *FIMD rotor*. More problematic is the determination of the modulus of elasticity (E) of the *FIMD rotor* (or at least its average value), which, as expected, differs from that of a homogeneous steel rotor due to the large number of joints or the amount of full-face bonding between the layers of the electrical sheets. This property was therefore determined experimentally, as shown in Fig. 9.27. The entire rotor was compressed by a specially manufactured mechanical press, and the compression was measured by laser triangulation displacement sensors (Fig. 9.28).

The compression force was varied between 70% and 100% of the value of the total assembly preload force of the eight clamping bolts, and force and displacement results from several series of measurements were averaged. The results are shown in Table 9.10.

Based on these measurements, a stiffness of approximately $9*10^5$ N/mm or a modulus of elasticity of 178,515 N/mm^2 could be determined. If Eq. 9.3 is applied, the natural bending frequency of the rotor is about 1850 Hz.

Of course, it must be noted that this is a highly simplified approximate calculation! Due to design restrictions, the modulus of elasticity could only be measured in the longitudinal direction, whereby it can be assumed that there is a strong anisotropy, which results in a deviation of the value in the radial direction!

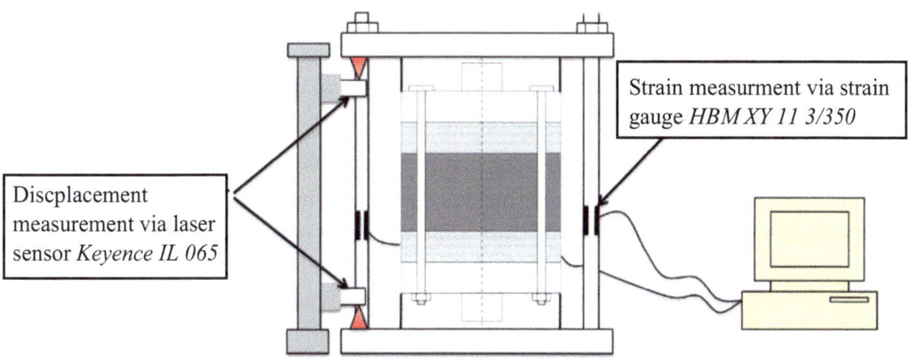

Fig. 9.27 Concept of the measurement of the FIMD rotor's modulus of elasticity

Fig. 9.28 FIMD rotor during compression measurement for the determination of the modulus of elasticity

Table 9.10 Results of stiffness measurement of the entire *FIMD rotor*

Percent of the mounting preload force	Compression force	Change of rotor length
70%	42,000 N	0.016 mm
100%	60,000 N	0.023 mm

2. **Numerical simulation**

A numerical simulation was carried out in this example by the engineering company *Dr.-Ing. Ernst Braun GmbH* (Martin-Luther-Straße 1, D-88400 Biberach) and was also based on some significant simplifications. The entire rotor was considered a homogeneous, isotropic body, and the bearing stiffnesses were assumed to be the same as the manufacturer's specifications of 430 N/μm without taking into account any mitigation by manufacturing tolerances in the bearing seat. However, a proper support of the bearing journals ends was considered. The simulation model is shown in Fig. 9.29.

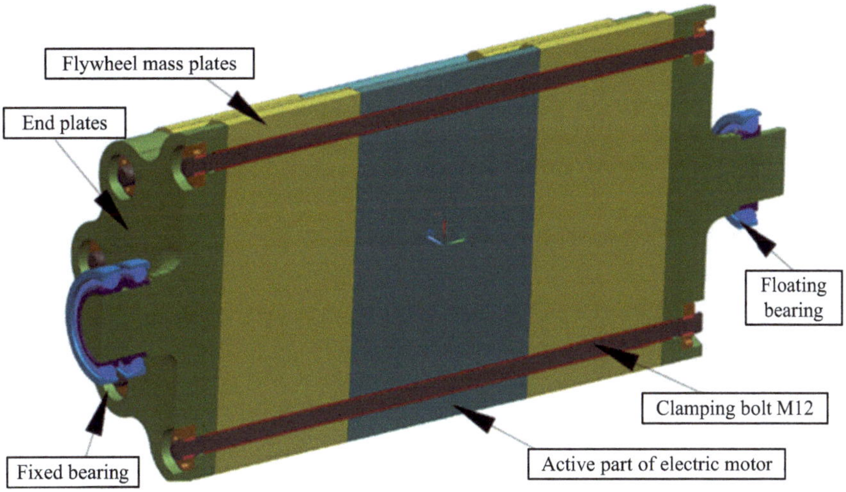

Fig. 9.29 Setup of the FEM simulation of the natural frequency. (Image rights: Dr.-Ing. Ernst Braun GmbH)

A natural frequency of the rotor—bearing system of about 265 Hz (corresponding to 15,900 rpm)—was determined.

3. **Empirical measurement using an impulse hammer**

This measurement was carried out with the kind support of Dr. Andreas Marn of the *Institute for Thermal Turbomachinery and Machine Dynamic* (*Institut für Thermische Turbomaschinen*) at *Graz University of Technology*. The rotor was freely suspended from an eyebolt and excited with a *roving hammer* (impulse hammer). A *B&K Type 4335 - 170,936* accelerometer was *successively* placed at different locations. The excitation by the impulse hammer was also carried out at various positions along the length of the rotor, whereby the active part of the electric machine had to be left out due to the risk of damage (potential short circuits of the electrical sheets on the outer circumference). The measurement setup is shown in Fig. 9.30.

The evaluation of the measurement data brought the following findings:

- At ~3200 Hz, a significant increase in amplitude can be seen at all measuring points (MP-1 to MP-4), which represents the first free bending eigenmode of the rotor (compare Fig. 9.31).
- The large number of sub- and superharmonic oscillations indicates nonlinear behavior, probably due to friction in the lamination stack.

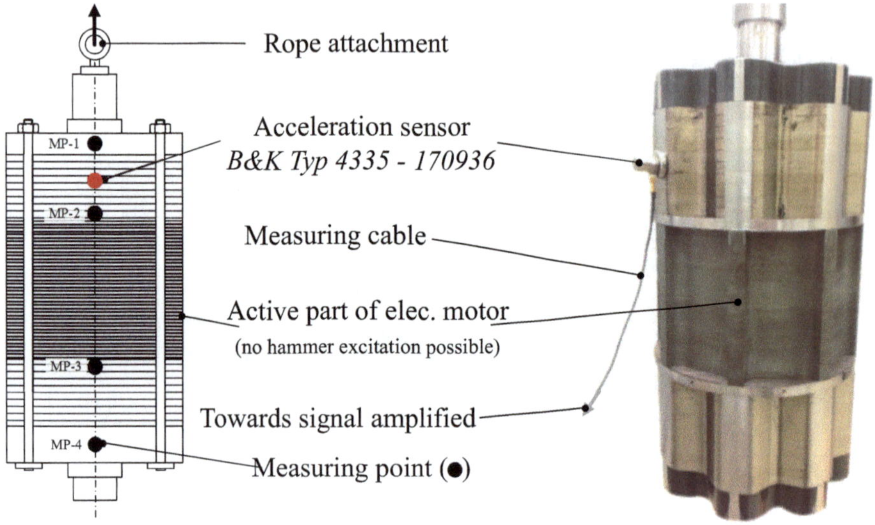

Fig. 9.30 Setup of the natural frequency measurement of the FIMD rotor

Interpretation of the Deviations of the Natural Frequency Results
A strong deviation of the results of the three methods for determining the natural frequency can be noticed. This deviation can have the following causes:

(a) Both, the measurement and the analytical calculation, show that the natural frequencies of the rotor itself (bending mode) are very high. This seems plausible considering the length-diameter ratio and material of the rotor.
(b) It can be assumed that the real bearing stiffnesses are many times lower than those specified by the manufacturer and are therefore decisive for the natural frequencies of the *system* in the operationally relevant speed spectrum of the flywheel.
(c) The real *total stiffness* (bearing plus bearing seat, end shield, and relevant housing parts) can differ from the value of the bearings alone by a factor of 10. (This corresponds to qualitative statements of the company *SKF* and own measurements/experience.)

It is therefore evident that the bearing arrangement plays a decisive role with regard to the machine-dynamic behavior of the entire FESS!

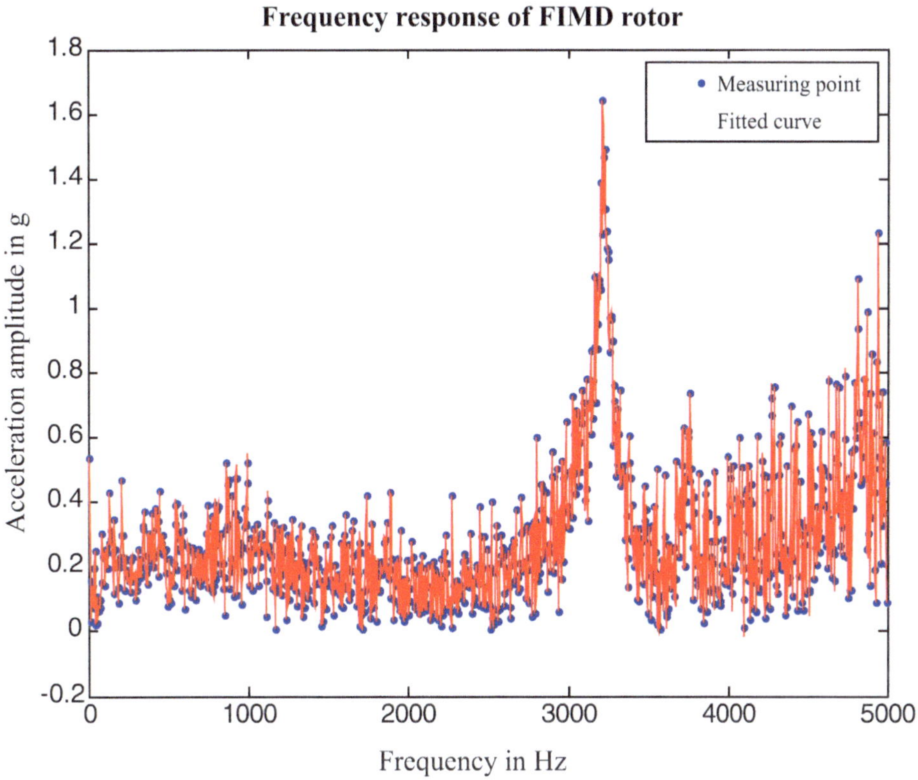

Fig. 9.31 Acceleration response of the FIMD rotor during the impulse hammer test at measuring point MP-2 (measurement supported by Dr. Andreas Marn, *Institute for Thermal Turbomachinery and Machine Dynamics* at Graz University of Technology). (Image right: Andreas Marn, Institute for Thermal Turbomachinery and Machine Dynamics of Graz University of Technology)

9.6.1.3 Influence of Bearing Stiffness on the Natural Frequency of the *FIMD Rotor System*

The stiffness of the rolling bearings, which in most cases is many orders of magnitude smaller than the stiffness of the rotor itself[2], depends on a large number of parameters. These are, among others:

1. Operating load
2. Bearing preload force
3. Geometric arrangement
4. Temperature

[2] Investigations have been carried out with compliant rotors (wound ropes or fiber bundles), but these concepts have not yet been brought to production maturity.

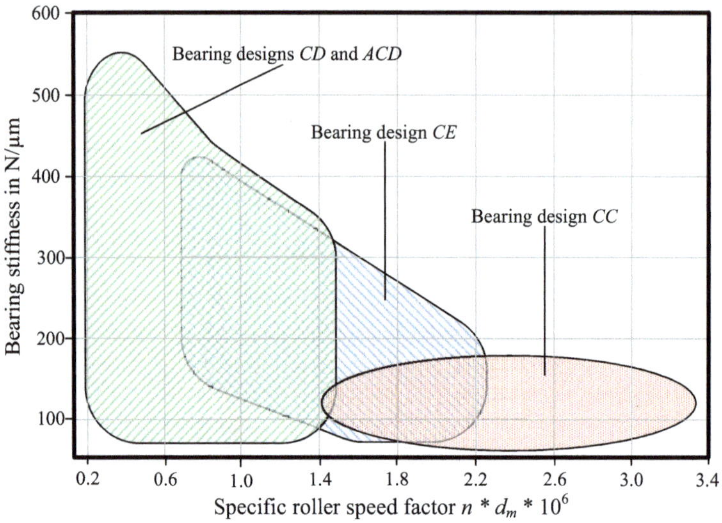

Fig. 9.32 Dependence between bearing stiffness and n^*d_m *factor* [12]. (Image rights: SKF)

5. Rotational speed
6. Running-in condition
7. Lubrication
8. Rolling element material
9. Geometric dimensioning and tolerancing (GDT) of bearing seat
10. Rolling element geometry

Of course, the surrounding housing also has an influence on the natural frequencies, which is not to be neglected, but these can only be estimated using complex numerical methods.

Furthermore, *SKF* indicates that the stiffness is a function of the n^*d_m *value*, a value which describes the average path speed of the rolling elements and is therefore a bearing type-specific variable. The effect of a reduction in stiffness with increasing n^*d_m *value* is shown in Fig. 9.32.

In order to demonstrate the significance of the influence of the bearing stiffness on the natural frequency behavior of the entire flywheel energy storage system, three representative scenarios were analytically recalculated on the basis of the linear single mass oscillator (aka harmonic oscillator). The scenarios are:

Case 1: Assumption of Maximum Bearing Stiffness

1. The stiffness of the bearings is based on manufacturer's specifications.
2. The duplex set is approximated by *a single* overall stiffness as the sum of the individual bearing stiffnesses.
3. This results in a stiffness of the fixed bearing of 640,000 N/mm.

Case 2: Reduction of the Bearing Stiffness Under Consideration of the *$n*d_m$ Value*

1. Reduction of the stiffness to about two thirds of its original value by considering the *ACD range* shown in Fig. 9.32.
2. Approximate bearing stiffness is 430,000 N/mm for the duplex set.

Case 3: Lowest Stiffness by Estimating the Influence of Bearing Seat and Periphery

1. Consideration of the mounting situation → bearing, end shield, cover, housing, and their respective contact stiffnesses are approximated and modeled as a series connection of springs!
2. Reduction of stiffness up to a factor of 10 based on experience.
3. This leads to a "worst-case" stiffness of the bearing arrangement of ~32,000 N/mm.

The results of this "best-case/worst-case assessment" are presented in Table 9.11.

The calculation was performed on the basis of the linear single-mass oscillator neglecting the influence of mass acceleration. Due to the symmetrical design it was assumed that each bearing side will support half of the rotor mass (42 kg). The results only serve to *estimate orders of magnitude* and to show that the natural frequencies caused by the resilience of the bearing are many times lower than the natural bending frequency of the rotor itself.

9.6.1.4 Commissioning and Problems of *FIMD Rotor Bearing System*

In addition to extensive temperature monitoring, the *FIMD flywheel* was equipped with capacitive displacement sensors and acceleration sensors on the end shield, both of which

Table 9.11 Results of the analytical *best-case/worst-case* estimation of bearing stiffnesses and resulting resonance speeds of the first eigenfrequency

Estimate of fixed bearing (*duplex set*)		**Radial stiffness in N/mm**	**Resonance speed in rpm**
Case 1	Manufacturer information	640,000	37,300
Case 2	Reduction by $n*d_m$ *value*	430,000	30,600
Case 3	Consideration of installation situation	130,000	16,800
Estimation of floating bearing (*single bearing*)		**Radial stiffness in N/mm**	**Resonance speed in rpm**
Case 1	Manufacturer information	320,000	26,300
Case 2	Reduction by $n*d_m$ *value*	210,000	21,300
Case 3	Consideration of installation situation	32,000	8300

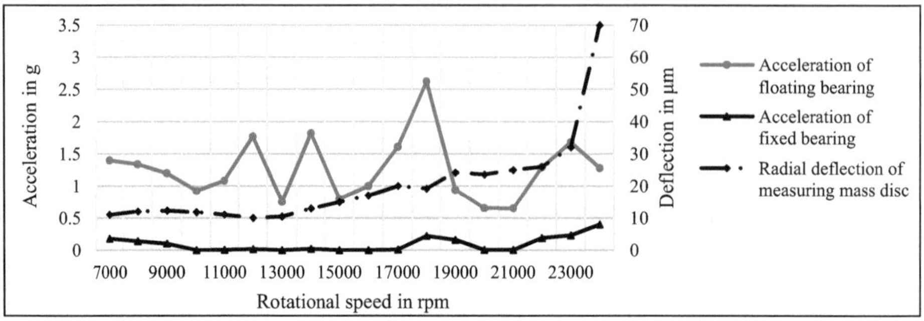

Fig. 9.33 Bearing accelerations and rotor vibrations during first commissioning of the FIMD flywheel system

are used for machine-dynamic monitoring (as shown in Sect. 7.5). During the first commissioning of the prototype the rotor speed was successively increased—as shown in the diagram in Fig. 9.33—the following observations were made:

1. At low speeds (<100 rpm), the capacitive distance signal of the circular "measuring mass disk" of the rotor shows an amplitude of about 12 µm. This value reflects the manufacturing-related out-of-roundness and eccentricity of the rotor.
2. Up to about 15,000 rpm, no significant rotor vibration could be detected. Between 17,000 rpm and 23,000 rpm, the deflection amplitude of the rotor increases significantly and reaches values up to 60 µm.
3. From 23,000 rpm onward, a sudden acoustic emission of the system, which increased sharply with rotor speed, indicated approaching resonance.
4. After switching off the system and a short break, an attempt was made to quickly pass through the presumed resonance at 23,500 rpm with high power (50 kW) of the electrical machine. However, the radial accelerations at the floating bearing became too high and bearing damage occurred (see Fig. 9.34).

9.6.1.5 Analysis of the *FIMD Rotor Bearing System*

The first real natural frequency of the system occurs at high angular speed (around 25,000 rpm), which means that—especially due to the high rotor mass of 84 kg—a lot of energy is dissipated when passing through the resonance. If the power provided by the motor generator is insufficient, it is not possible to overcome the resonance frequency, as the power of the electric machine is almost completely converted into vibration energy.

Looking at Fig. 9.33, it is clearly noticeable that the accelerations at the floating bearing are many times greater than those of the *back-to-back duplex set*. Since angular contact ball bearings are used in this case, it can be assumed that the axial preload of the floating bearing reached an inadmissibly low value due to the thermal expansion of the rotor, and

Fig. 9.34 Left: damaged floating bearing of the FIMD flywheel after first commissioning. Right: remaining Si3N4 rolling elements of the same bearing

the disk spring stack was no longer able to compensate for the change in length. (Bearing temperatures of 100–150 °C were measured.) The strong heating of the rolling bearings during operation has the following possible causes:

1. Too high internal preloading
2. Imbalance forces (high radial load)
3. Poor heat rejection due to grease lubrication
4. Operation of rolling bearings close to limiting speed
5. Viscous losses due to high lubricant quantity

Based on these experiences, it can be concluded that a reduction of the resonance speed as well as the radial bearing load by introduction of a resilient bearing seat and improved bearing cooling may represent a viable solution.

9.7 Resilient Bearing Seats for Rolling Bearings in FESS

The problem of high, unbalance-induced radial bearing loads described in Sect. 9.6.1.4 can be reduced by designing specific, resilient bearing seats. This way, the natural frequency of the system is shifted to lower speeds, which makes passing through the resonance a "low-energy process." The constructive design of a resilient bearing seat can be done in a variety of ways. Some options are described in [4], and possible design examples are shown in Fig. 9.35.

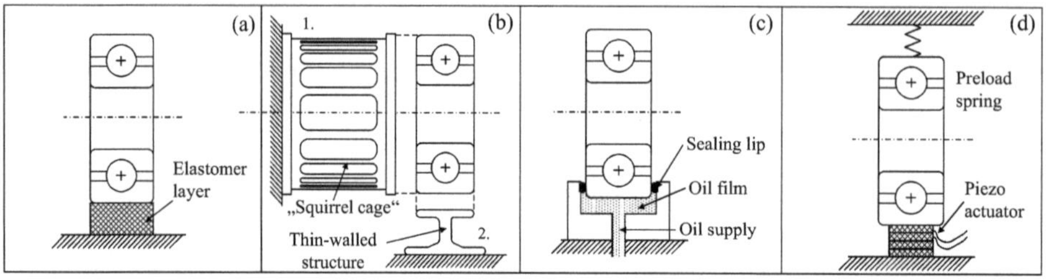

Fig. 9.35 Possible design versions of resilient bearing seats: (**a**) Simple elastomeric damper, (**b**) compliant metal structures ((1) squirrel cage or milled tube section or (2) thin-walled structure), (**c**) squeeze oil damper, (**d**) active piezo bearing tracking

9.7.1 Case Study *CMO Flywheel Energy Storage System*

The problems mentioned in Sect. 9.6.1.4 should be avoided by the design chosen for the *CMO flywheel.* The bearing concept and the essential design criteria are briefly outlined below.

Bearing Cooling Aspect

The bearing shield, which directly surrounds the rolling bearing, was provided with water cooling. In addition, an oil circulation lubrication system designed by *rosseta Technik GmbH* was implemented, which guarantees optimum heat rejection directly from the rolling element contact point.

Aspect of High-Frequency Resonance

The bearing shield was divided and a new design comprising two concentric elements was chosen. An elastomer layer is located between the *inner bearing shield,* which surrounds the rolling bearing outer ring, and the *outer bearing shield,* which is rigidly connected to the housing. This elastomer ensures low stiffness, yet sufficient internal damping, which reduces the resonance speed and hence mitigates bearing reaction forces.

While the gyroscopic reactions can only be eliminated by a suitable connection to the vehicle (see Sect. 9.5.4), imbalance forces caused by imperfect manufacturing can be reduced by appropriate bearing seats and supercritical rotor operation. The design of this resilient rotor connection must be such that the resonance lies below the operating range of the flywheel. The lowest operating speed at which the electric motor generators can still deliver constant power without requiring excessive currents is usually around 20% of the maximum speed, depending on the type of electric machine. The lower the resonance speed, the lower the energy and the easier it is to pass through it. Toward the lower end, however, the stiffness of the bearing seat is limited by the maximum permissible oscillation amplitude due to linear acceleration and gyroscopic reactions. Figure 9.36 shows rotor and end shield of the *CMO flywheel* after removal of the housing cover.

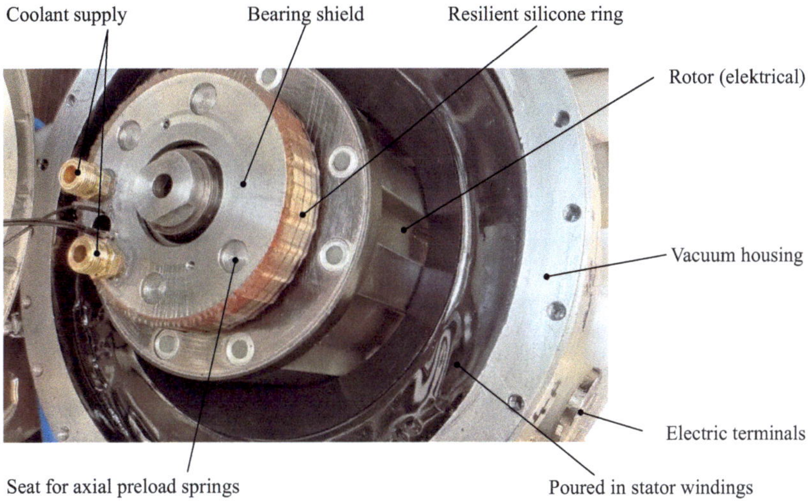

Fig. 9.36 Photo of bearing shield and rotor of the CMO flywheel

Whether flexible metallic structures or elastomers, such as the ring made of cast silicone shown in Fig. 9.38, are used depends on the desired damping characteristics. It must be mentioned that a *compromise* must always be found *between vibration insulation and vibration damping* [14]. Controllable or active measures such as the squeeze film damper, which are commonly used in turbomachinery or piezo actuators that even allow electrical bearing tracking, have also been considered for flywheel energy storage systems [15], but have so far been classified as inefficient solutions due to the great financial and design effort involved.

A low-cost solution used in the CMO flywheel energy storage system (see Sect. 7.4) is shown in Figs. 9.36, 9.37, and 9.38.

The mentioned lower speed limit of electromechanical flywheels allows to set the first natural frequency of the bearing-rotor system outside the FESS operating speed range. The reduction of the resonance frequency in combination with supercritical rotor operation in the electromechanically relevant speed range reduces the bearing loads and thus increases bearing life.

In order to characterize the suitability and properties of resilient bearing seats for FESS applications, it is necessary to investigate the system *bearing—bearing seat* in isolation due to the strong interdependencies in the whole energy storage device (magnetic forces, etc.).

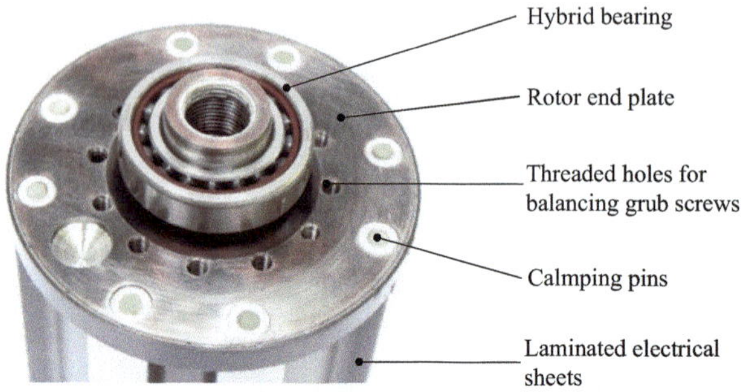

Fig. 9.37 CMO flywheel rotor with roller bearing

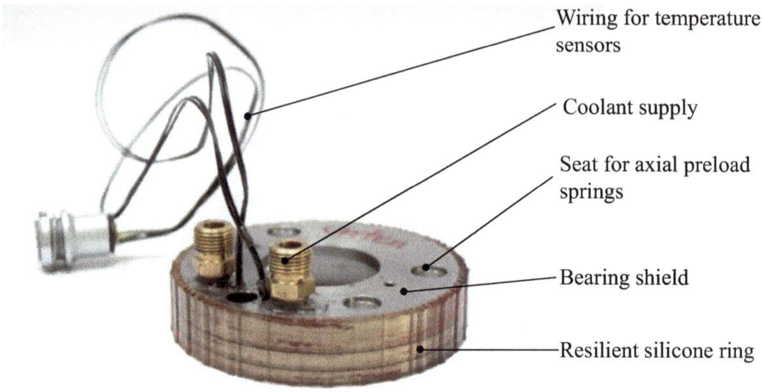

Fig. 9.38 Bearing shield of the CMO flywheel with flexible cast silicone ring

9.7.2 Investigation of Alternative Bearing Seat Concepts: Practical Example *LESS*

The practical examples of the *FIMD rotor* (see Sect. 9.6.1) and the *CMO flywheel* (see Sect. 9.7.1) have shown that failure of mechanical FESS bearings is primarily caused by the following two types of overload:

1. Thermal overload due to high speeds and lack of heat rejection (absence of convection cooling)
2. Mechanical overload when passing through resonances

While the thermal problem can be solved either by implementing a complex oil lubrication system or more simply by water-cooled bearing shields, as also shown in Fig. 9.36, the

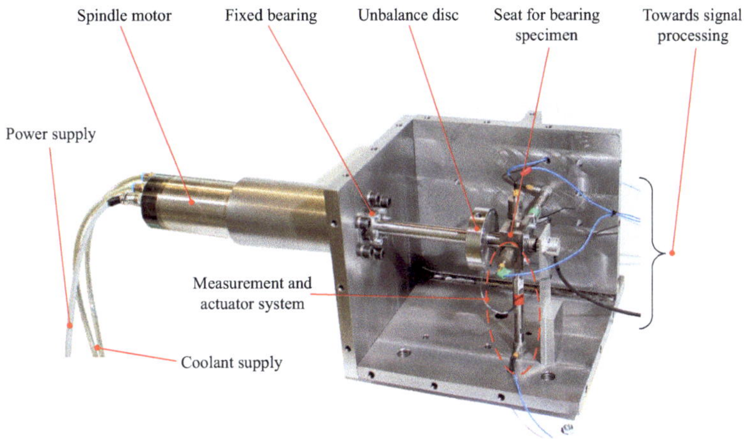

Fig. 9.39 Structure of the LESS test rig for investigation of active and passive bearing damping elements

resonance problem requires the modification of the entire machine dynamics behavior of the flywheel energy storage system.

In the Austrian FFG-funded project *Increased Service Life of Flywheel Energy Storage Systems* (*LESS*), the *Energy Aware Systems* Group at Graz University of Technology developed specific solutions for rolling bearings in flywheel energy storage systems. The main focus was on the investigation of active and passive measures for the manipulation of resilience and damping of the bearing seats. The concept of a special test rig, which was designed in the course of the project, is shown in Fig. 9.39.

The unit called "actuator and measuring system" consists of the following components:

- **Accelerometer:** Type *353B34* from *PCB Piezotronics*
- **Piezo force sensor:** *208C02 Multi-Purpose ICP* from *Pi*
  Optional, only in the case of *active* bearing tracking
- **Piezo actuators:** Type *P-842.20* with bending joints *P-176.50* and *P-176.60* from *Pi*

As can be seen in Fig. 9.40, the components are each installed in horizontal and vertical direction.

The investigation of an active bearing tracking, which allows active adjustment of the resonance frequency during operation, was carried out at the beginning of the project and—in short—brought the following findings:

1. In principle, the suitability of actively guiding bearings by means of piezo actuators and force feedback control could be proven.
2. The industrially available piezo actuators are only suitable for:
   (a) Small forces, i.e., they require a high balance quality of the flywheel

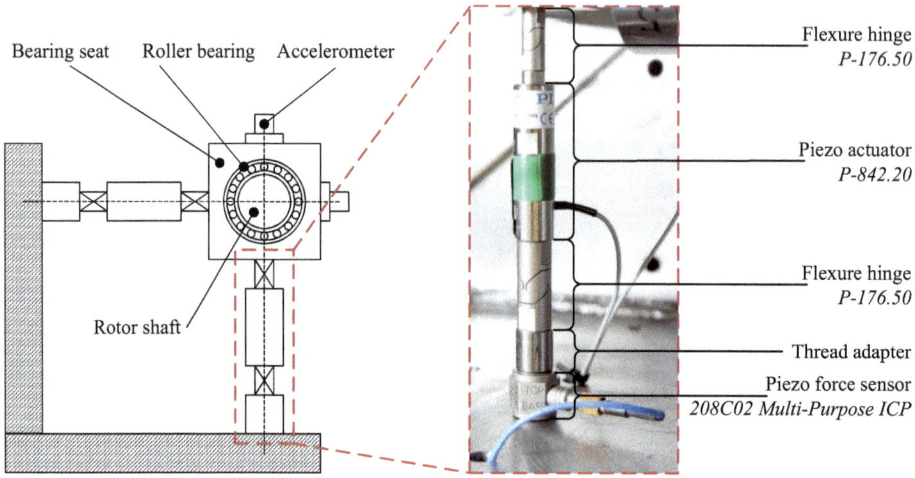

Fig. 9.40 Structure of the "actuator and measuring unit" of the LESS test bench

(b) Small displacements, i.e., the eccentricity must be very small

3. In order to actively dampen/control correspondingly high bearing reaction forces relevant for flywheel energy storage systems in the range of 0.1–5 kWh, correspondingly powerful charge amplifiers are required, which cause high costs and electrical losses.
4. The cost of a bearing seat consisting of piezo stack, amplifier, force measurement, and control system amounts to ~1500 Euro (even with low permissible bearing forces of less than 300 N).
5. Since the piezo stacks are very sensitive to tensile and bending stresses, compressive preloading by means of a spring package and careful handling are essential.

Due to these facts, the suitability of active, piezo-based bearing concepts for commercial flywheel energy storage systems in vehicles must be questioned. Especially the mechanical sensitivity of the system needs to be considered as a serious disadvantage, which can be accepted at most under laboratory conditions. More detailed information on the control system implementation of the project was published in [15].

9.7.2.1 Increasing Bearing Life Through the Use of Resilient Bearing Seats

The test rig shown in Fig. 9.39 allows the investigation of *two different bearing seat concepts*. On the one hand, the stiffness and damping of the bearing seat can be *actively* modified by means of piezo actuators, and on the other hand, various *flexible elements* such as elastomers or springs, i.e., *passive* solutions, can be evaluated. Figures 9.41 and 9.42 show practical possibilities of resilient FESS bearing seats, which were investigated at the

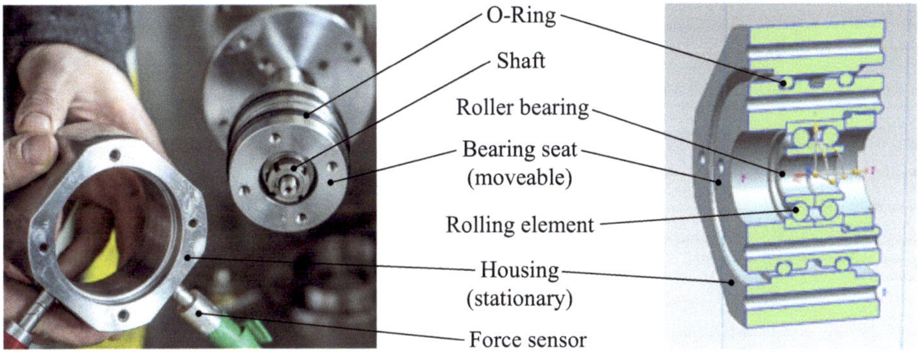

Fig. 9.41 Resilient bearing seat with elastomer elements (O-rings)

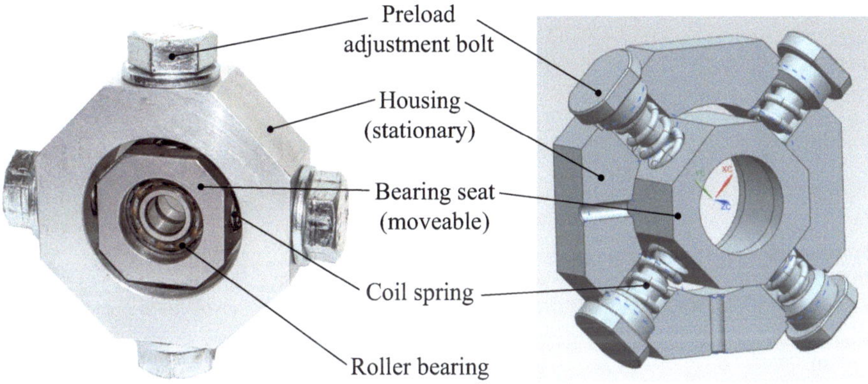

Fig. 9.42 Resilient beari lng seat with adjustable coil springs

Institute of Electrical Measurement and Sensor Systems of *Graz University of Technology* in the course of the *LESS project*.

In addition to a robustness superior to that of piezo actuators, these passive vibration isolation measures offer a significant advantage in terms of cost reduction. The unavoidable necessity to achieve a specific cost reduction of flywheel energy storage (€/kWh) has been mentioned several times in this book and justifies the investigation of passive low-cost bearing seats, despite the fact that the stiffness cannot be arbitrarily adjusted during operation.

Before the dynamic properties of the passive resilient bearing seats were checked on the *LESS test rig*, a quasi-static characterization was carried out, i.e., a force-displacement characteristic curve was recorded. The test setup used in this process is shown in Fig. 9.43, while actual measurement results are presented in Fig. 9.44.

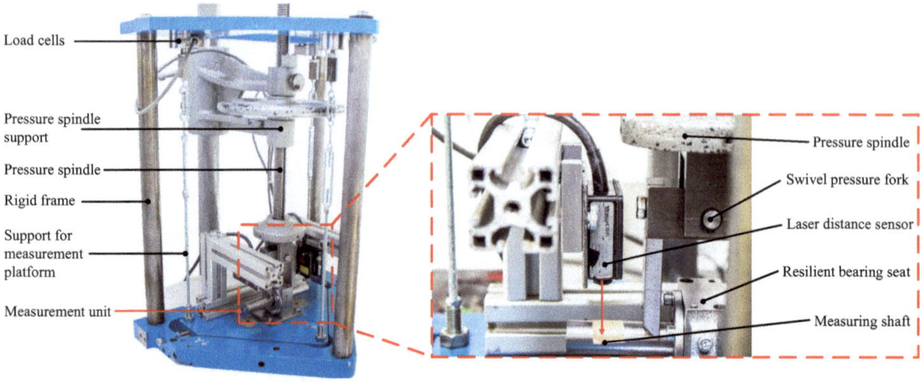

Fig. 9.43 Test setup for determining the stiffness of passive resilient bearing seats

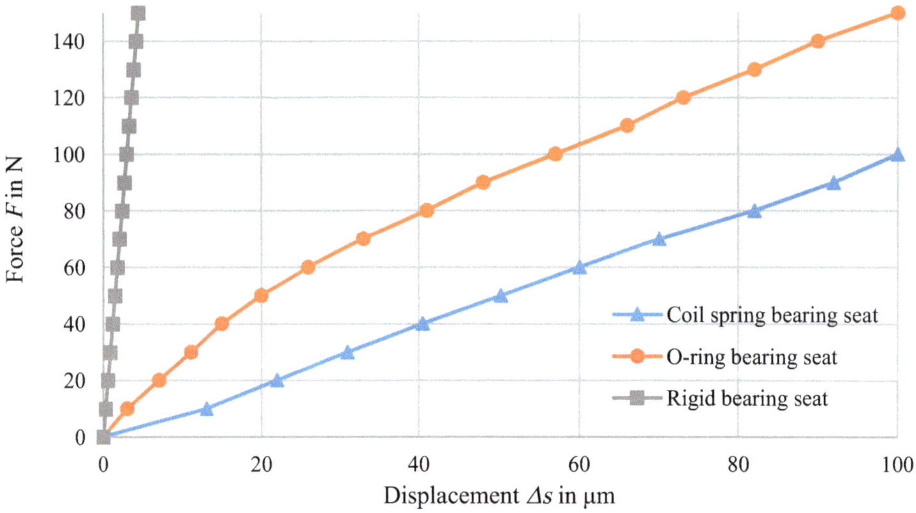

Fig. 9.44 Spring characteristics of two flexible and a rigid bearing seat

The easily recognizable degressivity of the spring characteristic curve of the bearing seat using O-rings is typical of the *pseudo-elasticity of elastomers*, which exhibit fluidlike behavior. While the "rigid" bearing seat has a measured resilience of 18.9 N/µm, the "2-O-ring bearing seat" reaches 1.6 N/µm, the "3-O-ring bearing seat" 2.4 N/µm, and the "spring bearing seat" about 1.0 N/µm. (For comparison: the stiffness of a rolling bearing with dimensions relevant to FESS lies in the range of several hundred *kilonewtons* per µm.)

During dynamic test bench investigations, the shaft with the unbalance disk was first accelerated to 3000 rpm, and this speed level was held for a few seconds to achieve a steady state. Then the speed was increased to 11,500 rpm, whereby the acceleration process only

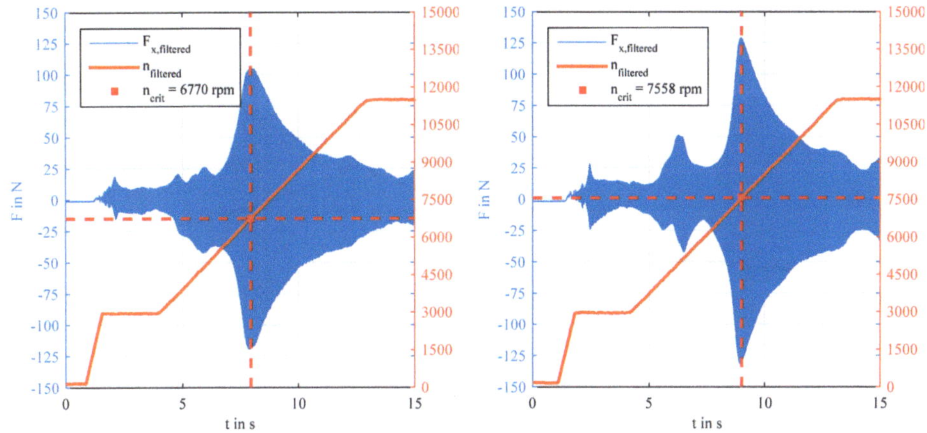

Fig. 9.45 Radial bearing load versus speed of the resilient bearing seats with 1.6 N/µm (**a**) and 2.4 N/µm (**b**). (Image rights: Michael Zisser)

lasted 5 seconds in order not to remain too long in the resonance, but pass through it rather quickly. As soon as the maximum oscillation amplitude had decayed again, the rotor ran supercritically. Figure 9.45 shows the resonance pass-through sequence of the 2- and *3-O-ring configurations*. It can be clearly seen that the more flexible bearing seat has lower maximum bearing loads (100 N vs 125 N).

The most important finding is that the unbalance-induced bearing load in supercritical operation decreases at higher speeds, whereas in a completely rigid bearing system, it increases with the second power of the rotational speed.

Figure 9.46 shows measurement results of bearing loads determined with the aid of the test setup shown in Fig. 9.39. Already at 10,000 rpm, a radial force reduction of 63% is achieved using a resilient bearing mount compared to the rigid bearing seat, which corresponds to a service life increase by a factor of 20.

9.7.2.2 Summary: Bearing Loads in FESS

Figure 9.47 summarizes all types of bearing loads occurring in flywheel energy storage systems, dividing them into four main categories, and lists their causes and possible solutions, which have been explained in detail in the previous chapters.

The following three aspects of FESS were picked out and analyzed in detail in this work:

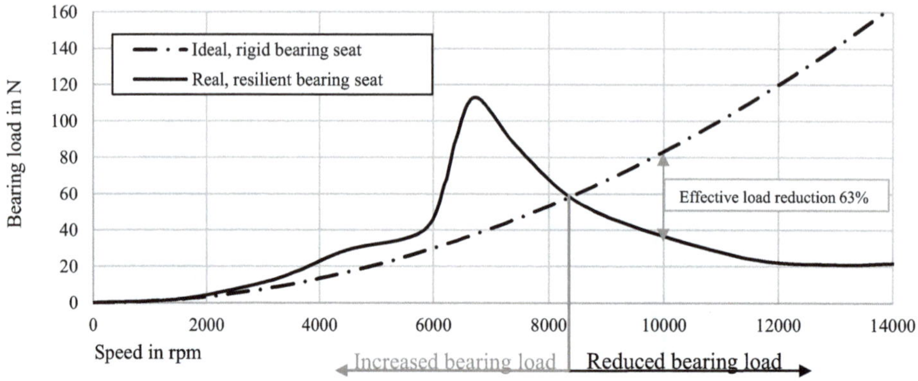

Fig. 9.46 Radial bearing loads with the same rotor imbalance, but rigid versus flexible bearing connection

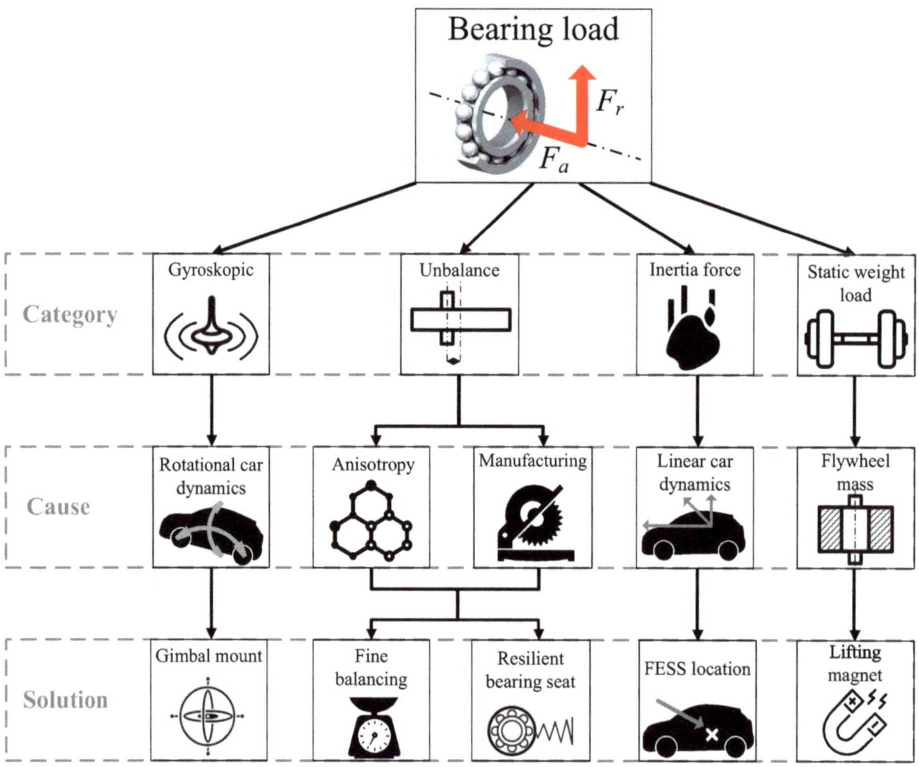

Fig. 9.47 Overview of bearing loads in energy storage flywheels

1. **Gyroscopic reaction forces**
   (a) It was shown that the FESS to vehicle mount has a considerable influence on the gyroscopic bearing loads. Ideally, a full gimbal is used as FESS suspension.
   (b) Another possibility for reducing gyroscopic bearing loads of FESS with a given energy content is to increase the rotor speed. An increase in speed at constant energy causes a reduction in rotor mass and thus in angular momentum. At this point, it must be remembered that the rotational speed influences the energy content to the second power, but it only has a linear effect on the gyroscopic torque.
2. **Imbalance forces**
   (a) A production-related eccentricity of the rotor causes mass imbalance forces, which can be reduced by dynamic balancing. Due to the limited accuracy of balancing machines, a certain "residual unbalance" is unavoidable and requires a solution, such as b.
   (b) A resilient bearing seat can reduce the unbalance-induced radial bearing loads at high speeds by ensuring supercritical rotor operation. Passive measures (using flexible, damping structures) have proven to be extremely effective compared to expensive and complex active measures (fast actuators for bearing tracking).
3. **Free inertial forces**
   (a) The free inertial forces of the rotor, which are caused by the linear acceleration of the vehicle and supported on the bearings, can only be eliminated by reducing the rotor mass (= speed increase) at a given energy content of the FESS.
4. **Static weight force**
   (a) In the case of a vertical axis of rotation—as it is common in almost all FESS applications—the weight force of the rotor on the bearings can be reduced by magnetic lifting, which enables a reduction of the axial bearing load and thus the loss torque. This solution is described in more detail in Sect. 10.3.1.

9.8 Thermal Properties of FESS Bearings

In addition to the mechanical loads described in Sects. 9.5 and 9.6, the bearings of a flywheel energy storage device are also subjected to *thermal loads*. Especially a rolling bearing experiences not only an increase in temperature due to internal friction (rolling friction of the rolling elements, cage friction, lubricant displacement dissipation, etc.) but also an external temperature input caused by the rotor. The rotor of the FESS can be significantly heated up by either electrical losses (eddy currents, hysteresis losses) or aerodynamic friction (windage losses) due to the high circumferential speeds. To counteract an impermissibly high operating temperature of the bearings, the bearing seats are usually water-cooled. Due to the vacuum atmosphere in the housing, however, the only way to cool the rotor is to dissipate the heat via the rolling bearings (see also Sect. 7.2.1). This is why the *thermal conductivity* of the bearings is of central importance. However, this

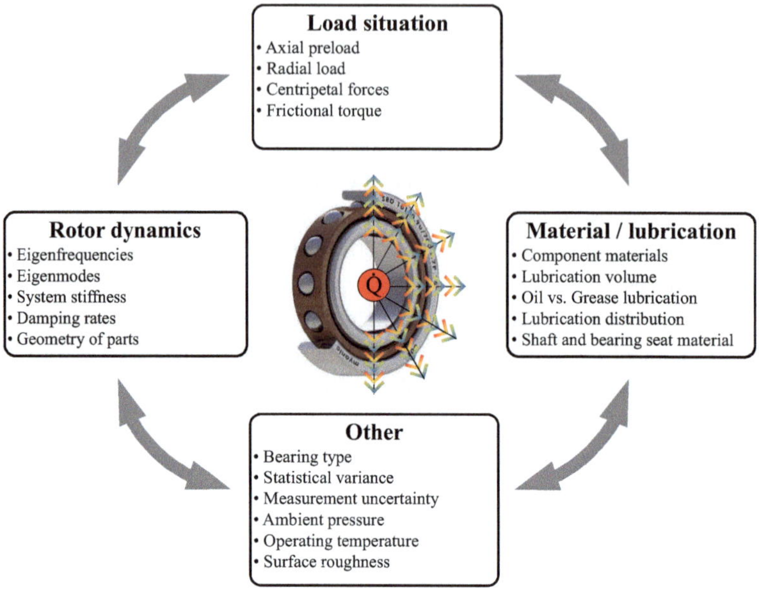

Fig. 9.48 Parameters influencing the thermal conductivity of rolling bearings

physical property depends on a large number of (operating) parameters and is also not specified by the manufacturer, in fact largely even unknown to many bearing companies. Figure 9.48 shows the main parameters influencing the thermal conductivity of rolling bearings.

9.8.1 Test Rig for Determining the Thermal Conductivity of Rolling Bearings

Although there are some analytical models that calculate the thermal conductivity of rolling bearings on the basis of a thermal resistance model (including the works of *Yovanovich* [16–18], *Bejan* [19], and *Baïri* [20]), some essential parameters, such as the influence of the lubricant, are not taken into account. The most reliable method is therefore the experimental determination of the thermal conductivity by measuring temperature gradients. Figure 9.49 shows a measuring principle in which a heat flux is conducted from a heated shaft via the rolling bearing to a cooled bearing seat. In this case, the bearing seat has a defined temperature measurement section, which is called the *spoke*, since its appearance is reminiscent of a cartwheel.

By measuring the temperature difference, neglecting convection and radiation (these assumptions are valid for most flywheel bearings) and assuming that the same amount of

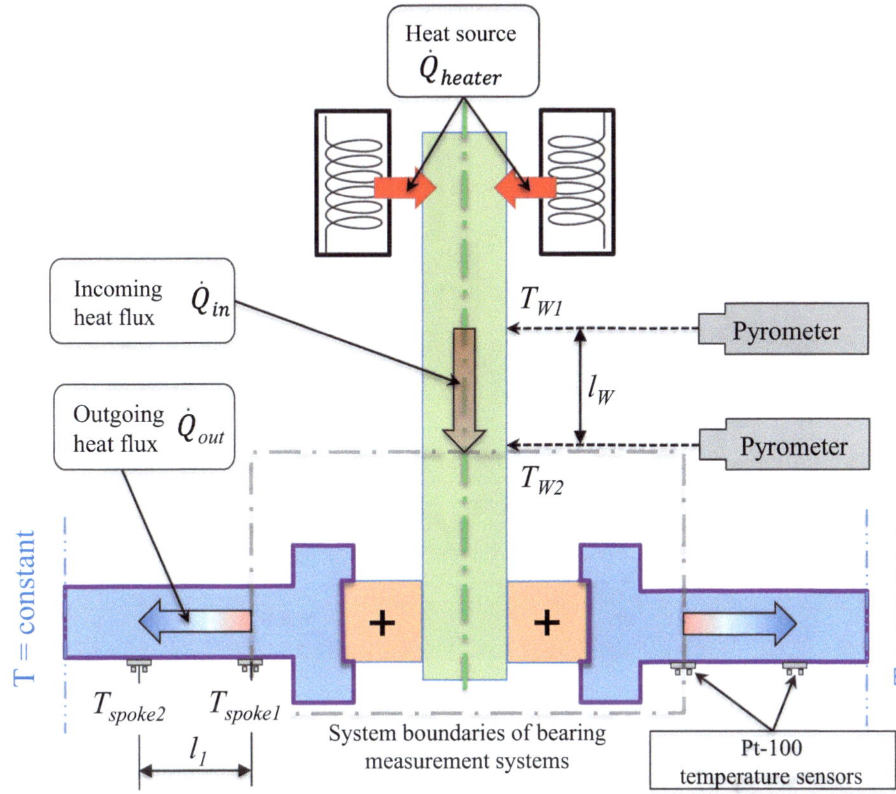

Fig. 9.49 Measuring principle for the indirect determination of the thermal conductivity of rolling bearings by measuring specific temperature gradients

frictional heat is generated on the inner and outer ring of the bearing, the *thermal conductivity* G_b can be determined as follows:

$$G_b = \frac{1}{2} * \frac{\dot{Q}_{in} + \dot{Q}_{out}}{T_{in} - T_{out}} \tag{9.14}$$

where the heat fluxes going into and out of the bearing can be calculated as follows (provided the *thermal conductivities* (λ_w, λ_{Spoke}) and material cross sections (A_W, A_{Spoke}) of shaft and web are known):

$$\dot{Q}_{in} = \frac{\lambda_W * A_W}{l_w} * (T_{W1} - T_{W2}) \tag{9.15}$$

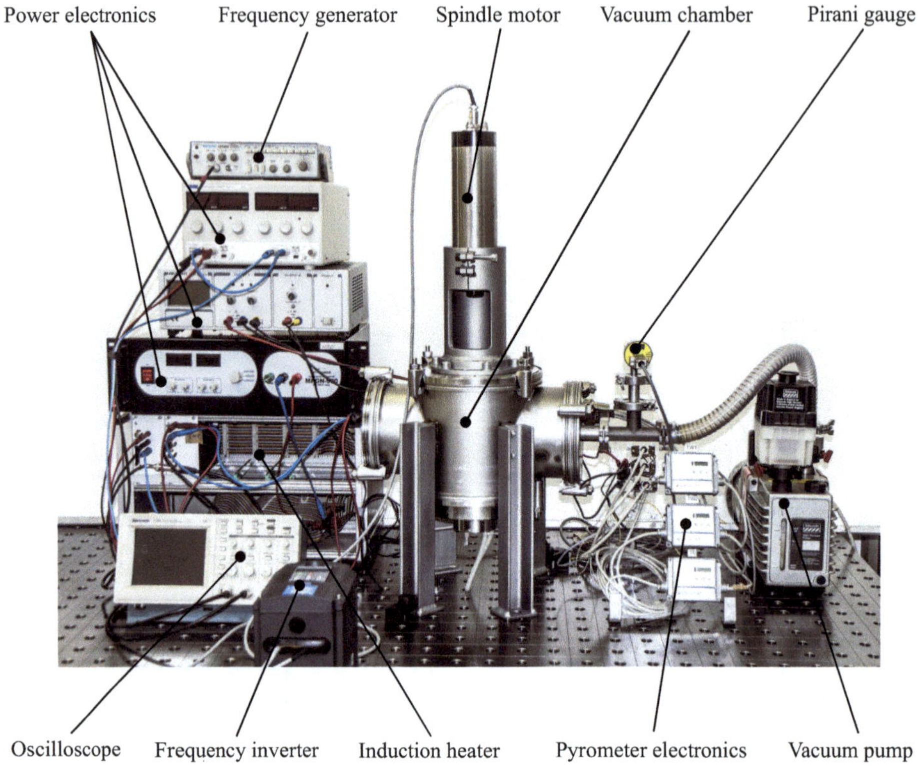

Fig. 9.50 Bearing heat conduction test bench with auxiliary units, *Energy Aware Systems* Group, Graz University of Technology

$$\dot{Q}_{Spoke} = \frac{\lambda_{Spoke} * A_{Spoke}}{l_{Spoke}} * \left(T_{Spoke1} - T_{Spoke2}\right) \tag{9.16}$$

Figure 9.50 shows the entire setup of the roller bearing heat conduction test rig at *the Energy Aware Systems* work group at Graz University of Technology, which operates according to the measuring principle described above.

Figure 9.51 shows an example of the measurement series of a rolling bearing type *71908 CEGA/HCP4A* compared with analytical calculations based on [19]. A striking influence of speed and preload on the *thermal conductivity* G_b of the bearing can be seen. It was also found out that the analytical calculation generally underestimates the real thermal conductivity, which may be due to the neglect of the lubricant influence.

A detailed description of the test rig and the underlying calculation methods can be found in [21].

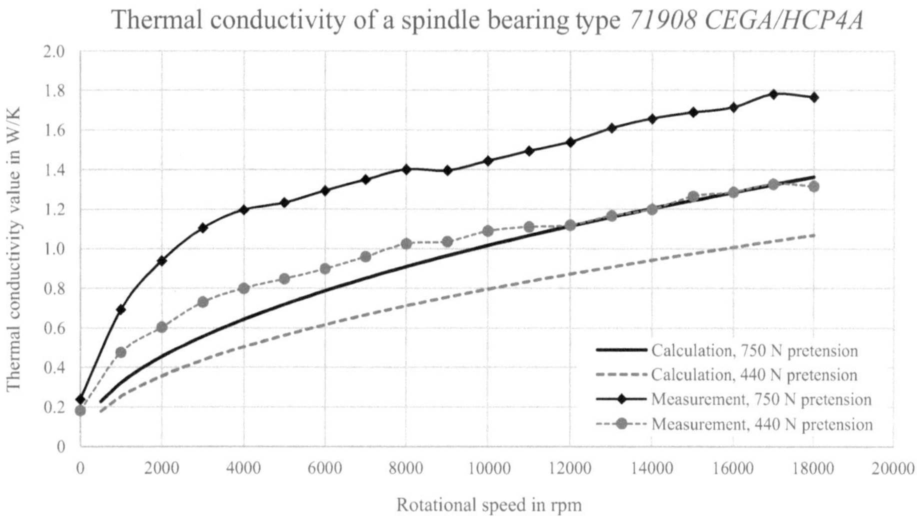

Fig. 9.51 Measurement results of the thermal conductivity of a rolling bearing *71908 CEGA/HCP4A* at different operating conditions and in comparison with analytical calculation. Grease filling capacity 0.5 cm^3, grease type *SKF LGLT-2*

References

1. F. Nelson (2007) Rotor Dynamics without Equations. International Journal of COMADEM, Nr. 10(3), pp. 2–10, Issue July 2007.
2. A. Buchroithner, A. Brandstätter and M. Recheis (2017) Determining Loads of Rolling Element Bearings in Mobile Flywheel Energy Storage Systems. IEEE Vehicular Technology Magazine, Volume: 12 Issue: 3, pp. 83-94. DOI: https://doi.org/10.1109/MVT.2017.2657804
3. J. Koyanagi (2011) Durability of filament-wound composite flywheel rotors. Mechanics of Time-Dependent Materials, Bd. 16, Nr. 1, pp. 71–83.
4. M. Recheis, A. Buchroithner, I. Andrasec, T. Gallien, B. Schweighofer, M. Bader and H. Wegleiter (2014) Improving kinetic energy storage for vehicles through the combination of rolling element and active magnetic bearings. IECON 2013 - 39th Annual Conference of the IEEE Industrial Electronics Society, Vienna, Austria. DOI: https://doi.org/10.1109/IECON.2013.6699884
5. K. Magnus (1971) Kreisel – Theorie und Anwendung, Springer-Verlag Berlin Heidelberg. DOI: https://doi.org/10.1007/978-3-642-52162-1
6. A. Brandstätter (2012) Mechanische Auslegung von Schwungrädern und Entwicklung eines Prüfstands zur Verifizierung der Eigenschaften für mobile Anwendungen. Institut für Maschinenelemente und Entwicklungsmethodik, Graz University of Technology, Austria.
7. M. Recheis (2017) Lifespan prolonging measures for bearings of mobile storage systems. Institute of Electrical Measurement and Measurement Signal Processing, TU Graz, Austria.
8. G. Bischof, K. Reisinger, T. Singraber and A. Summer (2014) Investigation of a passenger car's dynamic response due to a flywheel-based kinetic energy storage. International Journal of Vehicle

Mechanics and Mobility: Vehicle System Dynamics, Bd. 52, Nr. 2, pp. 201–217. https://doi.org/https://doi.org/10.1080/00423114.2013.869609

9. Parry People Movers Ltd. (2009) PPM Technology. Parry People Movers Ltd. Overend Road, Cradley Heath, West Midlands, B64 7DD, UK. http://www.parrypeoplemovers.com/technology.htm. [Accessed August 20th 2016].
10. Schenck-RoTec GmbH (2017) Warum ist Auswuchten so wichtig? The DÜRR Group, Bietigheim-Bissingen, Deutschland. https://schenck-worldwide.com/at-de1/why-balancing/index.php. [Accessed April 25th 2017].
11. H. Schneider (2013) Auswuchttechnik, VDI-Buch, Springer-Verlag Berlin Heidelberg. DOI: https://doi.org/10.1007/978-3-642-24914-3
12. SKF Gruppe (2014) SKF Hochgenauigkeitslager der Reihe „Super-precision bearings“. SKF GmbH, Gunnar-Wester-Straße 12, 97421 Schweinfurt, Deutschland.
13. K. Magnus, K. Popp and W. Sextro (2008) Schwingungen – Eine Einführung in die physikalischen Grundlagen und die theoretische Behandlung von Schwingungsproblemen. 8., überarbeitete Auflage. Vieweg+Teubner Verlag |GWV Fachverlage GmbH, Wiesbaden Germany.
14. P. Haidl, A. Buchroithner, M. Zisser, M. Bader, B. Schweighofer and H. Wegleiter (2016) Improved test rig for vibration control of a rotor bearing system. 23rd International Congress on Sound & Vibration (ICSV23), Athens, Greece.
15. M. Zisser, B. Schweighofer, H. Wegleiter, P. Haidl and M. Bader (2015) Test rig for active vibration control with piezoactuators. 22nd International Congress on Sound & Vibration (ICSV22), Florence, Italy.
16. M. Yovanovich (1970) Thermal Constriction Resistance Between Contacting Metallic Paraboloids: Application to Instrument Bearings. Proceedings of the 5th AIAA Thermodynamics Conference, pp. 337–358.
17. M. Yovanovich (1967) Thermal Contact Resistance across Elastically Deformed Spheres. Journal of Spacecraft and Rockets, vol. 1, pp. 119–122.
18. M. Yovanovich (1967) Analytical and Experimental Investigation on the Thermal Resistance of Angular Contact Instrument Bearings. Instrumentation Laboratory, pp. 1–85.
19. A. Bejan (1989) Theory of Rolling Contact Heat Transfer. Journal of Heat Transfer Vol. 111, pp. 257–263.
20. A. Baïri, N. Alilat, J. Bauzin, and N. Laraqi (2004) Three-dimensional stationary thermal behavior of a bearing ball. International Journal of Thermal Sciences Vol. 43, pp. 561–568.
21. A. Buchroithner, P. Haidl, H. Wegleiter, M. Simonyi and T. Murauer (2019) Design, operation and results of a low-cost test rig for investigation of thermal properties of rolling element bearings in vacuum. 18th European Space Mechanisms and Tribology Symposium (ESMATS2019), Munich, Germany, 18.-20. September 2019.

Stationary FESS for Modern Mobility 10

Flywheel energy storage systems, which typically offer high specific power output at moderate energy content and extremely high cycle life, are ideal for brake energy recuperation in vehicles used in dynamic driving cycles. Some specific problems when using FESS in automobiles, such as the gyroscopic bearing reaction forces described in the previous chapter or the crash safety issues discussed in Chapter 8, would be completely eliminated in stationary applications. If it is possible to minimize the rather high self-discharge rate of FESS and at the same time reduce costs, this concept could be a good alternative to chemical battery storage systems for storing renewable energy. The much-cited switch from conventional cars with internal combustion engines to what is likely to be a primarily electric transport system only makes sense if the electricity mix is largely made up of CO_2 neutral energy sources. In 2013, an average of 558 g CO_2 per kWh of electricity was still produced in the EU [1], which—if the charging and discharging efficiencies of electric vehicles are taken into account—is not much more environmentally friendly than energy production using a petrol engine, which generates around 690 g CO_2 per kWh [2].

In order to increase the effectiveness of private, decentralized PV systems and the degree of self-sufficiency, a flywheel energy storage system for solar energy was designed as part of a feasibility study at the *Institute for Electrical Measurement and Sensor Systems* at *Graz University of Technology*. A possible application that allows fast charging of electric vehicles with solar power even at night is shown in Fig. 10.1.

However, an important prerequisite for using a FESS as energy storage device for renewable energy is a low self-discharge, i.e., low torque loss of the bearing system.

A. Buchroithner, *Flywheel Energy Storage*,
https://doi.org/10.1007/978-3-658-35342-1_10

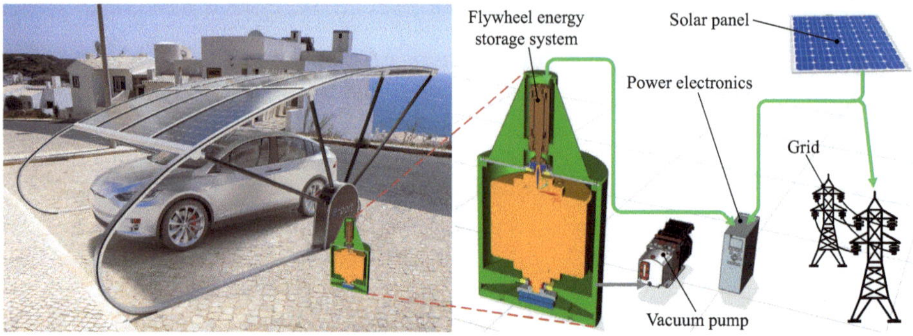

Fig. 10.1 Application of a FESS as a buffer storage for EV fast charging in combination with a solar cell-equipped carport. (Photo: Secar Technology GmbH)

10.1 Reduction of Torque Loss of FESS Bearings

10.1.1 Bearing Concepts for Stationary Flywheel Energy Storage Systems

In addition to the essential role regarding performance and energetic properties of the system, the bearing concept influences the costs of the FESS during the entire life cycle, as the following examples show:

(a) **Production and assembly costs**

Active magnetic bearings offer a particularly good suitability for high rotor speeds and the possibility of actively (during operation) changing stiffness and damping. However, the high costs must be mentioned as a serious disadvantage [3].

High-precision rolling bearings, which are suitable for the relevant speed ranges of flywheel rotors, not only require extremely precise manufacturing tolerances in the production of the bearing per se but also require exact manufacturing of the bearing seat and, consequently, good geometric dimensioning and tolerancing of the periphery. This is an aspect that also influences the system costs of the bearing (compare Fig. 10.2.).

(b) **Service costs**

While in FESS using active magnetic bearings only electronic components are subject to thermal aging, flywheels with rolling bearings are subject to wear. The renewal of the lubricant or, in some cases, of the entire rolling bearing is usually the only part that determines service costs.

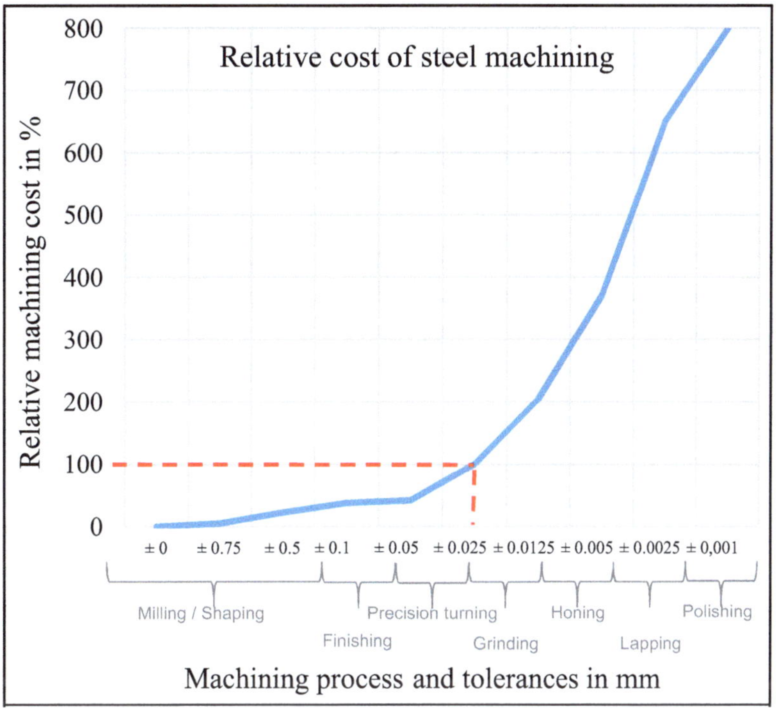

Fig. 10.2 Manufacturing costs via tolerance classes [4]. (Image rights: Springer Fachmedien Wiesbaden GmbH)

(c) **Operating costs**

In *UPS systems* where the flywheel is permanently running until a power failure occurs, the bearing losses are equal to the operating costs. If the *FESS* is used *to store renewable energy*, the bearing friction affects the self-discharge as well as the overall efficiency and subsequently the entire system profitability or payback period. Since aerodynamic losses can be minimized by evacuating the flywheel housing, bearing friction is one of the main causes of self-discharge and operating costs. (The low mechanical torque loss of active magnetic bearings is contrasted by electrical losses in the power electronics and ohmic losses in the coils.)

10.2 Loads and Friction Losses in Rolling Bearings for FESS Applications

The torque loss of rolling bearings is proportional to their speed and load, whereby the load-independent share depends on design aspects of the bearing (such as cage or possibly seal friction and rolling element centrifugal forces) and elasto-hydrodynamic lubrication effects. However, the majority of the torque loss is load-dependent, i.e., the rolling resistance increases proportionally to the axial and radial bearing load.

10.2.1 Bearing Loads of Stationary Flywheel Energy Storage Systems

Axial Bearing Loads

Instead of the gyroscopic loads in mobile FESS, the static weight load of the rotor dominates in the case of stationary flywheels. The significantly higher energy content normally required in stationary FESS requires larger rotor masses, which means that the fixed (or axial) bearing is subjected to a significantly higher load.

Radial Bearing Loads

The origin of all radial bearing loads in flywheel energy storage devices (with vertical axis of rotation) is found in unbalance-induced forces (see Sect. 9.6). These are an unavoidable product of the real manufacturing process of the rotor. And here again, a trade-off needs to be considered: Roller bearings are used instead of magnetic bearings to save costs. However, if a consistent low-cost approach is followed, the manufacturing tolerances must be kept within a reasonable range. Figure 10.2 shows the relative metalworking costs as a function of tolerance classes and different manufacturing methods.

10.2.2 Analytical Determination of Bearing Torque Loss

To estimate the torque loss of the rolling bearing system, three methods defined in the literature as industry standard were applied and compared. Table 10.1 provides an overview.

The actual calculation, no matter which of the three methods is used, always requires a good knowledge of the load situation. In the concrete case of an energy storage flywheel with 27 kg rotor mass (see Table 10.2), this means that in addition to the rotor weight, the eccentricity or unbalance must also be *exactly known*. The torque loss as a function of rotor speed is determined for the input values listed in Table 10.2 and is shown in Fig. 10.3.

Apart from a strong divergence of the torque loss calculation methods, it can be seen that a reduction of the degree of weight compensation from 0% to 90% does not result in a significant reduction of the calculated torque loss from a purely mathematical point of view, which—this can be forestalled—is in strong contradiction with the actual test bench results.

Table 10.1 Calculation methods for determining the bearing torque loss

Method	Formula	Description
FAG according to Palmgren [5]	$M_r = M_0 + M_1$ M_0 ... Speed-dependent torque M_1 ... Load-dependent torque	For this calculation, it is assumed that the frictional torque is composed of a load-dependent and a speed-dependent torque component
SKF simplified [6]	$M_r = 0,5\ \mu\ P\ d$ μ ... Constant coefficient of friction P ...Bearing load d ... Bearing diameter	This calculation applies to: Bearing load $P \approx 0.1\ C$ Good lubrication Normal operating conditions The frictional torque can be calculated using a constant coefficient of friction, the equivalent bearing load, and the bore diameter
SKF Expanded [6]	$M_r = \Phi_{ish} \cdot \Phi_{rs} \cdot$ $M_{rr} + M_{sl} + M_{seal} + M_{drag}$ Φ_{ish} ... Factor lubricant film thickness Φ_{rs} ... Influence of lubricant displacement M_{rr} ... Rolling friction torque M_{sl} ... Dynamic frictional torque M_{seal} ... Seal torque loss M_{drag} ... Viscous loss in the case of oil lubrication	This is a cause-dependent, physical-analytical calculation method. Sliding friction (also cage friction) as well as elasto-hydrodynamics of lubrication are also taken into account

Table 10.2 Input data for loss torque calculation, deep groove ball bearing *series 626*

Designation	Abbreviation	Value	Unit
Mass of the rotor	$m_{Flywheel}$	27	kg
Radius of the rotor	$r_{Flywheel}$	0.125	m
Eccentricity of the rotor	$e_{Flywheel}$[a]	10^{-6}	m
Speed of the rotor	$n_{Flywheel}$	20,000	min^{-1}
Degree of rotor weight compensation[b]	f_E	0.9	–
Resonance speed	n_{crit}	4500	min^{-1}
Radial load in supercritical operation[c]	$F_{rad\ overcritical}$	390	N

[a] Based on an assumed balancing quality of G = 6.3
[b] For magnetic weight compensation, see Sect. 10.3.1
[d] Determined by the eccentricity of the rotor and flexibility of the bearing seat

However, the reduction of the axial bearing load has a blatant influence on the mathematical estimation of bearing life, which is based on the classical method of *nominal bearing life calculation according to FAG* (*L_{10} method*), as shown in Fig. 10.4.

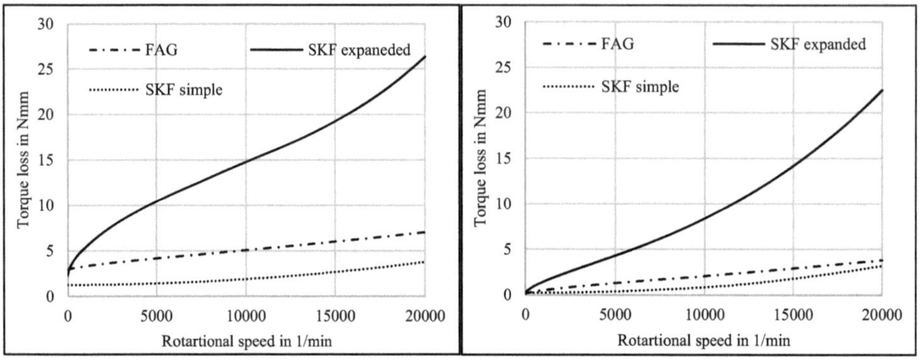

Fig. 10.3 Calculated torque loss of the 27 kg flywheel supported by deep groove ball bearings series 626 at 0% (left) and 90% rotor weight compensation (right) considering supercritical rotor operation. (Image rights: Clemens Voglhuber)

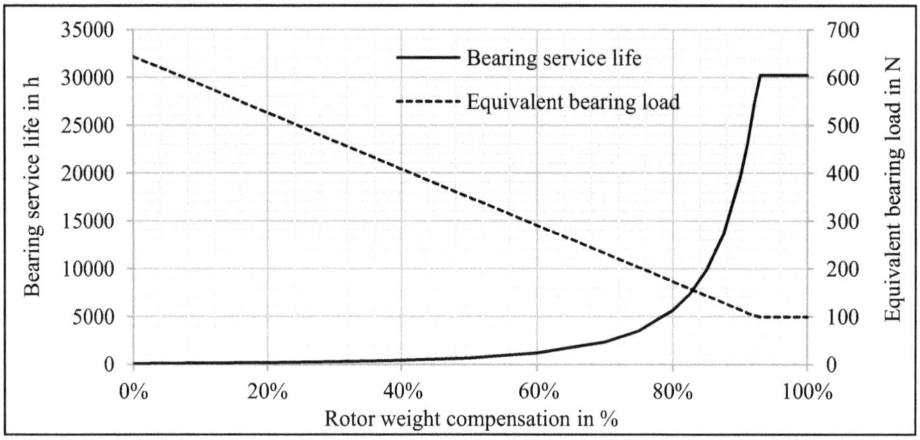

Fig. 10.4 Influence of the degree of magnetic rotor weight compensation on the service life of a deep groove ball bearing BR 626 at a constant speed of n = 10,000 min^{-1} (Image right: Clemens Voglhuber)

10.3 Bearing Load Reduction for Energy Storage Flywheels with Roller Bearings

10.3.1 Reduction of Axial Loads

As a method for reducing the axial bearing load, a *magnetic weight compensation* system was designed. In the first instance, the simplest topology of a flywheel energy storage system was considered, namely, a rotating, cylindrical disk made of steel without a motor

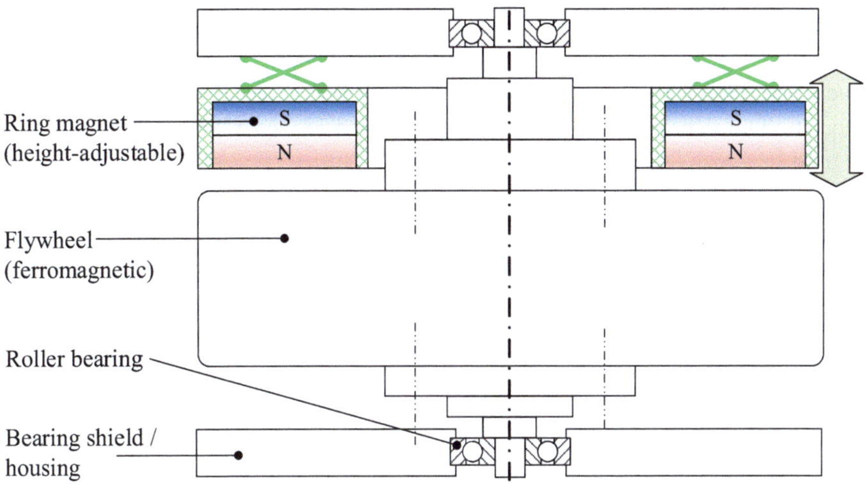

Fig. 10.5 Option 1: attracting arrangement with ferrite ring magnet

generator for electrical energy conversion. The solutions were evaluated within the scope of a master's thesis at Graz University of Technology [7].

Lifting magnets have been used several times in FESS [8], but the power losses still amounted to several hundred watts, which in the case of active linear solenoids is due to the electrical losses. Permanent magnets, on the other hand, were mainly used in a segmented design, which leads to inhomogeneity of the magnetic field and subsequently generates eddy current losses in moving (iron) parts. These losses inevitably occur when the magnetic field changes. The greater the time-related change in the field and the greater the conductivity of the (rotor) material in the field, the greater the eddy currents. On the one hand, these eddy currents cause losses via the ohmic resistance, and, on the other hand, they in turn generate a magnetic field that is directed against the direction of relative movement and thus slows down the rotor. Under certain circumstances, these effects may even cancel out the positive effect of weight compensation. Possible causes for a changing magnetic field are

- Anisotropy of the (ferromagnetic) rotor material
- Inhomogeneity of the magnetic field of the permanent magnet(s)
- Unsteadiness of the rotor surface within the magnetic field

10.3.1.1 Option 1: Attracting Arrangement with Hard Ferrite Ring

From a design point of view, this option offers the simplest solution for weight compensation. A ring magnet is placed above the steel rotor and reduces the axial roller bearing load through its attracting (lifting) force, as shown in Fig. 10.5.

Advantages

- Due to the possible expansion of the magnet almost to the outer rotor diameter, the magnetic flux density plays a subordinate role and low-cost hard ferrite magnets can be used.
- Good availability of ring magnets with bore diameters adequate for the flywheel shaft (>50 mm).
- Due to their ferromagnetic properties, steel rotors do not require a second, co-rotating magnet.

Disadvantages

All disadvantages of this arrangement have to do with eddy current losses. However, since the aim is to develop a bearing concept with the lowest possible losses, special attention must be paid to this issue.

- The good electrical conductivity of a (steel) rotor favors the propagation of eddy currents.
- Insufficient running tolerances of the rotor (axial run-out) cause a periodically changing distance between rotor and magnet and thus a change in flux density.
- The surface of the rotor facing the magnet must not show any inhomogeneities (holes, balancing grooves, sliding blocks, etc.).

The homogeneity of the magnetic flux of a hard ferrite magnet with 220 mm outer diameter and 80 mm bore was measured as shown in Fig. 10.6 using a *Teslameter 904 T*. The results are shown in Fig. 10.7 and show fluctuations of up to 20%. The SmCo disk magnet, which can be seen on the left of the picture, performs significantly better than the hard ferrite ring.

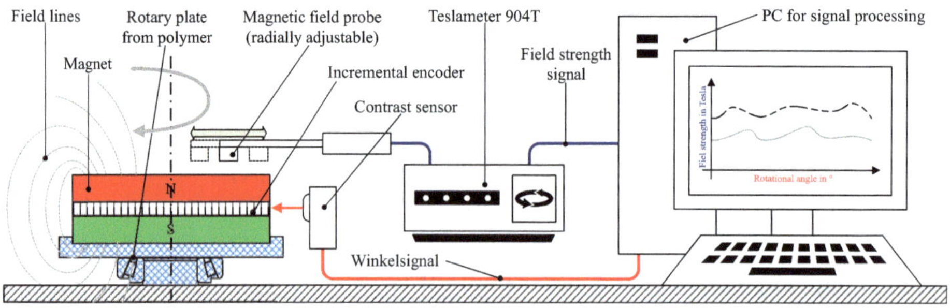

Fig. 10.6 Measurement setup for measuring the homogeneity of the magnetic field of a linear solenoid

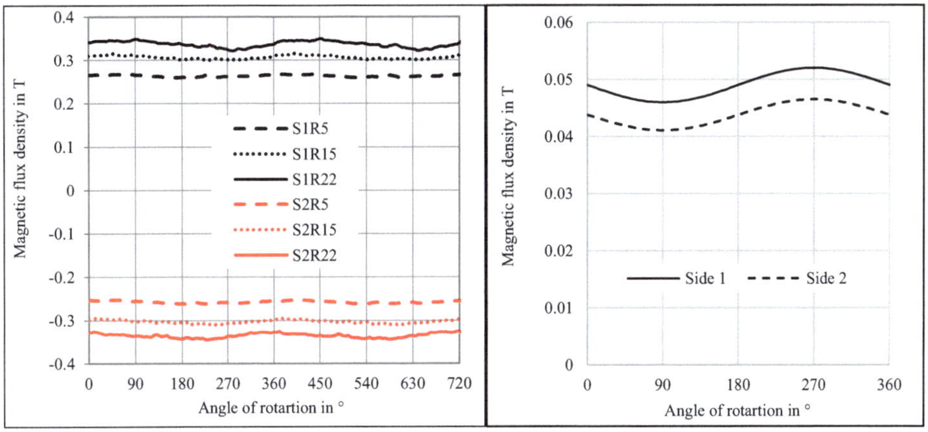

Fig. 10.7 Magnetic flux density of two magnet types. Left: SmCo disk magnet with 40 mm diameter. Right: hard ferrite magnet with 220 mm diameter and 80 mm bore

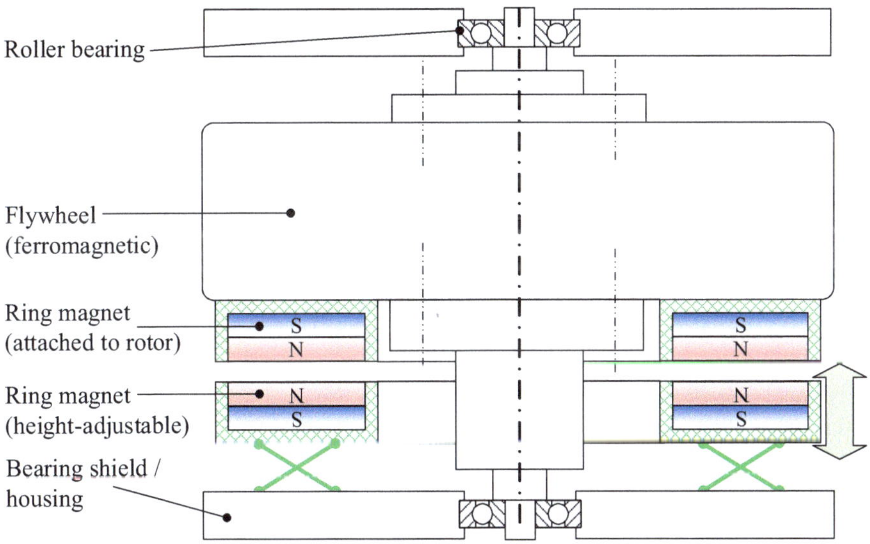

Fig. 10.8 Option 2a: repelling arrangement of two ferrite rings

10.3.1.2 Option 2: Two Magnets in Repelling Arrangement

The strong fluctuations in the magnetic flux density of the magnets used in option 1 would—in order to keep the torque loss due to eddy currents low—require a ferromagnetic, but electrically poorly conductive rotor material. However, since low-cost rotors would have to be made of structural or heat-treatable steel, the only way to reduce losses is to fit a second, repulsive ferrite ring, since this has poor electrical conductivity.

However, the configuration shown in Fig. 10.8 has the following problems:

- The high angular velocities usual for FESS cause high centrifugal forces in the rotor material. The tensile strength of hard ferrite is only about 50 MPa, almost a factor of 20 lower than conventional heat-treatable steels.
- In the case of repelling magnet arrangements, there is a risk of demagnetization. If hard ferrite is used, this detrimental effect may already occur at room temperature.

10.3.1.3 Option 2b: Repelling Arrangement of Two SmCo Disk Magnets

On the basis of the problems identified in Option 2a, the following solution was designed:

- The centrifugal force-induced stresses were mitigated by reducing the magnet diameter.
- Magnets with higher flux density (neodymium-iron or samarium-cobalt) were chosen.
- The magnets were designed without bore hole to further reduce the maximum centrifugal force stresses (Fig. 10.9).

Since energy storage flywheels usually run in a vacuum and there is no convective cooling, it was necessary to select magnets that are temperature-resistant with respect to their magnetic properties. In this case samarium-cobalt disk magnets were chosen. Nevertheless, the temperature dependence of the magnetic force requires the air gap to be adjustable. Figure 10.10 shows the magnetization characteristics of the selected SmCo magnet, and Fig. 10.11 a simulation of the magnetic field in *COMSOL*.

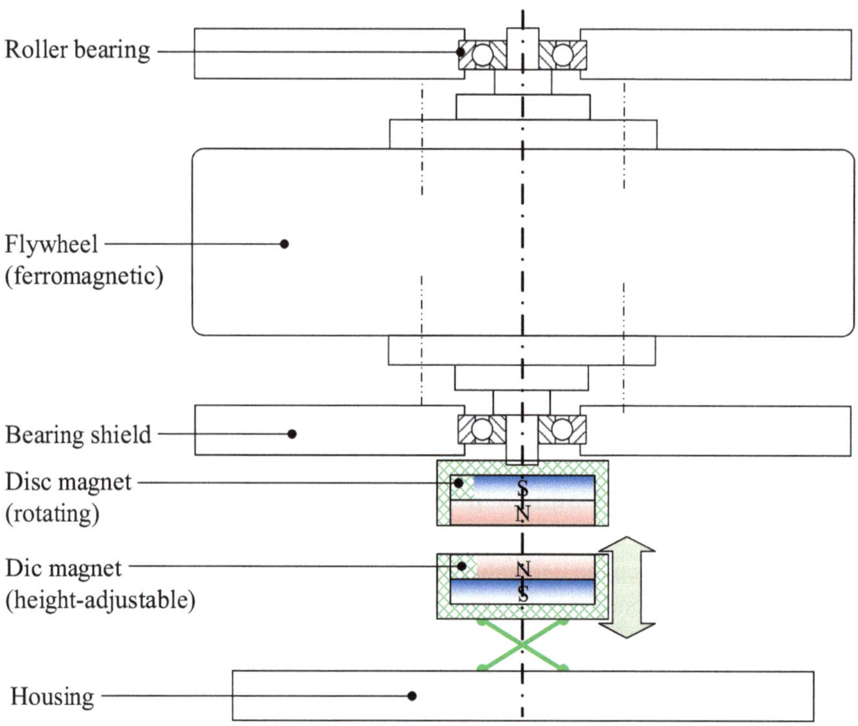

Fig. 10.9 Option 2b: repelling arrangement of two SmCo disk magnets

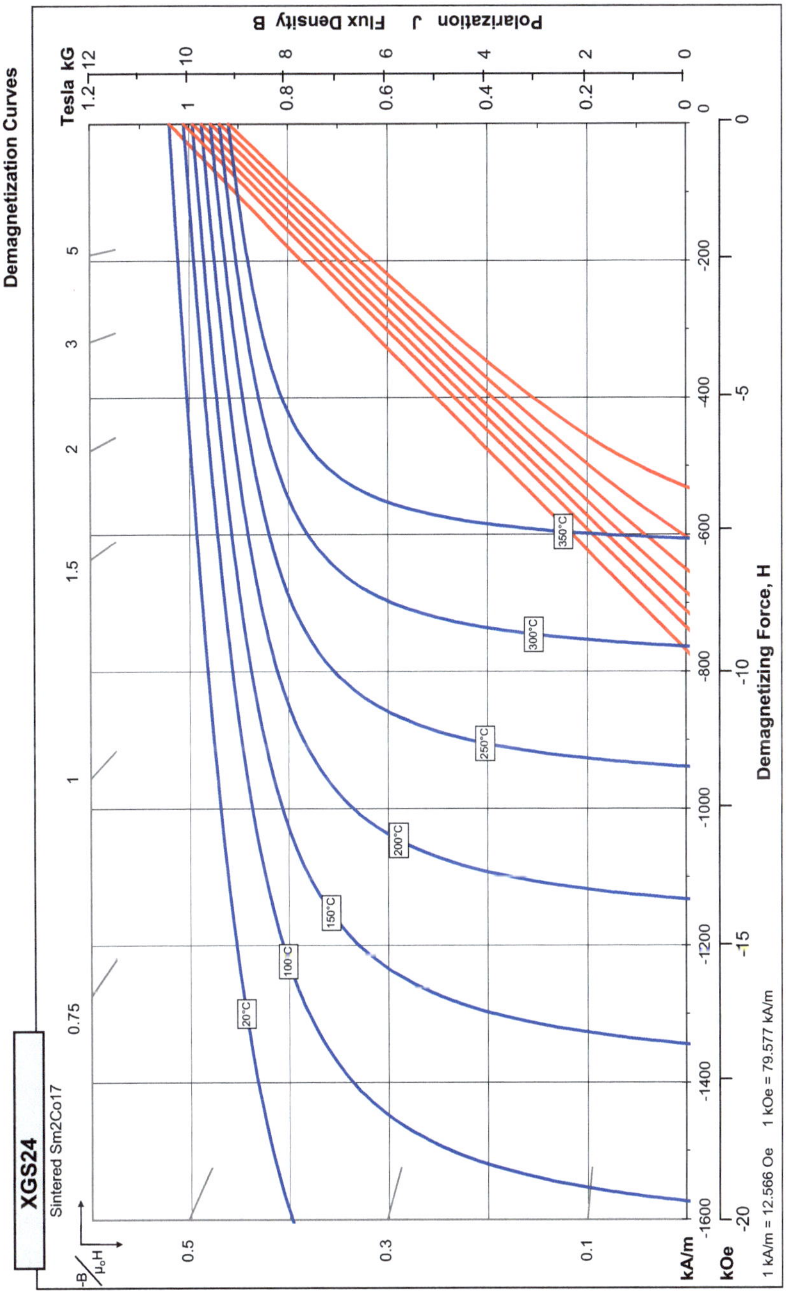

Fig. 10.10 Demagnetization characteristics of the Sm2Co17 magnet (*XGS24*) [9]. (Image rights: BVI Magnet GmbH)

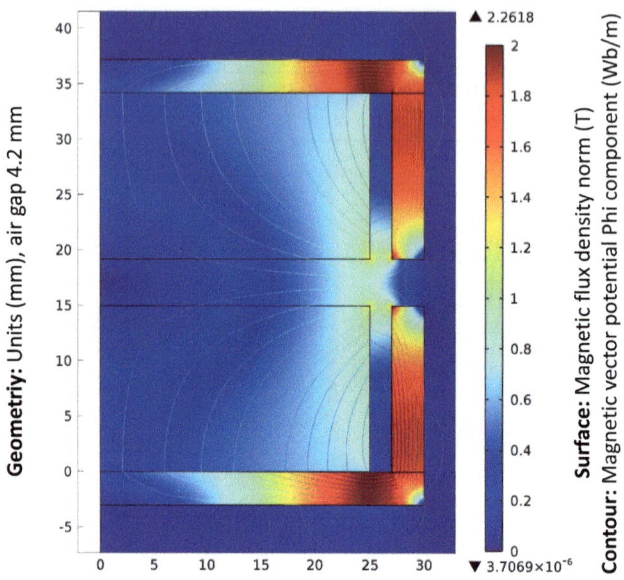

Fig. 10.11 Magnetic field simulation of the repelling magnetic discs of option 2b in Fig. 10.9

10.4 Reduction of Radial Bearing Loads

In order to increase the service life of rolling bearings in flywheels, active and passive measures for vibration isolation and damping were investigated as described in Sect. 9.7.2 [10]. In this context, mainly piezo actuators for active bearing tracking were compared with cost-effective, flexible structures, whereby the latter achieved satisfactory results [11]. Due to *Earnshaw's theorem*, a bearing arrangement based exclusively on permanent magnets cannot be implemented in a stable manner [12], which is why either active magnetic bearings (AMBs) or, as in this case, rolling bearings for radial guidance are still necessary. A resilient seat in the end shield of these bearings allows supercritical rotor operation.

10.4.1 Cast Silicone Bearing Seat

The bearing seat shown in Fig. 10.12 is designed as follows: The rolling bearing has been pressed into an aluminum bearing sleeve (i.e. *inner* bearing shield), which has structural grooves on the outer circumference. The gap to the *outer* bearing shield was filled with cast silicone (SHa 30). The easily recognizable degressivity of the force-deflection characteristic curve of the bearing seat (compare Fig. 10.13) is typical for the hyper-elasticity of elastomers, which exhibit fluidlike behavior.

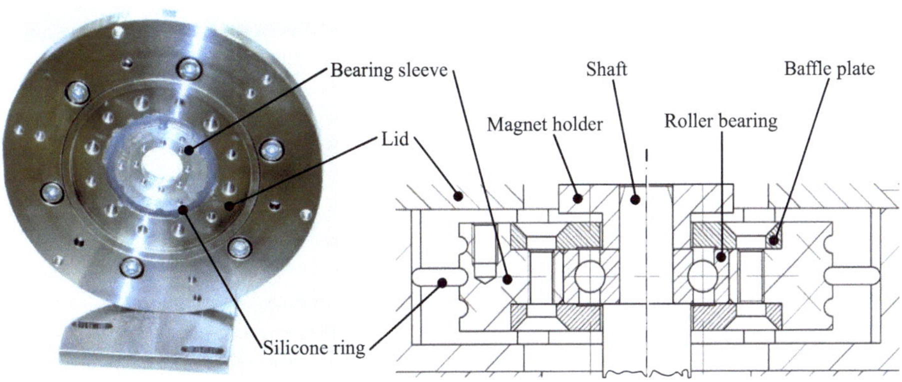

Fig. 10.12 Cast silicone end shield of the test flywheel's bearings

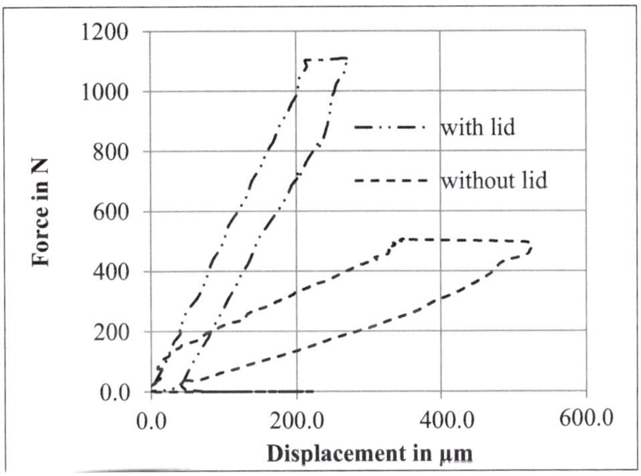

Fig. 10.13 Force-displacement diagram of the silicone bearing seat

As can be seen in Fig. 10.13, the resilience of the bearing seat is significantly reduced by attaching the covers, as these prevent the displacement of the elastic material. The combination of the measures for reducing the axial and radial loads allows significant downsizing of the ball bearings, which in turn further reduces the effective diameter of the frictional force and thus the torque loss (see Fig. 10.14).

The combination of measures to reduce the axial and radial loads, consisting of a lifting magnet and a flexible bearing seat, allows significant rolling bearing downsizing, which in turn further reduces the effective diameter of the frictional force and thus the torque loss.

In order to quantify the torque loss of the bearing arrangement, a so-called spin-down test rig was built and put into operation. The gradient of the run-out curve (speed over time)

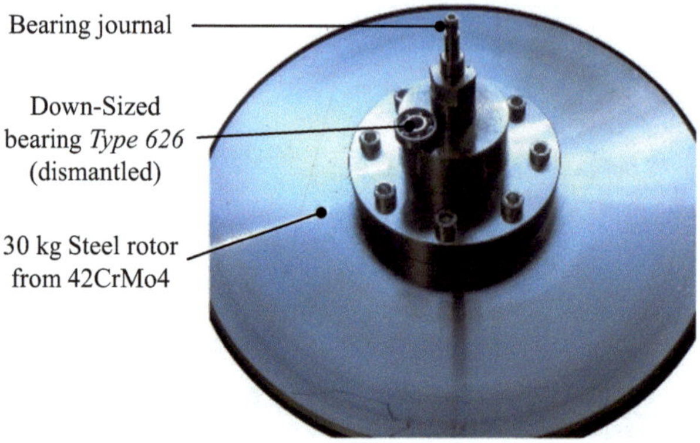

Fig. 10.14 Flywheel shaft assembly and deep groove ball bearing series 626 in size comparison

is a measure for the friction losses of the system, which can be calculated with the help of the conservation of angular momentum theorem stated in Eq. 10.1:

$$M_{loss} = J_{Rotor} \cdot \frac{\partial \omega}{\partial t} \tag{10.1}$$

The adjustment of the degree of rotor weight compensation by axial infeed of the SmCo magnet is carried out via an electromechanical linear actuator. The actual axial bearing load is measured using the setup shown in Fig. 10.15. The test rig includes the following measurement equipment:

- Measurement of the actual axial bearing load/preload (bearing shield in differential design with strain gauge bending beams, as shown Fig. 10.16)
- Measurement of rotational speed (laser contrast sensor)
- Measurement of the atmospheric pressure (Pirani vacuum probe)
- Measurement of various temperatures (Pt-100 temperature sensors)
- Measurement of the acceleration at the bearing seat (piezo accelerometers)
- Measurement of the vibration amplitude of the shaft (laser triangulation sensor)
- Measurement of the axial run-out of the flywheel (laser triangulation sensor)

In order to eliminate the aerodynamic torque loss, the test rig design shown in Fig. 10.17 was integrated into a vacuum chamber. The torque for accelerating the flywheel is applied by means of an asynchronous motor and magnetic coupling that acts through a membrane made of fiberglass composite. The maximum speed is 24,000 rpm.

10.4.1.1 Results

The aim of the empirical investigation was to prove the functionality and performance of a low-loss low-cost flywheel bearing concept. Contrary to the calculations carried out in Sect. 10.2.2, the effectiveness of the magnetic weight compensation must be described as astonishingly high, **since a reduction of the loss torque of about 80% could be achieved**

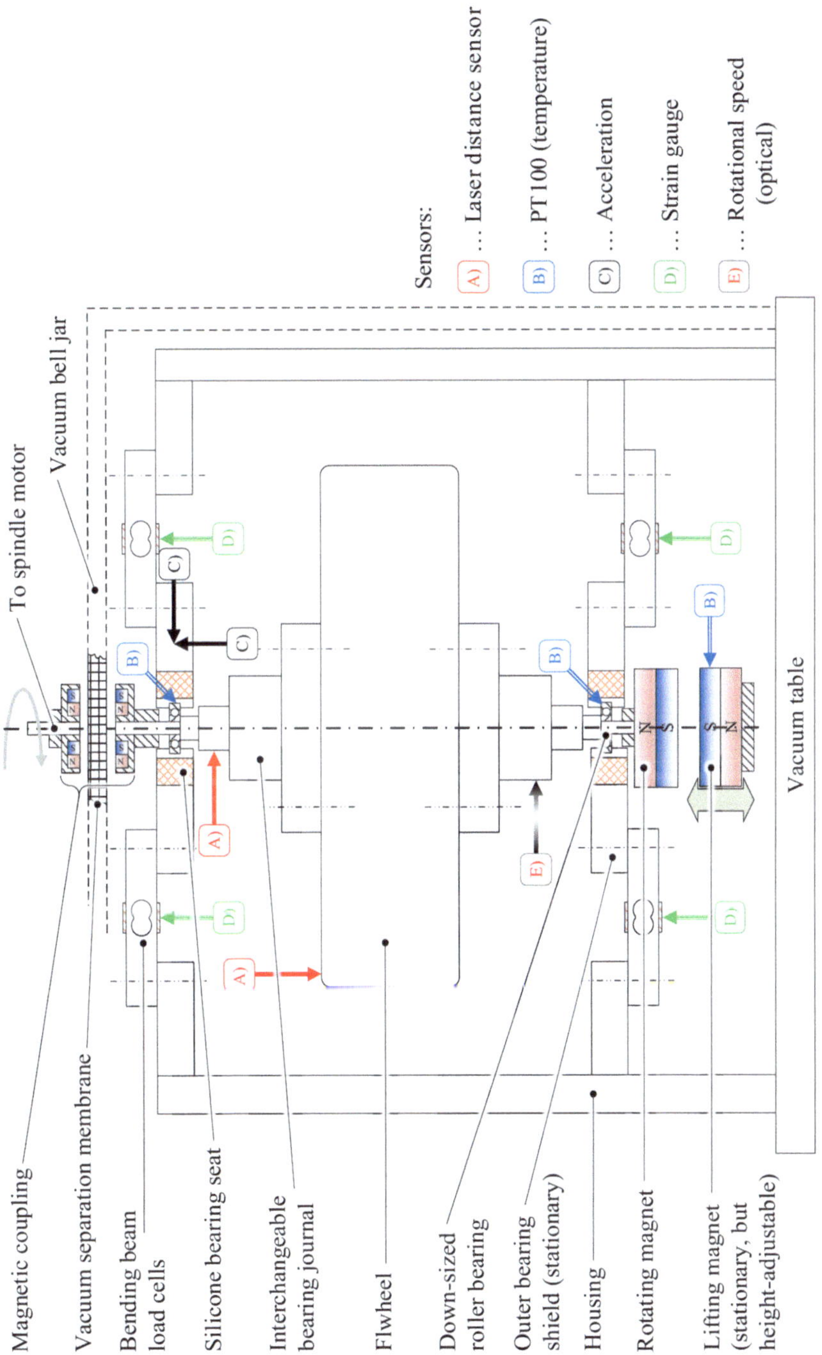

Fig. 10.15 Structure of the spin-down test bench with marking of the measuring points

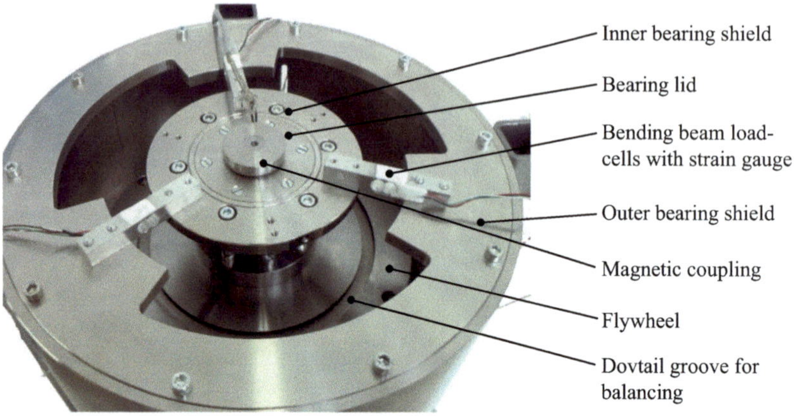

Fig. 10.16 Bearing shield assembly with "spoke wheel" made of load cells (bending beam design) to determine the axial bearing loads

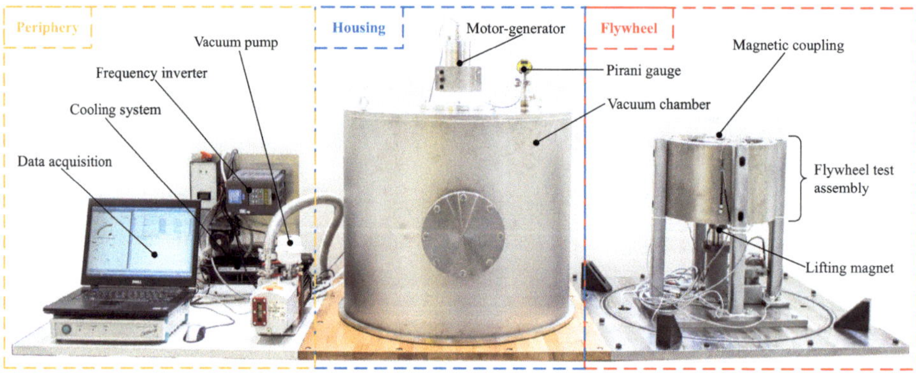

Fig. 10.17 Complete test bench setup for loss torque measurement (spin-down test rig)

in principle. Figure 10.18 shows the absolute torque loss of the flywheel at 1000 rpm and ambient pressure, whereby the share of aerodynamic losses (windage) is marked by a blue, semitransparent area.

However, low torque loss over the *entire speed range* is decisive for the efficiency and self-discharge of the flywheel energy storage system, which makes machine dynamics considerations even more important.

Figure 10.19 shows a 5000 rpm flywheel spin-down curve, with an obvious discontinuity at about 2500 rpm. The reason for this can be found in passing through the resonance

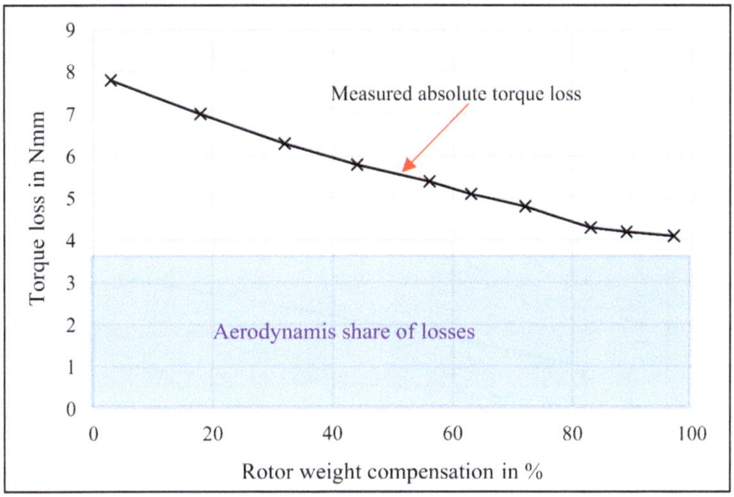

Fig. 10.18 Determined torque loss of the ball bearing arrangement over degree of magnetic weight compensation at 1000 rpm. (Image rights: Clemens Voglhuber)

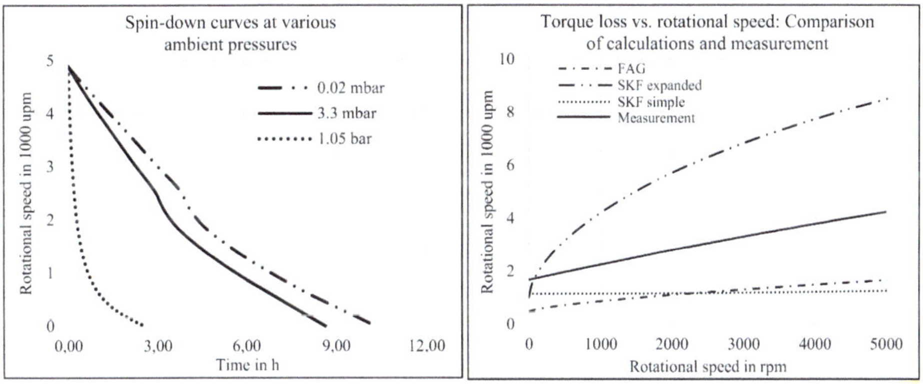

Fig. 10.19 Spin-down tests of the flywheel at various ambient pressures. (Image rights: Clemens Voglhuber)

frequency of the system, which is confirmed by a measurement of the vibration amplitude of the flywheel shaft (compare Fig. 10.21).

A comparison of the torque loss measurement and calculation in Fig. 10.20 shows that the simple methods of *SKF* and *FAG* underestimate the friction losses by a factor of about 2 to 4, while the *expanded SKF method* assigns too much importance to the fluid dynamics components of higher power.

The asymmetry of the diagram in Fig. 10.21 (left) can be explained by gyroscopic effects and a manufacturing-related anisotropy of the cast silicone (air inclusions). Figure 10.21 (right) shows the influence of an unbalance of 25 g attached to the flywheel at a

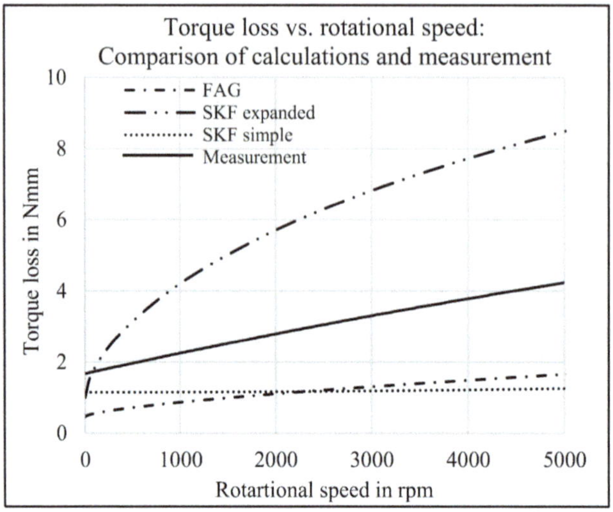

Fig. 10.20 Comparison of the measured torque loss of the flywheel bearings with analytical calculation methods. (Image rights: Clemens Voglhuber)

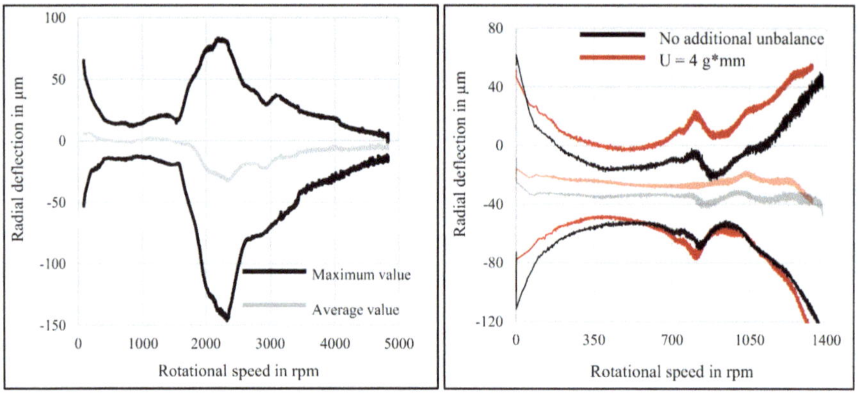

Fig. 10.21 Amplitude of the flywheel shaft of the spin-down test bench over speed. (Image right: Clemens Voglhuber)

radius of 100 mm (corresponds to U = 4 g*mm). The influence of unbalance can clearly be seen as it causes significantly higher shaft oscillation. Due to the large deflections of the flywheel, higher speeds than 1500 rpm were not possible in this test.

10.5 *FlyGrid*: Flywheel Energy Storage for EV Fast Charging and Grid Integration

The theoretical potential of flywheel energy storage technology, for both mobile and stationary applications, is now clear to every reader of this book. This potential must be given even more importance, especially in the context of the energy revolution. The transition from vehicles with combustion engines to purely electric mobility is considered one of the most important steps in the course of decarbonization. It is equally important to achieve the climate protection goals as it is to achieve political and economic independence. While the strong predicted growth in the number of electric vehicles can be described as a thoroughly positive development, it will result in a number of new challenges for energy suppliers, network operators, and vehicle and charging station manufacturers and, ultimately, for the customers.

10.5.1 Developments in Electric Mobility

In particular, the ever-increasing EV charging powers, combined with an increasing supply from volatile sources, result in an enormous electricity network load, which can cause instabilities and—in the worst case—even blackouts. Nevertheless, a development toward fast charging (100 kW and more) can be considered absolutely necessary to combat EV customer's range anxiety. Mobility experts consider the lack of a suitable quick-charging infrastructure to be the greatest threat to e-mobility. In order to largely avoid costly network expansion and yet provide a high-performance, nationwide network of fast-charging stations, new, innovative solutions must be found that not only guarantee 100% customer satisfaction but also facilitate the integration of renewable energy sources.

10.5.2 Aims of the *FlyGrid* Project

In the FFG-funded Austrian *FlyGrid* project, a high-performance flywheel energy storage system is integrated into an innovative, fully automatic EV charging station. As a result, even when connected to a conventional low-voltage distribution network (400 V, 50 Hz), high charging powers can be achieved while simultaneously allowing load averaging in the network. The system is designed to integrate local volatile sources—such as PV modules on a carport—and thus contributes to increasing the share of renewable energy for transportation.

Superior cycle life of the flywheel energy storage system, the possibility of feeding high power back into the grid, and easy transportability as a mobile "fast-charging container" (e.g., for electrified construction machinery) are further characteristics of the *FlyGrid* concept. This results in a wide range of possible applications, which are of high relevance not only to vehicle fleets and public transport but also for network operators, as shown in

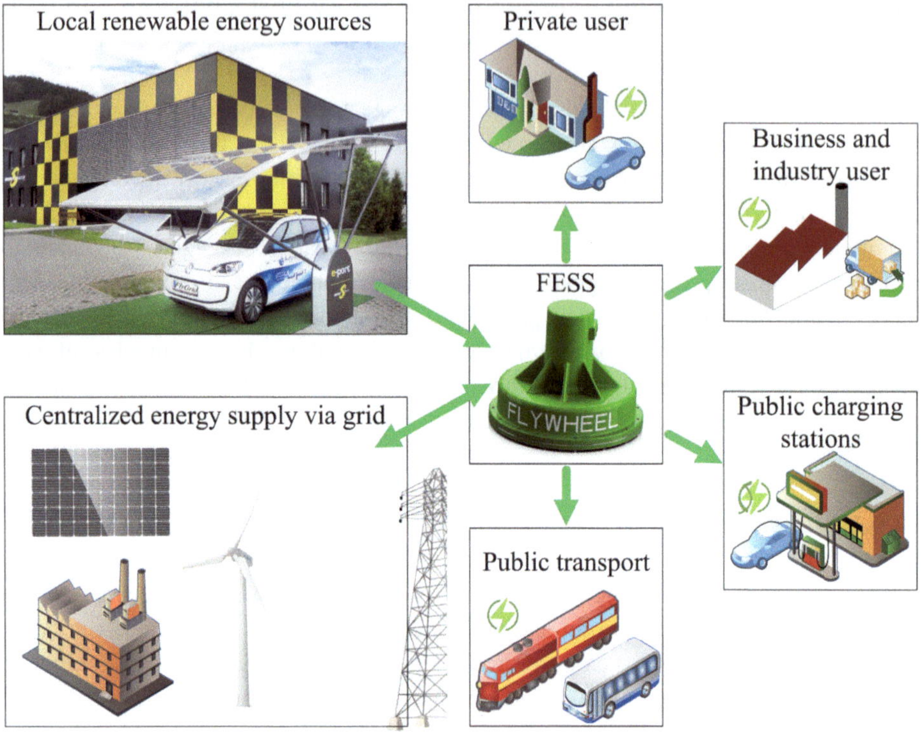

Fig. 10.22 Use cases of the *FlyGrid* energy storage system

Fig. 10.22. *FlyGrid* is a disruptive technology that can be produced in Central Europe or other Western countries for that matter, through which the following overriding goals can be achieved with a high socioeconomic impact:

- Reduction of charging time of EVs and higher market penetration
- Higher customer satisfaction through improved EV charging network
- Avoidance of cost-intensive network expansion
- Improved integration of renewable sources for the supply of electric mobility
- Improved grid stability and power quality
- Portable fast-charging solution for electrical construction machinery or events

10.5.3 Core Element Flywheel Energy Storage

At the center of the system and the research task is the electromechanical flywheel energy storage unit. A possible design of the device is shown in Fig. 10.23. A vacuum pump, a cooling system, and the frequency converter are the usual peripheral components, as

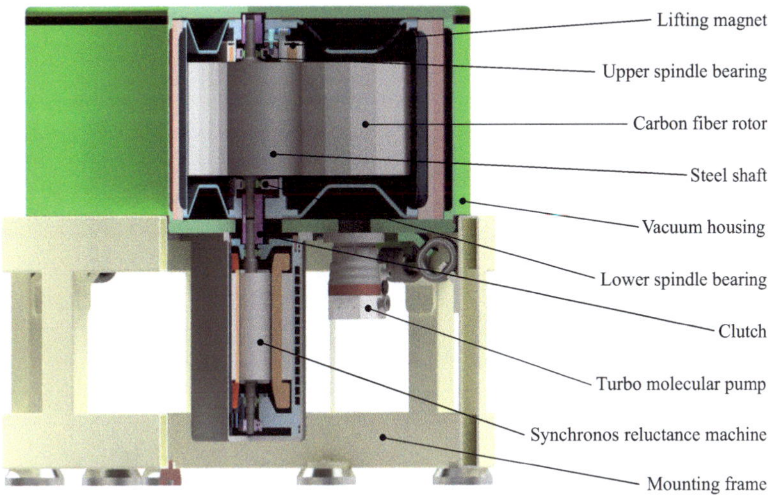

Fig. 10.23 Possible layout of the *FlyGrid flywheel system*

described in Sect. 2.2.3. Compared to batteries, this concept has some decisive advantages for the planned application [13]:

- High number of charge cycles (long service life)
- No capacity fade due to aging
- High power density
- Simple determination of the state of charge (SoC)
- Deep discharge possible (no transport requirements/problems)
- No toxic or rare raw materials required, unproblematic recycling

Chapter 7 has shown that the specific energy of the system is defined by the ratio of the permissible centrifugal force stresses to the density of the rotor material. Rotors made of carbon fiber composite have a high tensile strength at low density, which is why they can achieve high specific energies. Figure 10.24 shows maximum achievable specific energies based on different rotor materials. The theoretical potential of carbon nanotubes is higher than that of fossil fuels in terms of energy content. Of course, under the current state of the art, this must still be considered a distant future goal.

In order to improve the system properties, the development activities related to the FESS focus on optimizing the bearing concept and increasing the energy density of the flywheel rotor by utilizing the potential of high-performance materials.

The *overall FlyGrid system*, from the grid to the vehicle, is shown schematically in Fig. 10.25.

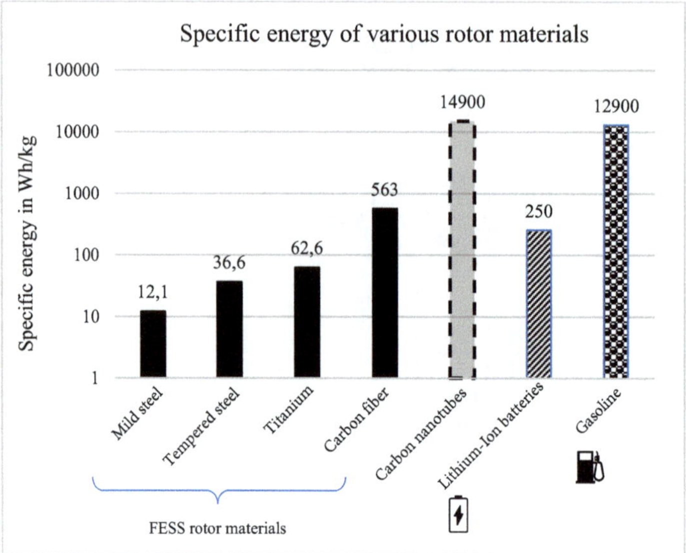

Fig. 10.24 Potential of the specific energy of flywheel rotors. Note the logarithmic scale

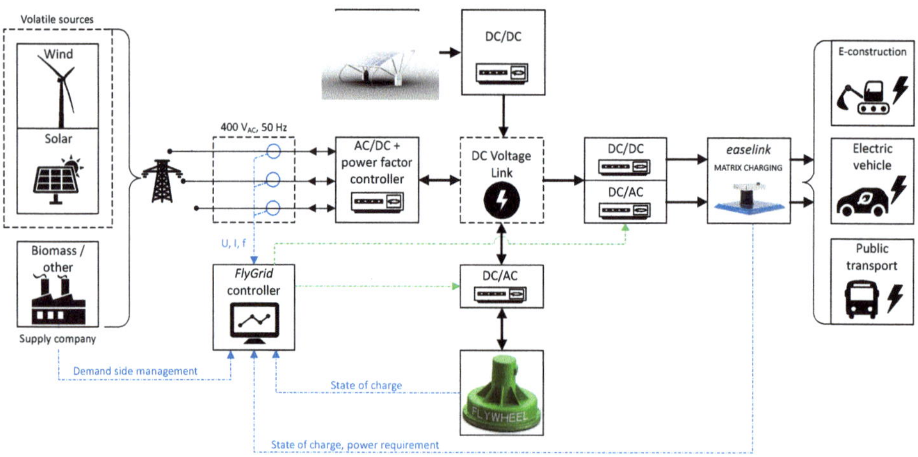

Fig. 10.25 Representation of the overall system of *FlyGrid*: from the energy source to the vehicle

References

1. EEA – European Environment Agency (2015) Overview of electricity production and use in Europe. Kongens Nytorv 6, 1050 Kopenhagen, Denmark.
2. International Energy Agency (2012) CO_2 Emissions from Fuel Combustion – Documentation for Beyond 2020 Files. International Energy Agency, Paris.

3. P. Bühler (1995) Hochintegrierte Magnetlager-Systeme. ETH Zürich, Switzerland.
4. G. Halevi (2003) Process and Operation Planning, Springer Netherlands. DOI: https://doi.org/10.1007/978-94-017-0259-1
5. A. Palmgren (1957) Neue Untersuchungen über Energieverluste in Wälzlagern. VDI-Berichte, Band 20. Verein Deutscher Ingenieure.
6. SKF Gruppe (2008) Hauptkatalog 2008. Hauptverwaltung der SKF Gruppe, SE-415, 15 Göteborg, Sweden.
7. C. Voglhuber (2016) Entwicklung und Inbetriebnahme eines Prüfstands zur Bestimmung des Verlustmoments eines passiv magnetisch entlasteten Schwungrades. Institut für Maschinenelemente und Entwicklungsmethodik, Graz University of Technology, Austria.
8. EMEA Active Power Solutions Ltd. (2015) CleanSource® 750HD UPS. EMEA Active Power Solutions Ltd., Lauriston Business Park, Pitchill, Evesham, UK.
9. BVI Magnet GmbH (2016) BVI Magnet GmbH, Schönaustr.77, 44227 Dortmund, Deutschland. http://www.bvi-magnete.de/index.php. [Accessed April 18th 2016].
10. M. Zisser, P. Haidl, B. Schweighofer, H. Wegleiter and M. Bader (2015) Test Rig for Active Vibration Control with Piezo-Actuators. The 22nd International Conference on Sound and Vibration (ICSV22), Florence, Italy, 2015.
11. P. Haidl, A. Buchroithner, M. Bader, M. Zisser, B. Schweighofer and H. Wegleiter (2016) Improved test rig design for vibration control of a rotor bearing system. 23rd International Congress on Sound & Vibration (ICSV23), Athens, Greece.
12. M. Lang (2003) Berechnung und Optimierung von passiven permanentmangetischen Lagern für rotierende Maschinen, Fakultät V - Verkehrs- und Maschinensysteme der Technischen Universität Berlin, Deutschland. https://doi.org/10.14279/depositonce-739
13. A. Buchroithner, H. Wegleiter und B. Schweighofer (2018) Flywheel Energy Storage Systems Compared to Competing Technologies for Grid Load Mitigation in EV Fast-Charging Applications. IEEE 27th International Symposium on Industrial Electronics (ISIE 2018), Cairns, Australia. DOI: https://doi.org/10.1109/ISIE.2018.8433740,

11 Summary and Outlook

The book *Flywheel Energy Storage in Automotive Engineering* pursues a consistently holistic approach to the topic. In times of CO_2-induced global warming and constantly rising energy prices, it is essential to put even technical details of the hybrid powertrain—such as, for instance, the energy storage system—in a global context.

It is thus the analysis of the *supersystem* consisting of vehicle, customer, and environment that not only defines the essential target properties of mobile flywheel energy storage systems but also questions the usefulness of this application per se. The results of the *supersystem analysis*, the so-called threshold properties, not only represent generally valid development goals, but indirectly define which components of the *subsystem* require specific optimization. While components that appear critical at first glance, such as the electric motor generator or vacuum components, can be acquired off-the-shelf with satisfactory performance from other technological sectors, the *rotor*, *bearing*, and *housing* turned out to be FESS-specific key elements.

The special operating conditions of mobile flywheel energy storage systems (high rotational speeds, vacuum, gyroscopic reactions, etc.) and the resulting *strong system-internal interdependencies* do not permit isolated optimization of individual (sub) components, but require another system analysis, this time of the FESS *subsystem*. Especially due to the high-speed characteristic of flywheel energy storage systems, the rotor and bearing design are closely interlaced by machine dynamics. Likewise, the burst containment must be adapted to the structure and material of the rotor to guarantee safety in the event of vehicle crash or rotor failure.

In this book, approaches for the development of cost-optimized solutions for *rotor*, *housing*, and *bearings* were presented, since these three components were identified as critical in the system analysis. The functionality and validity of the solutions were empirically proven, either by prototypes or component test rigs, and consequently the next

A. Buchroithner, *Flywheel Energy Storage*,
https://doi.org/10.1007/978-3-658-35342-1_11

Table 11.1 Summary of problems and solutions concerning the critical FESS components *rotor*, *housing*, and *bearings*

Rotor			
Initial problems	Costs	Energy density	Safety
Solution approach	Multi-disk design → Cost reduction through steel rotor instead of high-strength fiber composite, energy density increase through thin sheets (higher specific strength), better safety through deformation work in sheets, and light rotor fragments		
Still unsolved problems	Balance quality and machine dynamics	Bearing loads	Low-energy density of steel
Outlook	Investigation of innovative steel and composite rotor structures, low cost magnetic bearings, and/or hybrid bearings in resilient seats		
Housing			
Initial problems	Costs	Energy density	Safety
Solution approach	Cost-effective burst containment of ductile structural steel, optimized design ideally exploiting the material strength and avoiding oversizing		
Still unsolved problems	Statistically significant number of tests required	Variation of material quality requires safety margin	
Outlook	Investigation of possible, lighter concepts by combining modern materials.		
Bearings			
Initial problems	Costs	Torque loss	Service life
Solution approach	Reduction of radial bearing loads by resilient bearing seats and supercritical rotor operation. Reduction of torque loss through rolling bearing downsizing and permanent magnetic weight compensation		
Still unsolved problems	Eddy current losses	Strength of the thin shaft ends (bearing journals) in case of a vehicle crash	
Outlook	Examination of circulating oil lubrication and dry-running bearings		

necessary steps in the development process were defined. Table 11.1 summarizes the optimization of the FESS *subsystem*.

Even if not all component-specific challenges could be completely solved in the first instance, the concepts developed and described in the scope of this book make it possible to achieve a consistent cost reduction compared to the current state of the art. By achieving a price that is comparable to (or even lower) than that of competing technologies, it is possible for FESS to enter suitable niches in the market and thus gain more experience in the field. Consequently, further cost reduction through optimized manufacturing processes will be possible due to an increase in the number of units produced.

The fact that it has not been possible up to now to completely improve all relevant properties of FESS to those of the ideal reference energy storage by optimization in the *subsystem* (component improvement) could be countered with a further iteration of the optimization in the *supersystem.*

Although the implementation of low-cost flywheel systems with steel rotor and rolling bearings is possible—as the prototypes *CMO* and *FIMD* described in this book showed—they are only suitable for highly dynamic load cycles and applications with moderate energy requirements (e.g., hybrid commercial vehicles in public transport). The just mentioned reduction of the self-discharge by passive magnetic rotor weight compensation and supercritical rotor operation in resilient bearing seats (see Sect. 9.7 or Fig. 11.1) proved to be a feasible way, but prefer stationary applications due to the now fragile rotor shaft. However, in the course of the energy revolution, stationary energy storage systems are becoming increasingly important for storing renewable energy and supporting the power grid. Especially with higher penetration rates of electromobility, buffer storage is

Fig. 11.1 Looking to the future: a low-cost/low-loss flywheel energy storage system on a vacuum test rig at Graz University of Technology. (Image rights: Barbara Krobath)

indispensable. Improvements in the field of composite materials (CFRP) will enable even higher energy contents of FESS in the future. And while chemical batteries usually require raw material resources from South America or Asia which are sometimes limited resources, flywheel energy storage systems can represent an important step toward the independence of Europe—or the Western world for that matter—as a market economy.

The proposed solutions therefore provide a basis for the possible imminent commercialization of FESS technology. First successful implementations of these solution approaches in the form of prototypes or component test benches indicate that the direction taken has not only brought about a significant improvement in terms of effective energy storage properties, but is also an indicator of further development potential that can and will be exploited in the future.

It has been shown that the optimization of *a single* component in the *subsystem* cannot take place in isolation, but that all interactions and interdependencies *(horizontal* and *vertical*; see Sect. 6.2.2) must be taken into account:

- A change in the rotor topology causes a change in the burst behavior and therefore requires a different housing/burst containment architecture.
- A change in the bearing stiffness influences the resonance speed and consequently limits the speed range available for the electric machine.

The list of these relationships and interdependencies is long and complex, and representations such as those in Sects. 7.3, 8.1, or 9.2 (Paradigms of Rotor, Housing, or Bearing Design) are an attempt to greatly simplify and only approximate reality. But this general approach can be regarded as a universally valid dogma of mechanical engineering.

MIX
Papier aus verantwortungsvollen Quellen
Paper from responsible sources
FSC® C105338

If you have any concerns about our products,
you can contact us on
ProductSafety@springernature.com

In case Publisher is established outside the EU,
the EU authorized representative is:
Springer Nature Customer Service Center GmbH
Europaplatz 3, 69115 Heidelberg, Germany

Printed by Libri Plureos GmbH
in Hamburg, Germany